Lecture Notes in Biomathematics

Managing Editor: S. Levin

22

Francesco M. Scudo
James R. Ziegler

The Golden Age
of Theoretical Ecology: 1923–1940

A Collection of Works by V. Volterra, V. A. Kostitzin,
A. J. Lotka and A. N. Kolmogoroff

Springer-Verlag
Berlin Heidelberg New York 1978

Authors

Francesco M. Scudo
Laboratorio di Genetica
Biochimica ed Evoluzionistica
Via S. Epifanio, 14
27100 Pavia/Italy

James R. Ziegler
Associate in the
Agricultural Experiment Station
Division of Biological Control
University of California, Berkeley
1050 San Pablo Avenue
Albany, California 94706/USA

AMS Subject Classifications (1970): 92-01, 92-02, 92-03, 92 A 05, 92 A 10

ISBN 978-3-540-08769-4   ISBN 978-3-642-50151-7 (eBook)
DOI 10.1007/978-3-642-50151-7

2141/3140-543210

PREFACE

This is the part of any book where the authors usually discuss why they wrote it. We hope, however, that the text will justify itself. In fact, any well-trained ecologist will immediately grasp the significance of these seminal works. We have therefore tried to keep our interpretive comments to a minimum.

Students of "modern" theoretical ecology will want to contrast the papers in this collection with their modern derivatives. We believe that those who do so will be surprised, if not amazed, by the ecological sophistication and intellectual power of the earlier works. They will stand as a challenge to those who study them, and we hope, provide a standard for the quality of their work.

By presenting this collection of works, most of them not easily available and/or for the first time in English, we hope to help them attain the high level of recognition they deserve. We are also enabling readers not sufficiently familiar with Italian to acquire enough of a background to properly follow the works in French not presented here by including Volterra's "Variazioni e fluttuazioni del numero d'individui in specie animali conviventi" (1927), still available in the original edition.

The idea for a cooperative project of this kind was born in the summer of 1973 while the two of us were colleagues at the same New England university. It became a commitment one year later, and we decided to carry out the major portion of our work in Italy during the summer months of 1975.

We could not have chosen a location more conducive to our task, immersed as we were in an atmosphere of historical perspective and isolated from the usual mundane distractions. Somehow, after ten weeks, and perhaps 600 hours apiece of translating and typing, we finished a preliminary draft.

We first worked in a small cottage in the vineyard village of Mazzolino, southeast of Pavia, in the hills overlooking the valley of the Po and within sight of the Alps. Cool nights, long walks through the countryside, and the local Barbera cleared our heads and prepared us for another day at the typewriter.

When the heat receded, we would pack our office-sized electric Olivetti in the trunk of a small red Citroen, tie the typing table and chair to the roof, and descend into Pavia to an apartment in the medieval section of town near the University. Whenever the heat became unbearable, we gathered our families and typewriter and retreated to the hills, usually to an isolated farmhouse near the village of Fortunago in the northern Apennines, about half-way between Pavia and Bobbio.

The next two years saw the manuscript edited, revised, typed, and retyped by typists at Cornell and U.C. Berkeley. We communicated by mail between the United States and Italy. Service was slow and irregular. Letters were lost or misdirected,

R. COMITATO TALASSOGRAFICO ITALIANO

Istituito con la legge 13 luglio 1910, N. 442 modificata con la legge 5 giugno 1913, N. 599.

MEMORIA CXXXI

Variazioni e fluttuazioni del numero d'individui in specie animali conviventi

per il Prof. VITO VOLTERRA

VENEZIA
PREMIATE OFFICINE GRAFICHE C. FERRARI
1927 - Anno V E. F.

Plate I

Title page of Volterra's "Variazioni e fluttuazioni del numero d'individual in specie animali conviventi", 1927.

et permet de ne pas subdiviser les groupes raciaux en sous-groupes sexuels.

1. *Structure cellulaire*. — Soit donc $(p_1, p_2, \ldots, p_n)$ une population composée initialement de représentants de deux ou plusieurs races de la même espèce et contenant au bout de quelques générations toutes les races que peuvent donner les croisements $(p_h \times p_k)$.

Admettons, d'après la théorie chromosomique de l'hérédité, qu'avant sa maturation une cellule sexuelle contienne deux jeux de chromosomes-porteurs des caractères. Considérons un caractère α (un couple de caractères d'après la terminologie habituelle) qui peut se présenter tantôt sous forme A, tantôt sous celle a et qui est localisé dans un des chromosomes non sexuels. Comme, dans la cellule, ce caractère est porté par deux chromosomes, trois combinaisons sont possibles : il peut arriver que dans les deux chromosomes le caractère α figure sous forme A; dans ce cas, par rapport à α, la cellule sera *homozygote du type* $\varepsilon_1 = (A, A)$; dans le cas où α figure sous forme a, la cellule sera homozygote du type $\varepsilon_2 = (a, a)$; enfin, dans le cas mixte, la cellule sera hétérozygote du type $\varepsilon_3 = (A, a)$.

On peut donc caractériser la structure chromosomique d'une cellule sexuelle avant ses divisions réductionnelles par le symbole

(3) $$E = (\varepsilon_{i_1}^1, \varepsilon_{i_2}^2, \varepsilon_{i_3}^3, \ldots, \varepsilon_{i_m}^m),$$

les indices supérieurs donnant le numérotage des caractères et les indices inférieurs, égaux soit à 1, soit à 2, soit à 3, indiquant l'aspect de ces caractères.

2. *Divisions de maturation et combinaisons ultérieures*. — Dans ces divisions, la cellule perd au hasard la moitié de ses chromosomes en n'en conservant qu'un de chaque type. Le croisement, en combinant les cellules réduites, reconstitue des cellules à deux jeux de chromosomes. On peut exprimer cette opération et ses résultats par les relations symboliques suivantes :

(4) $$\begin{cases} \varepsilon_1 \times \varepsilon_1 = (\varepsilon_1); & \varepsilon_2 \times \varepsilon_2 = (\varepsilon_2); & \varepsilon_3 \times \varepsilon_3 = \frac{1}{4}(\varepsilon_1) + \frac{1}{4}(\varepsilon_2) + \frac{1}{2}(\varepsilon_3); \\ \varepsilon_1 \times \varepsilon_2 = (\varepsilon_3); & \varepsilon_1 \times \varepsilon_3 = \frac{1}{2}(\varepsilon_1) + \frac{1}{2}(\varepsilon_3); & \varepsilon_2 \times \varepsilon_3 = \frac{1}{2}(\varepsilon_2) + \frac{1}{2}(\varepsilon_3). \end{cases}$$

Ces relations expriment les lois de Mendel et donnent la répartition probable des produits de tous les croisements dans le cas le plus simple d'un caractère. Dans le cas de $m > 1$, cet algorithme doit être appliqué indépendamment à chaque caractère avec multiplication symbolique et groupement ultérieur des termes. Pour en donner une idée, considérons le cas de structures ne différant que par deux caractères indépendants. Le

Plate II

Original of Kostitzin's "Tensor notation" for Mendelian segregations in models for natural selection in diploid interbreeding organisms, 1938b.

and an occasional "air-mail" parcel was sent by boat. These aggravations, however, were overcome. All we really lost was time, gaining, meanwhile, in terms of several improvements.

We could not even have begun, however, without the support and encouragement by a number of key people, especially by Professors Samuel Karlin and the Volterra family. We are also deeply indebted to Academician A.N. Kolmogoroff, to the Volterra family, and to all other holders of copyrights, for their permissions to publish the works included in this volume. These include the publishers of the following journals: Acta Biotheoretica, Journal of the Washington Academy of Sciences, the American Journal of Epidemiology, Human Biology, and the Proc. Edinburgh Math. Soc., the Hermann & Cie and Armand Colin publishing companies. Among the latter we are particularly grateful to the Academie des Sciences of Paris for having granted us exceptional conditions for our translations and to the Laboratorio della Biologia del Mare of the Italian National Research Council in Venice for having forfeited its copyrights on Volterra's "Variazioni e fluttuazioni del numero d'individui in specie animali conviventi" (1927). Mrs. Luisa D'Ancona, Professors A. Falaschi, S.D. Jayakar, C. Petit, M. Prenant, R. Rossetti, Yu. Shmiglevsky, and V.A. Volkov were of great help at various stages in the preparation of this work.

The texts by Volterra are largely based on the 1962 re-edition by Dr. Elena Freda, who corrected a number of misprints and skilfully made up for some redundancies in Volterra's early publications. Our research in Italy was sponsored by the Laboratorio di Genetica Biochimica ed Evoluzionistica of the Italian National Research Council in Pavia.

We are grateful to Prof. Simon Levin and Mrs. Alice Peters, who, in their connection with Springer-Verlag, provided help and infinite patience throughout the lengthy preparation of the final manuscript. We wish also to express our appreciation and thanks to Nora Lee in Berkeley and Darlene Lynch in Ithaca for their superb jobs of typing this most difficult material.

Above all others, however, we are deeply indebted to our wives who gallantly endured the consequences of our private commitment. Finally many thanks in advance to the readers who might have an understanding attitude towards the many shortcomings of our efforts and will point them out to us.

I

'ORDRE d'idées qui a donné naissance à la théorie moderne de l'évolution organique est de nature essentiellement quantitative. On sait que Darwin autant que Wallace a été mené à l'énoncé du principe de la survie du plus apte en réfléchissant sur le problème abordé par Malthus : comment le nombre des êtres vivants est-il maintenu entre les bornes que nous observons effectivement? Par un hasard assez singulier, cet aspect du problème de l'évolution, qui, semble-t-il, aurait dû être le premier à attirer l'attention des biologistes prédisposés à l'analyse mathématique, n'a reçu leur pensée sérieuse que tout récemment. Une certaine école, il est vrai, profitant entre autres des travaux géniaux de Gregor Mendel, s'est bien occupée depuis des années de l'analyse biométrique appliquée aux questions de survie et de reproduction dans leur relation avec le problème de l'évolution organique. Mais, dans leurs recherches, les disciples de cette école se sont bornés presque entièrement à la discussion des caractéristiques d'une seule espèce, et des conséquences, en ce qui touche sa survie, de ces caractéristiques. L'interaction des diverses espèces entre elles et avec leur milieu ambiant a reçu tout au plus une considération passagère et incidente de la part de ces auteurs. Les écologistes, par contre, se sont contentés presque entièrement d'études empiriques sur ce sujet.

Il reste donc, dans la science biométrique, une lacune béante. Aujourd'hui nous voyons un groupe encore peu nombreux occupé à la combler. Nous assistons, semble-t-il, à la naissance de toute

Plate III

Opening page of Lotka's I volume of the "Théorie analytique des associations biologiques", 1934, a small sample of a work we regret not having been able to include. It is noteworthy for being equally critical toward the two major deviations from Darwinian theories, the complete neglect of ecology by "neo-Darwinists", and the often irrelevant "empiricism" of most contemporary "ecologists".

CONTENTS

HISTORICAL INTRODUCTION

In the early decades of this century the theories of ecology and of evolution proposed by Darwin and Wallace were being extensively tested along two independent lines. On the one hand the rediscovery of Mendel's theory led to a great flourishing of investigations on the mechanisms of inheritance and on the power of selection, mostly in artificial settings. On the other hand patterns in the distribution and abundance of organisms were being studied in a quantitative manner, with explicit reference to Darwin's theory for their causation.

In both of these areas the need for explicit analytical reasoning became soon apparent. Following the classic contributions by Hardy and Weinberg in the first decade of this century various systems of mating were treated in terms of recursions by Jennings, Robins, Wentworth and Remick. Much at the same time Ross was laying the foundation of the mathematical theory of epidemics in terms of differential equations.

After World War I mathematical models soon acquired a pivotal position in the development of both these areas. The mathematical theory of population genetics was developed in terms of frequencies and mostly in discrete time by Fisher, Haldane, and Wright. Much at the same time, and independently of each other, Lotka in the United States and Volterra in Italy began to concentrate their mathematical skills on the problem of the variations and fluctuations in the numbers of individuals and species.

Lotka's main previous interest was physical chemistry, with special emphasis on the oscillatory behavior of chemical reactions. He had also dealt with demographic problems (1907, 1911a), and with thermodynamic aspects of evolutionary theory (e.g., 1911b, 1915). His earlier interests in mathematical approaches to malaria epidemics (1912, 1919) matured with a series of five remarkable papers (cf., Part III).

Volterra's previous interests were mostly in mechanics, including irreversible phenomena and elasticity. The latter had led him to develop the theory of functionals and integro-differential equations, for which he is best known. He had devoted his best efforts during World War I to military technology. His previous interests in quantitative aspects of biology (e.g., 1901, 1906) were suddenly awakened in 1925 by a curious phenomenon pointed out to him by D'Ancona (cf., Scudo, 1971). Corresponding to severe limitations on fishing in the Upper Adriatic during the war, the proportion of predator species had greatly increased. The virtual cessation of fishing seemed to be the cause, and much the same effect had already been predicted by Darwin with reference to hunting. Volterra began to investigate analytical models to corroborate these suggestions. Lotka had also briefly investigated an analogous problem posed by agriculture, but in a quite different theoretical context (1924, Chapter 22). Still unaware of Lotka's work, Volterra began to investigate the effect of external perturbations of preexisting natural equilibria using the same quadratic differential equation model for a predator and a prey which Lotka had also considered (1924). Volterra soon extended this treatment of "conservative" predator-prey interactions to the case of any number of species. He soon realized that such models, while being of great mathematical interest, had severe limitations as a representation of nature. The simplest acceptable formulations of predator-prey interactions also had to incorporate a measure of efficiency in food transfer, and competition, which Volterra also treated on his own (cf., Part II).

Both Lotka and Volterra realized, from the very beginning of their ecological investigations, that to achieve some degree of realism delayed effects had to be explicitly taken into account. Lotka introduced them in his joint paper with Sharpe (1923) as incubation lags in the differential equations describing malarial epidemics. Volterra set up integro-differential models to account for the effect of the prey on the fertility of the predator, and for the accumulation of toxic catabolites on the growth of bacterial populations (cf., Parts III and I respectively).

Volterra's fame attracted a number of European mathematicians to approaches of this kind. One of the first among them, and certainly the most successful, was V. A. Kostitzin. Although very little is known of his life and his non-biological works (Scudo and Ziegler, 1976), his previous works were mostly on geophysics and integro-differential equations. His interest in theoretical ecology seems also to have been stimulated by a curious case of parasitism studied by Pérez (1928, 1931), and involving the multiple parasitic infestation of a crab by the rhizocephalous barnacle, <u>Chlorogaster</u>. As junior author to his wife, a parasitologist, Kostitzin published a numerical study of Pérez's data (Kostitzin and Kostitzin, 1931). Kostitzin was soon led to develop, much in the same vein as Lotka and Volterra, a very general theoretical framework which incorporated parasitism as a mathematical specialization of the more general case of symbiosis (cf., Part III). His interests both in geophysics and in biological evolution led him to propose one of the first, and so far one of the most comprehensive treatments of phenomena involving the two disciplines (cf., Part V).

Even more original, however, is Kostitzin's approach to genetics and evolution. He was the first to realize (1936a, 1937, 1938a) that differential equation models analogous to Volterra's are particularly suitable to describe natural selection due to competition and predation on diploid interbreeding populations. He also discussed very thoroughly the implications of his models in the general framework of the "neo-Darwinian" theory of evolution. It is interesting to contrast the deductions from his preliminary results with many of the problems suggested by considerations of gene frequencies (cf., Part I and IV).

Prior to World War II, mathematical ecology had reached a level of conceptual and analytical sophistication which had no parallel in any other branch of mathematical biology, and which has not been matched since. There is a refreshing diversity of philosophical attitudes, choice of natural problems, and of mathematical technique among the contributors to this volume. Lotka tends to concentrate on detailed quantitative studies of relatively simple systems, for large collections of values of the parameters. Kostitzin, and even more Volterra and Kolmogoroff, show a preference for more general analytical problems. Volterra and Kolmogoroff sought primarily general qualitative results, and Volterra limited quantitative considerations to simple cases, by the way of illustration. Kostitzin tends to analyze large collections of special cases as a means of approaching problems which are not amenable to analysis in their more general form. There are also systematic differences in more specific technical aspects, and in the set-up of special problems. Volterra, for instance, sought to incorporate the effect of

seasonal changes through periodically variable vital coefficients.
Kostitzin approached the same problem by explicitly combining into recursions the effects of continuous change during the seasons (cf.,
Part I). This approach by Kostitzin, together with the classical one
by Nägeli (1874), and the one by Nicholson and Bailey (1935), are the
major examples of discrete-time treatments in ecology prior to World
War II.

Most of the works we have considered here are discussed in greater
or less detail in monographs by D'Ancona (1942, 1954). Some of these
works are also beautifully summarized in a review by Rescigno and
Richardson (1973), and in the textbook by Poluektov et al. (1974). But
for these and a few other rare exceptions, most of the treatments discussed here have had little direct or indirect influence on the more
recent ecological literature. In most of the recent literature references to Volterra and Lotka are limited to their earliest treatments of
conservative interactions between predators and prey, or, at best, to
an early popularization by Volterra (1926, translated into French,
English and Russian). Not even Lotka's mature, comprehensive works
(1934, 1939 in French), have been translated, to the best of our knowledge, into any other major language. Volterra's "Leçons" (1931 b) are
now available in Russian, and what is probably the most comprehensive
biography of Volterra is also in this language (Poliscink, 1977). Also
to the best of our knowledge none of Kostitzin's works have yet appeared
in any other major language except the "Biologie Mathématique" in
English (1937 b), although a translation of his work into Russian is
being undertaken.

Lotka's two volumes on the "Théorie Analytique des Associations
Biologiques" (1934, 1939) would deserve much more thorough comments than
those we will be able to provide here. The first volume is a short,
largely "philosophical" essay on life, including energetic and evolutionary aspects, in a sense a more mature "distillate" of his famous
"Mathematical Biology" (cf. Plate 3 and the Introduction to Part II).

The second volume is confined to demography proper, a subject we shall not enter in any detail since it is very thoroughly covered in the reader edited by Smith and Keyfitz (1977). Incidentally it is also interesting to follow in this reader how Lotka's preoccupations with his own priorities, also noticeble in his paper (1932, cf., Part II), explodes into a fullblown controversy with Kuczynsi.

PART I

LOGISTIC APPROACHES

INTRODUCTION

We have chosen to begin with logistic treatments since they are the simplest, and historically the first to be widely used. The simplicity is achieved by imposing an artificial dichotomy which discriminates between the population itself and the rest of the world, i.e., the "environment." Because the rest of the world is only reflected in certain "demographic" or "vital" coefficients of a single population, the logistic approach is particularly useful in dealing with special "internal" population phenomena. Here we present some of the logistic treatments developed by Volterra and Kostitzin quite late in their biological careers.

The first paper, by Volterra, deals, for the logistic, with a very general and theoretically interesting question. Many differential equation models in ecology can be derived on the assumption that they minimize certain biologically interesting quantities. In this context the "canonical" framework of classical mechanics applies to ecological models as well.

The second paper, also by Volterra, deals with the effects of low density on the growth of a cross-breeding population. The simplest, biologically meaningful representation of the process is given by a cubic differential equation. Here the quadratic term represents the sexual encounters of individuals, and the cubic term represents a variety of limiting factors. With compulsory cross-breeding there is a minimum density below which the population tends to extinction. Starting at higher densities the population tends toward a stable equilibrium.

The same problem is dealt with in a different way by Kostitzin in the third paper. The equation he derives has the logistic as the limiting case for high population densities. Its behavior is again characterized by a stable equilibrium, and a threshold below which local extinction occurs. This would be intuitively expected from any reasonable model for this effect. In the same paper Kostitzin deals with the limitations of the logistic treatment and the complex, multiple, and yet restricted meanings of the "vital coefficients" in the Verhulst-Pearl equation. He also treats explicitly the case in which the vital coefficients are periodical functions of time, representing the seasonal cycle.

In his "Biologie mathématique" (1937) Kostitzin also considered various other logistic treatments, both in terms of more general analytical forms, and of more specific biological situations. Among the latter Kostitzin took into account the different limiting effects of space and of nutrients. He also considered the case in which nutrients are not renewed. He also dealt with the effects of emigration and immigration, and with different life stages, overlapping and not, of metamorphosing animals.

The fourth paper in this section, by Kostitzin, deals with some of the effects of "internal" variation in a population with an annual life cycle. The main problem is much the same as in the very first mathematical treatment of evolution by Nägeli (1874)--that of the stable coexistence of different varieties of the same species. Kostitzin's case is much simpler than Nägeli's, and mathematically more elegant. Rather than the perennial plants considered by Nägeli, Kostitzin deals with an insect having an annual life cycle and a molt between the immature and adult life stages. The choice of an insect is particularly appropriate since for many insects partial reproductive isolation ("segregation") can result from purely behavioral preferences for different host species. In Kostitzin's model, "segregation" is explicitly introduced through a difference in the onset of the adult molt. The biological context is, however, much the same as Nägeli's, and neither model specifically incorporates Mendelian variation.

This treatment by Kostitzin is a rare example of a discrete-time model derived from a detailed consideration of both continuous and discontinuous processes throughout the annual cycle. Dealing with a collection of special cases he works out the conditions under which the two "segregants" can stably coexist. He points out that the particular isolating mechanism he has explicitly considered is just one among many which could lead to similar results. He also points out that "physiological" mechanisms of splitting into two forms with partial reproductive isolation are the ideal preconditions for genetic divergence, and, ultimately, for speciation.

The last three papers in this section are motivated by detailed observations of bacterial growth in a confined medium, both in the case of a single species and

of two species in competition. In such conditions a logistic-like growth is
followed by a very rapid decline. In the first note Volterra interprets this
pattern as being a result of the toxic effects of catabolites which accumulate in
the medium. This effect is represented by integro-differential equations, which
are solved by Volterra for a special case. In the following paper Kostitzin
provides more general solutions to Volterra's equations. The method is extended
to cover situations in which the census methods do not discriminate between living
and dead cells, and to include the intoxicating effects of accumulated dead cells.
The last paper, co-authored by Volterra and Kostitzin, analyzes an extensive
collection of data in the light of intoxication models. The authors point out the
existence of upper limits to the rate of fission and to other "vital coefficients",
as the initial concentration of nutrients increases.

Kostitzin continued to be interested in the dynamics of bacterial populations
even after his interest in general ecological problems began to wane around 1940.
As late as 1956 he provided a clever analytical solution to a quite general model
of bacterial growth in a chemostat. To our knowledge this is the last contribution
by Kostitzin to ecology.

Calculus of variations and the logistic curve

V. Volterra

1. I have been able to show that the equations of the struggle for existence depend on a question of Calculus of Variations, or more precisely on a problem of minimum.

In order to obtain this result, I have replaced the notion of *population* by that of *quantity of life*.[1] In this manner I have also obtained some results by which dynamics is brought into relation to problems of the struggle for existence.

2. The quantity of life X and the population N of a species are connected by the relations

$$X = \int_0^t N\,dt \quad , \quad N = \frac{dX}{dt}$$

where N is to be considered as a function of the time. All depends on a theorem of Calculus of Variations that I have already used in my Memoir above cited. I shall now give the enunciation, the demonstration, and some applications of this theorem.

3. THEOREM. *The differential equation*

(I) $$\frac{dz}{dt} = f(z) = a(z-a_1)(z-a_2)\ldots(z-a_n) \quad ,$$

where $f(z)$ is a rational and integral polynomial of degree n, can be obtained by equating to zero the first variation of

$$P = \int_0^T F(t)\,dt \quad ; \quad F = \sum_1^n m_i \left(a_i - \frac{dX}{dt} \right) \ell n \left(a_i - \frac{dX}{dt} \right) + KX \quad ,$$

by putting $dX/dt = z$ *and by making a convenient choice of the constants* m_i , K .

In fact we shall have

$$(II) \qquad \delta P = \int_0^T \delta F dt = \int_0^T \left\{ -\sum_i \left[m_i \frac{d\delta X}{dt} \ln\left(a_i - \frac{dX}{dt}\right) - m_i \frac{d\delta X}{dt} \right] + K\delta X \right\} dt$$

and with an integration by parts, neglecting the terms out of the integral, this takes the form

$$(III) \qquad \delta P = \int_0^T \delta X \left[\sum_i \frac{-m_i \dfrac{d^2X}{dt^2}}{a_i - \dfrac{dX}{dt}} + K \right] dt \ .$$

If $\delta P = 0$ for every value of δX then we shall have

$$\sum_i \frac{m_i \dfrac{d^2X}{dt^2}}{a_i - \dfrac{dX}{dt}} = K \ ,$$

from which we deduce, by putting $dX/dt = z$,

$$(IV) \qquad Q(z)\frac{dz}{dt} = K(a_1-z)(a_2-z)\ldots(a_n-z) \ .$$

4. $Q(z)$ is a polynomial of degree $n-1$, linear and homogeneous with respect to $m_1, m_2, \ldots, m_n$.

We can write

$$(V) \qquad Q(z) = A_1 z^{n-1} + A_2 z^{n-2} + \ldots A_n$$

where

$$A_i = m_1 p_{1i} + m_2 p_{2i} + \ldots + m_n p_{ni} \; ,$$

A_i and $p_{1i}, p_{2i}, \ldots, p_{ni}$ being of degree $i-1$ with respect to $a_1, a_2, \ldots, a_n$.

If we write

(VI) $\qquad\qquad A_1 = A_2 = \ldots = A_{n-1} = 0 \; , \quad A_n = A$

(A constant), we have n equations, linear with respect to $m_1, m_2, \ldots, m_n$.

The determinant of these equations is

$$D = \begin{vmatrix} p_{11}, & p_{12}, & \ldots, & p_{1n} \\ p_{21}, & p_{22}, & \ldots, & p_{2n} \\ \cdots\cdots\cdots\cdots\cdots\cdots \\ p_{n1}, & p_{n2}, & \ldots, & p_{nn} \end{vmatrix}$$

Considering that $p_{1i}, p_{2i}, \ldots, p_{ni}$ are of degree $i - 1$ with respect to $a_1, a_2, \ldots, a_n$ we shall see that D is of degree

$$1 + 2 + \ldots + (n-1) = \frac{n(n-1)}{2} \; .$$

Moreover by changing i with s we produce an exchange between columns i and s ; hence D admits the factors $a_i - a_s$. They are in number of $n(n-1)/2$. We conclude that D will be equal to a constant multiplied by all the factors $a_i - a_s$, which can be obtained by means of the combinations two by two of the indexes $1, 2, \ldots, n$. Hence D will not be zero if $a_1, a_2, \ldots, a_n$ are all different among them.

Therefore we can determine $m_1, m_2, \ldots, m_n$ as solutions of equations (VI). Then we have $Q(z) = A$ and

$$\frac{dz}{dt} = a(a_1-z)(a_2-z)\ldots(a_n-z)$$

where

$$a = K/A .$$

5. We have demonstrated the theorem in the case where $f(z)$ does not have multiple roots. The case of multiple roots can be treated by the usual procedure supposing that two or more roots become indefinitely near among them.

We shall consider only a particular example. Let us suppose a case having three roots, and that $a_2 = a_3$.

We can write

$$\text{(VII)} \quad F = m_1 \left(a_1 - \frac{dX}{dt} \right) \ell n \left(a_1 - \frac{dX}{dt} \right) + m_2 \left(a_2 - \frac{dX}{dt} \right) \ell n \left(a_2 - \frac{dX}{dt} \right) +$$

$$+ \frac{m_3}{\varepsilon} \left[\left(a_2 + \varepsilon - \frac{dX}{dt} \right) \ell n \left(a_2 + \varepsilon - \frac{dX}{dt} \right) - \left(a_2 - \frac{dX}{dt} \right) \ell n \left(a_2 - \frac{dX}{dt} \right) \right] .$$

As ε approaches zero, F approaches

$$m_1 \left(a_1 - \frac{dX}{dt} \right) \ell n \left(a_1 - \frac{dX}{dt} \right) + m_2 \left(a_2 - \frac{dX}{dt} \right) \ell n \left(a_2 - \frac{dX}{dt} \right) + m_3 \, \ell n \left(a_2 - \frac{dX}{dt} \right) + m_3 \, ,$$

and if we take $\delta P = 0$ we shall obtain the equations

$$\text{(VIII)} \quad m_1 a_2^2 + m_2 a_1 a_2 + m_3 a_1 = A$$

$$\text{(IX)} \quad 2m_1 a_2 + m_2 (a_1 + a_2) + m_3 = 0$$

$$\text{(X)} \quad m_1 + m_2 = 0$$

whose determinant D is $- (a_1 - a_2)^2$, and therefore is not zero if

$$a_1 \gtrless a_2 \quad .$$

6. Starting from the expression (II) for δP we obtain for the second variation

$$\int_0^T \Sigma_i \frac{m_i \left(\frac{d\delta X}{dt}\right)^2}{a_i - \frac{dX}{dt}} \, dt \ .$$

If $m_i > 0$ and $a_i > dX/dt$ ($i = 1, 2, \ldots, n$) the preceding expression is positive and then we shall have a minimum.

7. In a lecture published in 1937[2] I spoke about the variational principle in mathematical biology. Having received that Memoir, Prof. L. Amoroso asked me this question: Is it possible to reduce the equation of Verhulst-Pearl to the Eulerian form in the sense indicated in my Memoir and to reduce consequently the movement of human population to a principle of minimum?

I have been able to answer this interesting question by means of the preceding theorem showing that it is also applicable to the Verhulst-Pearl case.

8. In fact the Verhulst-Pearl differential equation can be written

$$\frac{dN}{dt} = N(\varepsilon - \lambda N) \ .$$

Now let us proceed as in Section 2 above. Letting

$$N = \frac{dX}{dt}$$

the integral whose variation we must consider can be written

$$\int_0^T \left(m_1 \frac{dX}{dt} \ln \frac{dX}{dt} + m_2 \left(\varepsilon - \lambda \frac{dX}{dt} \right) \ln \left(\varepsilon - \lambda \frac{dX}{dt} \right) + KX \right) dt$$

and its first variation will be

$$(XI) \qquad \int_0^T \left[\left\{ m_1 \ln \frac{dX}{dt} + m_1 - m_2\lambda \ln\left(\varepsilon - \lambda \frac{dX}{dt}\right) - m_2\lambda \right\} \frac{d\delta X}{dt} + K\delta X \right] dt \ .$$

Integrating by parts and neglecting the terms out of the integral we shall have

$$\int_0^T \left[\left(-\frac{m_1}{\frac{dX}{dt}} - \frac{m_2\lambda^2}{\varepsilon - \lambda\frac{dX}{dt}} \right) \frac{d^2X}{dt^2} + K \right] \delta X dt \ .$$

Equating to zero this first variation and assuming $m_1 = \lambda m_2 > 0$, $K = m_1$ we reach the equation

$$(XII) \qquad \frac{d^2X}{dt^2} = \frac{dX}{dt}\left(\varepsilon - \lambda\frac{dX}{dt}\right) ,$$

that is

$$(XIII) \qquad \frac{dN}{dt} = N(\varepsilon - \lambda N) ,$$

which is Pearl's equation.

If we calculate the second variation starting from equation (XI) we find

$$(XIV) \qquad \int_0^T \left(\frac{m_1\left(\frac{d\delta X}{dt}\right)^2}{\frac{dX}{dt}} + \frac{m_2\lambda^2\left(\frac{d\delta X}{dt}\right)^2}{\varepsilon - \lambda\frac{dX}{dt}} \right) dt \ .$$

Now since

$$0 < \frac{dX}{dt} = N < \frac{\varepsilon}{\lambda}$$

we can say that the second variation is positive and hence that we have a minimum.

9. In a preceding paper[3] I have extended Pearl's equation, taking into account beside the *mortality term* and *Pearl's term* also a *birth term;* I have thus found an equation [see (3), §8, equation (I')] which can be written in the form (I). Therefore the theorem enunciated in Section 2 of the present paper is applicable to the cases treated in the earlier above mentioned paper.

Footnotes:

[1]Principes de Biologie Mathématique: Part I, §1, N. 2; Acta Biotheoretica vol. III, part I, 1937.

[2]Applications des Mathématiques à la Biologie, L'enseignement Mathématique, vol. 36, 1937, pp. 297-330.

[3]Population growth, equilibria, and extinction under specified breeding conditions: a development and extension of the theory of the logistic curve, Human Biology, vol. 10, 1938, pp. 1-11.

Population growth, equilibria, and extinction under specified breeding conditions:
a development and extension of the theory of the logistic curve

V. Volterra

Professor D'Ancona has asked me if it is possible to state the mating conditions under which a population living under defined environmental conditions will decrease and vanish or,on the contrary,continue to grow steadily larger.

1. We can give a criterion to distinguish these two cases and settle a critical value of the population such that for a smaller value the population exhausts itself and, for a larger one, it increases.

2. Let N be the population and ε the mortality coefficient. If we suppose that the ratio of the males to the females remains constant, the number of of the males will be αN and the number of the females βN, α and β being two constants such that $\alpha + \beta = 1$. The number of the meetings of the two sexes in a unit of time will be prportional to $\alpha N \cdot \beta N = \alpha \beta N^2$; and if the birth of m individuals corresponds to n meetings, the number of the births in a unit of time can be expressed by:

$$K\alpha\beta \ \frac{m}{n} \ N^2 \ = \ \lambda N^2$$

K and λ being two positive constants. It results from this that in a time dt we have:

$$dN \ = \ - \ \varepsilon N \ dt + \lambda N^2 \ dt$$

that is:

$$\frac{dN}{dt} \ = \ (-\varepsilon + \lambda N)N \ .$$

3. This equation is identical to the equation of Verhulst-Pearl, except for the signs of the coefficients which are interchanged.

Thus if we take the Verhulst-Pearl integral with the necessary sign-interchange, we get the formula:

$$N = \frac{-\varepsilon N_0 e^{-\varepsilon t}}{-\varepsilon - N_0 \lambda (e^{-\varepsilon t} - 1)} = \frac{\varepsilon N_0}{(\varepsilon - N_0 \lambda) e^{\varepsilon t} + N_0 \lambda} = N_0 \frac{1}{(1-h) e^{\varepsilon t} + h} \quad ,$$

where N_0 is the initial population at the epoch $t = 0$ and $h = N_0 \lambda / \varepsilon$.

4. We can then distinguish three cases:

(1) $\qquad\qquad\qquad \varepsilon - \lambda N_0 = 0 \quad$ (i.e. $h = 1$)

(2) $\qquad\qquad\qquad \varepsilon - \lambda N_0 > 0 \quad$ (i.e. $h < 1$)

(3) $\qquad\qquad\qquad \varepsilon - \lambda N_0 < 0 \quad$ (i.e. $h > 1$) .

In the first case the population N *remains always constant* and equal to $\varepsilon/\lambda = N_0$. In the second case, when t increases indefinitely, the denominator increases indefinitely and N tends to zero, and therefore the population tends to vanish. In the third case, the denominator decreases and becomes zero when

$$e^{\varepsilon t} = \frac{N_0 \lambda}{N_0 \lambda - \varepsilon}$$

i.e. at the epoch

$$t = \frac{1}{\varepsilon} \ln \frac{N_0 \lambda}{N_0 \lambda - \varepsilon} \quad ,$$

we have $N = \infty$.

Of course this occurrence cannot happen; other causes lessen the increase of the population.

Thus a critical value $N_0 = \varepsilon/\lambda$ *exists above which the population grows continuously but below which it decreases and vanishes. When* $N_0 = \varepsilon/\lambda$ *the population remains stationary.*

5. The case of the vanishing can be verified, but of course the case in which the population increases indefinitely and becomes infinite after a finite

time is not a realizable one. There must exist some cause which lessens the endless increase.

In the case considered by Pearl which leads to the logistic curve, it is supposed that with the increase of the population the means of subsistence become rarefied and a negative term proportional to the already existing population must be added in the increase coefficient. It would be possible to think that in our case too Pearl's term $- \mu N$ (μ being positive) must be added in the increase coefficient $- \varepsilon + \lambda N$. In this manner the latter would become $- \varepsilon + (\lambda-\mu)N$. But, if $\lambda-\mu$ is positive we find again the case we have already treated; otherwise, $\lambda-\mu$ is negative and then we have a rapid decrease of N until N becomes zero.

6. We can consider all the cases in which the increase coefficient is linear in N . The equation takes then the form

(I)
$$\frac{dN}{dt} = (a + bN)N ,$$

and we can distinguish four cases:

(1) $a > 0 , \quad b < 0$

(2) $a < 0 , \quad b > 0$

(3) $a > 0 , \quad b > 0$

(4) $a < 0 , \quad b < 0 .$

The integral of equation (I) is

$$e^t = \left(\frac{N}{N_0} \cdot \frac{a+bN_0}{a+bN} \right)^{1/a} , \quad \text{i.e.} \quad N = \frac{N_0 a}{(a+bN_0)e^{-at} - N_0 b} .$$

The first case is Pearl's and leads to the logistic curve.

The second case is the case we have just discussed.

The third leads to $N = \infty$, for

$$t = \frac{1}{a} \ell n \left(\frac{a + bN_0}{bN_0} \right) .$$

But an infinite increase after a finite time is impossible.

The fourth case leads to $N = 0$ for $t = \infty$; and is thus the extinction case.

7. We have seen that by the allowance of a restraint similar to Pearl's, i.e. a decrease proportional to the increase of the population, we add nothing essentially new to the study already made.

The failure is due to the fact that the increase coefficient is a linear function of the population. But it is obvious that, by examining the question more profoundly, we can succeed by logic and natural means in establishing for the said coefficient a second degree polynomial. In this manner the resulting demographical phenomenon becomes entirely altered.

In fact we supposed that to n meetings between individuals of different sexes would correspond m births, but the hypothesis of a constant ratio between the births and the meetings can be admitted only as a first approximation.[1] If we wish to consider the possible cases with greater accuracy, it is necessary to suppose that this ratio can be influenced by the population itself, and not solely by the rarefaction of the means of subsistence which results from the increase of the population. As a first approximation we can suppose, as it is done in analogous cases, that this ratio decreases linearly with the population. The birth number in a unit of time will be expressed by:

$$K\alpha\beta \; \frac{(m - \rho N)}{n} \; N^2$$

ρ being positive; i.e. by

$$(\lambda - \gamma N)N^2$$

λ and γ being positive.

8. Thus if we take into account the *mortality coefficient* - ϵ , Pearl's *term* - μN, and the *birth term* $(\lambda - \gamma N)N$, we shall obtain the increase coefficient:

$$- \epsilon + (\lambda - \mu)N - \gamma N^2 \; ,$$

from which it results that (I) is replaced by

$$\frac{dN}{dt} = (-\epsilon + (\lambda - \mu)N - N^2)N$$

or

(I')
$$\frac{dN}{dt} = - (c - bN + aN^2)N$$

where a and c are positive.

We will suppose too that b is positive, keeping the case of b negative apart for the present.

Let us suppose that the roots of the equation

(II)
$$aN^2 - bN + c = 0$$

are real: they will be positive. Let us suppose first that they are unequal and call them α and β $(\alpha > \beta)$. The equation (I') will be

$$\frac{dN}{dt} = - a(N-\alpha)(N-\beta)N$$

or as well

$$- adt = \frac{dN}{(N-\alpha)(N-\beta)N}$$

and integrated

$$e^{-c(\alpha-\beta)t} = \left(\frac{N}{N_0}\right)^{\alpha-\beta} \left(\frac{N-\alpha}{N_0-\alpha}\right)^{\beta} \left(\frac{N-\beta}{N_0-\beta}\right)^{-\alpha} \quad .$$

N_0 is the value of N when $t = 0$ and we suppose N_0 different from $0, \alpha, \beta$.

We have further

$$\frac{d^2N}{dt^2} = - aX \frac{dN}{dt}$$

where

$$X = 3N^2 - 2(\alpha+\beta)N + \alpha\beta .$$

But

$$X_{N=\alpha} > 0, \ X_{N=\beta} < 0, \ X_{N=0} > 0 ,$$

so the equation $X = 0$ has two positive roots: one γ lying between α and β, and the other δ smaller than β, i.e. we have

$$\alpha > \gamma > \beta > \delta > 0 ,$$

and

$$\frac{d^2N}{dt^2} = - 3a(N-\gamma)(N-\delta) \frac{dN}{dt} .$$

9. I. If $N_0 > \alpha$, we have $dN/dt < 0$, $d^2N/dt^2 > 0$. N will decrease tending towards α and the curve $N(t)$ will always turn its concavity upwards. It has the straight line $N = \alpha$ for an asymptote (Figure A).

II. If $\alpha > N_0 > \gamma$, we have $dN/dt > 0$, $d^2N/dt^2 < 0$. N will increase tending towards the value α and the curve will always turn its concavity downwards, its asymptote being the straight line $N = \alpha$ (Figure B).

III. If $\gamma > N_0 > \beta$, we have $dN/dt > 0$, N will increase towards the α value and the curve $N(t)$ will have the straight line $N = \alpha$ as an asymptote. When N shifts between N_0 and γ we have $d^2N/dt^2 > 0$ and the curve $N(t)$ turns its concavity upwards; for $N = \gamma$, the curve presents an inflexion and as

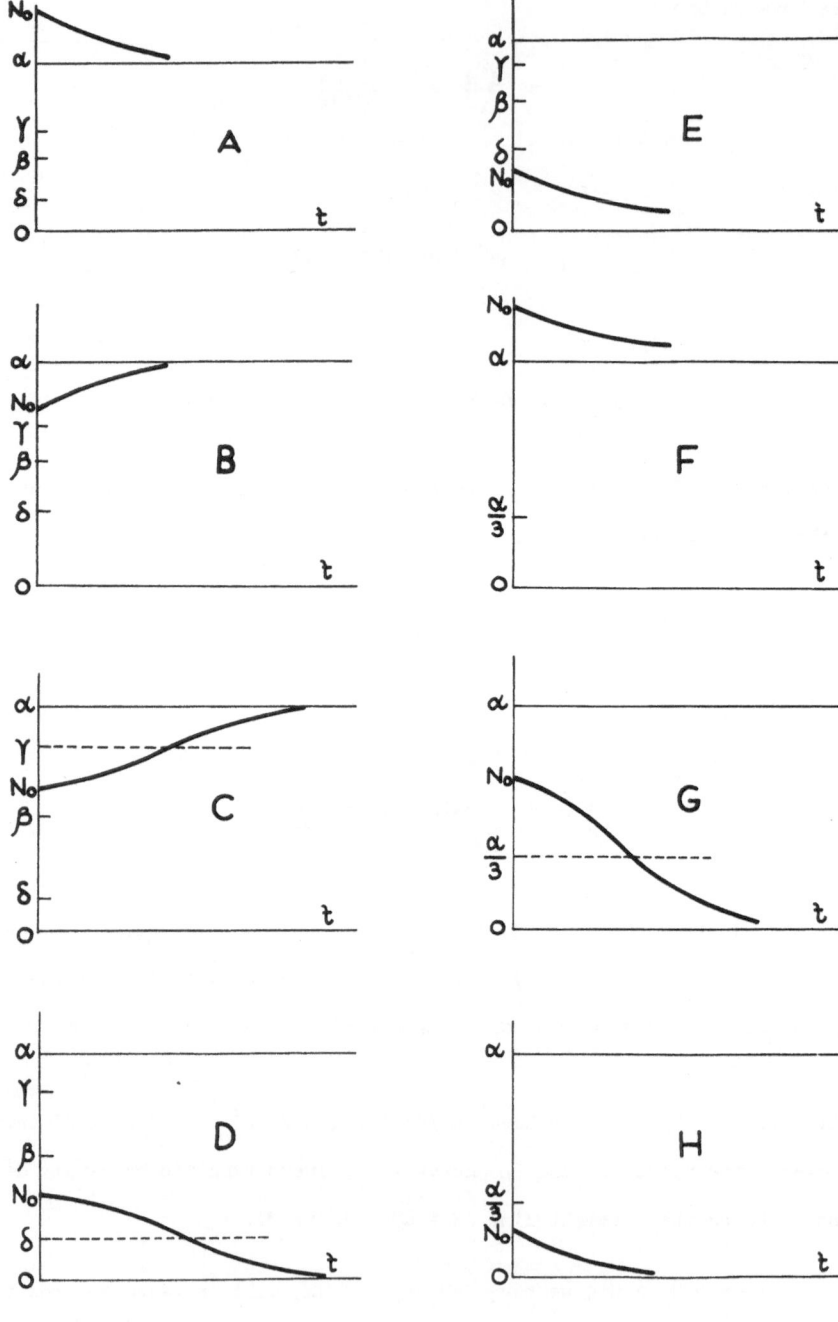

N shifts between γ and α , we have $d^2N/dt^2 < 0$, and the curve turns its convexity upwards. The curve $N(t)$ will have the same trend as the logistic curve of VERHULST-PEARL (Figure C).

IV. If $\beta > N_0 > \delta$, we have $dN/dt < 0$, N will decrease tending towards 0 and the curve $N(t)$ will have the axis $N = 0$ as an asymptote. When N shifts between N_0 and δ , we shall have $d^2N/dt^2 < 0$ and the curve $N(t)$ will turn its convexity upwards; for $N = \delta$, the curve will have an inflexion and when N shifts between δ and 0 , d^2N/dt^2 will be > 0 and the curve will turn its concavity upwards (Figure D). The species will decline and ultimately perish.

V. If $\delta > N_0$ we shall have $dN/dt < 0$; N will decrease and tend towards zero and the curve $N(t)$ will have as an asymptote the axis $N = 0$. We shall have $d^2N/dt^2 > 0$ and the curve $N(t)$ will turn its concavity upwards (Figure E). There will be an exhaustion in this case also.

10. The populations $N = \alpha$, and $N = \beta$ *are equilibrium populations* because they correspond to $dN/dt = 0$ and therefore to $N = $ constant. But for $N = \alpha$ *the equilibrium is stable,* while for $N = \beta$ *the equilibrium is unstable,* because if N departs a little from the α value, it returns to α indefinitely, while if N departs from β , it gets away from β further and further as it tends towards α or zero.

11. Let us now consider the case of a double root equal to α . We shall then have

$$\frac{dN}{dt} = - a(\alpha-N)^2N ,$$

whence it results that N decreases. The integration gives

$$e^{-at} = e^{\frac{N-N_0}{\alpha(\alpha-N)(\alpha-N_0)}} \cdot \left(\frac{N}{N_0} \frac{\alpha-N_0}{\alpha-N}\right)^{1/\alpha^2}$$

where N_0 is the initial value of N for $t = 0$.

We have then

$$\frac{d^2N}{dt^2} = - a(N-\alpha) (3N-\alpha) \frac{dN}{dt} \quad .$$

I. If $N_0 > \alpha$, we shall have $dN/dt < 0$, and $d^2N/dt^2 > 0$, therefore N will decrease tending indefinitely towards α . The straight line $N = \alpha$ will be an asymptote of the curve $N(t)$ which will always turn its concavity upwards (Figure F).

II. If $\alpha/3 < N_0 < \alpha$ we shall have $dN/dt < 0$; N will decrease indefinitely tending towards 0 . When N shifts between N_0 and $\alpha/3$, d^2N/dt^2 will be < 0 and the curve $N(t)$ will have its convexity turned upwards. For $N = \alpha/3$, the curve $N(t)$ will have an inflexion and when N shifts between $\alpha/3$ and 0 , d^2N/dt^2 will be > 0 and the curve will turn its concavity upwards. The axis $N = 0$ will be an asymptote (Figure G).

III. If $N_0 < \alpha/3$ we shall have $dN/dt < 0$ and $d^2N/dt^2 > 0$. N will decrease and tend towards 0 . The curve $N(t)$ will turn its concavity upwards. The axis $N = 0$ will be an asymptote (Figure H).

In the case of two roots equal to α , the population $N = \alpha$ will be an equilibrium population. It will be stable for a disturbance which tends to increase it and unstable for one which tends to decrease it.

12. In the case when equation (II) has imaginary roots, the integral of (I') will be

$$e^{-t} = \left(\frac{N^2(aN_0^2-bN_0+c)}{N_0^2(aN^2-bN+c)} \right)^{1/2c} e^{\frac{b}{c\sqrt{K}} arctg} \frac{2a(N-N_0)}{\sqrt{K}\left(1 + \frac{(2aN-b)(2aN_0-b)}{K}\right)}$$

where N_0 is the initial value of N for $t = 0$ and $K = 4 ac - b^2 > 0$.

It immediately appears that dN/dt < 0 and N decreases and tends towards
0 for t = ∞ . The curve N(t) has the axis N = 0 as an asymptote and it is
easy to distinguish the cases where there is an inflection or not. It is easy to
recognize that if b < 0 , N tends towards 0 for t = ∞ .

In short:

*If the initial population is an equilibrium population, the population
remains constant. In all other circumstances the species will disappear, except
when: (a) two equilibrium populations exist and the initial amount of the population is greater than the stable equilibrium population or contained between the
stable equilibrium population and the unstable equilibrium population. In both
cases, the population tends towards the stable population and in the case where
the population lies between the two equilibrium populations and is close enough
to the unstable one, it can tend towards the stable population by changing the
trend of its increase; (b) only one equilibrium population exists smaller than the
initial population. In this case the population tends towards the equilibrium
population.*

FOOTNOTE

[1] Compare R. Pearl and S. L. Parker, Proc. Nat. Ac. of Sci., vol. 8 (1922).

The logistic law and its generalizations

V. A. Kostitzin

1. In this paper I shall consider certain generalizations of the biological and statistical law, which is improperly called the "logistic law". In its simplest form the logistic differential equation

$$(1) \qquad p' = np - mp - hp^2 = \varepsilon p - hp^2$$

represents the balance of a dense population, living in a habitat relatively rich in food, and in relatively stable conditions. This is a law of statistic origin, and its coefficients represent some averages both with respect to the population and with respect to time. This remark is very important: one often forgets that the coefficients, n representing natality, m representing mortality, and h representing a limiting factor, competition, are never constant. They are functions of time, t, and by giving the differential equation of the process in the form (1) with constant coefficients, one is forced to consider these coefficients as averages with respect to t.

2. Another aspect which is often forgotten is the role of the universe, external with respect to the population under study. This role appears implicitly in equations of the type (1), and in each term one could point to that part which comes from emigration, immigration, to the action of other populations in the same biotope, etc. One must then make sure each time that all such effects can in fact be expressed by terms of this order.

Suppose, for instance, that equation (1) expresses very well the growth of a certain population in a large biotope. This does not mean that if the biotope is subdivided into two parts one will have the same success applying equation (1), even with modified coefficients, to each sub-biotope separately, or to different groups which comprise its overall population. Nevertheless, one can cite in the demographic literature several examples of such errors.

3. <u>Variable vital coefficients.</u> In the case of variable vital co-efficients the solution of equation (1) can be given in the following form

$$(2) \qquad \frac{p_0}{p} = p_0 \int_0^t h(s) \, e^{-\int_s^t \varepsilon(u)du} \, ds + e^{-\int_0^t \varepsilon(u)du}$$

in which p_0 is the initial value of p . Suppose that one has observations during a time T , that during this time the coefficients do not change much, and that one introduces the averages

$$(3) \qquad \bar{\varepsilon} = \frac{1}{T} \int_0^T \varepsilon(u)du \, , \qquad \bar{h} = \frac{1}{T} \int_0^T h(u)du \, .$$

Equation (2) becomes

$$(4) \qquad \frac{p_0}{p} = e^{-\bar{\varepsilon}t} + \frac{p_0 \bar{h}}{\bar{\varepsilon}} (1 - e^{-\bar{\varepsilon}t}) \quad .$$

Can this formula be applied to the first half of the time interval $(0, T)$? It can if the functions $\varepsilon(u)$, $h(u)$ do not vary in a systematic manner; it cannot if the variations are systematic. And one must say that they often are.

4. <u>Periodic vital coefficients.</u> Consider a population which, through matings, births, etc., taking place at certain times of the year, has a seasonal regime. One must then apply formula (2) in which the coefficients $\varepsilon(u)$, $h(u)$ are periodic of period a . Let us, then, set

$$(5) \qquad \varepsilon(a+u) = \varepsilon(u) \, , \quad h(a+u) = h(u)$$

$$(6) \qquad E = \int_0^a \varepsilon(u)du$$

and introduce the annual cycle in time:

$$(7) \qquad\qquad t = na + z \quad (0 < z < a) .$$

Under this regime

$$(8) \qquad \int_0^t h(s)\, e^{-\int_s^t \varepsilon(u)du}\, ds = \int_0^z h(s)\, e^{-\int_s^z \varepsilon(u)du}\, ds$$

$$+ \frac{1-e^{-nE}}{e^E-1} \int_0^a h(s)\, e^{-\int_s^z \varepsilon(u)du}\, ds$$

and, therefore,

$$(9) \qquad \frac{p_0}{p(na+z)} = p_0 \int_0^z h(s)e^{-\int_s^z \varepsilon(u)du}\, ds$$

$$+ \frac{p_0(1-e^{-nE})}{e^E-1} \int_0^a h(s)e^{-\int_s^z \varepsilon(u)du}\, ds + e^{-nE-\int_0^z \varepsilon(u)du} .$$

Two very different cases can occur: 1) E is negative, and the population tends to disappear; 2) E is positive, and the population tends, at the limit, toward the periodic regime

$$(10) \qquad\qquad \lim_{n\to\infty} p(na+z) = q(z) .$$

The function q(z) is defined by the relation:

(11)
$$\frac{1}{q(s)} = \int_0^z h(s)\, e^{-\int_s^z \varepsilon(u)du}\, ds + \frac{1}{e^E - 1} \int_0^a h(s) e^{-\int_s^z \varepsilon(u)du}\, ds \; ;$$

which does not depend on the initial value, p_0 . One can express $p(t)$ through the intermediary $q(z)$ in the following way

(12)
$$\frac{p_0}{p(t)} = \frac{p_0}{p(na+z)}$$

$$= \frac{p_0}{q(z)} + e^{-nE - \int_0^z \varepsilon(u)du} \left[1 - \frac{p_0}{e^E - 1} \int_0^a h(s) e^{\int_0^s \varepsilon(u)du}\, ds \right] .$$

This formula makes it possible to compare the values of $p(t)$ at different phases of the cycle.

     5. <u>The effect of population density</u>.   Consider a population which is sparce, or which occupies a habitat with few resources. One can generalize equation (1) in a way which represents the growth of a population no matter what its density is. The term np in this equation theoretically represents the contribution of normal natality, proportional to the population size, p . The proportionality might seem obvious, but the reasoning upon which it is based breaks down when p is sufficiently small.

     Suppose, in fact, that the density of the population is very low, and that fruitful encounters between females and males follow time-consuming searches by both. It is clear that the principle of encounters should be applied to this very factor, and that the term expressing the corresponding contribution must be proportional to p^2 rather than to p .

     It is then natrual to look for an analytical form which would hold in all such cases. Let us consider the probability of fertilization, $\pi(p)$, and give to equation (1) the form

$$(13) \qquad p' = \sigma p \pi(p) - mp - hp^2 .$$

Through considerations analogous to the kinetic theory of gas (see e.g., Pearl, 1932), one can suppose that the number of favorable chances would be proportional to p , and that the number of unfavorable chances is constant. One finds in this way

$$(14) \qquad \pi(p) = \frac{\alpha p}{p+\beta}$$

and equation (13) takes the form

$$(15) \qquad p' = \frac{np^2}{p+\beta} - mp - hp^2 .$$

For sufficiently large p equation (15) approaches the logistic equation (1). For small p this reduces to

$$(16) \qquad p' = \tau p^2 - mp ,$$

which has been studied by Volterra (1938) and by myself (1937). The integration of (15) is not difficult, and I shall not carry it out here. This equation, however, has a property which is not found in the logistic equation. If the initial value p_0 is below a certain limit, $\theta = \frac{m}{\tau}$, the population does not survive. If, on the contrary, p_0 is larger than θ , the population tends toward a positive equilibrium value, which does not depend on p_0 . In this way a considerable ensemble of biological and demographic observations can be accounted for.

References

Kostitzin, V. A. Biologie mathématique. Paris: A. Colin, 1937. Pages 61-63.

Pearl, R. The influence of density of population upon egg production in Drosophila melanogaster. J. exp. Zoll. 63, 57 (1932).

Volterra, V. Population growth, equilibria, and extinction under specified breeding conditions: A development and extension of the thoery of the logistic curve. Human Biology 10, 1 (1938).

The physiological segregation and the variation of species

V. A. Kostitzin

1. The idea of natural selection is often countered with the fact that
different animal varieties coexist in a stable manner. Up to now observations
have not been able to solve this problem for the lack of sufficient time. The
same observations could always support different, and often contradictory interpretations. In such a situation mathematical reasoning has certain sure advantages: It is the only way to follow, and to work out the often distant consequences of the various assertions.

We shall examine the conditions for the coexistence of two varieties of
animals which differ only in the onset of sexual maturity. This difference must
be sufficiently large to greatly reduce or eliminate natural crossings, even if
artificial or accidental crossings are still possible. Suppose, for simplicity,
that one is dealing with insects which pass through only two phases: 1) an
immature phase (egg, larvae, nymphs), and 2) an adult phase (imagoes). This is
not at all an *ad hoc* hypothesis. To the contrary, our results will hold *a fortiori*
if the species has a more complex structure. We suppose, on the other hand, that
the imagoes disappear before winter. This hypothesis could also be removed with
no drawback other than that of more cumbersome computations.

2. To begin with let us recall the case of a single undifferentiated
group. Let p_n be the effective adult population in the n-th year (n-th generation), q_n the effective population of young resulting from it, t the duration
of the annual cycle measured at the time of appearance of each generation, τ
the time of appearance of the imagoes, $\sigma > \tau$ the time of oviposition, $\lambda = e^{\rho}$
the rate of oviposition, m the coefficient of adult mortality, μ the
coefficient of immature mortality, and h a limiting coefficient of the adults.
From the equations

$$(1) \qquad p_n' = - mp_n - hp_n^2 \qquad (\tau < t < \sigma) ,$$

$$(2) \qquad q_n' = - \mu q_n \qquad (\sigma < t < 1 + \tau) .$$

if follows that

$$(3) \qquad \frac{1}{p_n(t)} = \frac{1}{p_n(\tau)} e^{m(t-\tau)} + \frac{h}{m} (e^{m(t-\tau)} - 1) \qquad (\tau < t < \sigma) ,$$

$$(4) \qquad q_n(t) = q_n(\sigma) e^{-\mu(t-\sigma)} \qquad (\sigma < t < 1 + \tau) .$$

Now, oviposition is proportional to the corresponding effective adult population:

$$(5) \qquad q_n(\sigma) = e^{\rho} p_n(\sigma) ,$$

and the initial number of adults can be assumed to be the same as that of the surviving immatures produced by the previous generation:

$$(6) \qquad p_{n+1}(\tau) = q_n(1+\tau) ;$$

Setting $t = 1 + \tau$ in equation (4) one has:

$$(7) \qquad q_n(1+\tau) = q_n(\sigma) e^{-\mu(1+\tau-\sigma)} ,$$

and $p_n(\sigma)$ is given by (3) for $t = \sigma$:

$$(8) \qquad p_n(\sigma) = \frac{p_n(\tau) \, e^{-m(\sigma-\tau)}}{1 + \frac{h}{m} \, p_n(\tau) [1 - e^{-m(\sigma-\tau)}]} .$$

Combining these results the effective adult population is connected over two consecutive generations by

$$(9) \qquad \frac{p_{n+1}(\tau)}{p_n(\tau)} = \frac{e^{\rho-\mu(1+\tau-\sigma)-m(\sigma-\tau)}}{1 + \frac{h}{m} p_n(\tau)\left[1-e^{-m(\sigma-\tau)}\right]} \quad .$$

To begin with notice that the expression

$$(10) \qquad E = \rho - \mu(1 +\tau-\sigma) - m(\sigma-\tau)$$

plays here a role analogous to the rate of increase, ε, in the logistic equation, and in the familiar Lotka-Volterra equations. When E is negative the population tends to disappear. When E is positive the process tends to a finite limit value

$$(11) \qquad p_\infty(\tau) = \frac{m(e^E-1)}{h\left[1-e^{-m(\sigma-\tau)}\right]} \quad ,$$

so that the population is periodically renewed, and, in this sense, is at equilibrium. For this to occur it is necessary that the coefficient λ be considerably larger than unity:

$$(12) \qquad \lambda > e^{\mu(1+\tau-\sigma)+m(\sigma-\tau)} \gg 1 \ ,$$

that is to say, that each female must lay a large number of eggs. This result is very interesting. We have built a mathematical model which represents the life cycle of a group of animals, and the model shows that the animals can only survive if they produce a large number of eggs. Reason dictates, and nature obeys. Notice that it is very easy to experimentally verify the inequality (12).

3. <u>The case of two varieties</u>. Consider now the case of two varieties, and set

p_{n1}, p_{n2} — the effective number of adults with early and late maturity respectively,

q_{n1}, q_{n2} — the numbers of immatures,

σ_1, σ_2 — the times of oviposition,

m_1, m_2 — the adult mortalities,

μ_1, μ_2 — the immature mortalities,

h_1, h_2 — the limiting coefficients of the adults,

$\lambda_1 = e^{\rho_1}$,

$\lambda_2 = e^{\rho_2}$, — the proportionality coefficients of oviposition

τ — the time at which imagoes appear, common to both varities.

It might seem surprising that, in dealing with two varieties which only differ in the timing of sexual maturity,   we have introduced different vital coefficients. The reasons are fully explained in a previous paper of mine (1). It is enough to notice that oviposition almost always decreases the vitality of the imago, and that, because of the temporal shift between the two varities, it must be assumed that the coefficients of mortality and self limitation also differ. For the same reason the coefficients of proportionality $\lambda_1 = e^{\rho_1}$, $\lambda_2 = e^{\rho_2}$ must also differ.

The problem is described by the equations

(13)
$$\begin{cases} p'_{n1} = -m_1 p_{n1} - h_1 p_{n1} p_n , \\[2mm] p'_{n2} = -m_2 p_{n2} - h_2 p_{n2} p_n , \end{cases}$$

(14)
$$q'_{n1} = -\mu_1 q_{n1} , \quad q'_{n2} = -\mu_2 q_{n2}$$

with

(15)
$$p_n = p_{n1} + p_{n2}$$

Equations (13) give

(16)
$$\begin{cases} P_{n1}(t) = p_{n1}(\tau)\, e^{-m_1(t-\tau)-h_1 P_n(t)} \quad , \\[3mm] P_{n2}(t) = p_{n2}(\tau)\, e^{-m_2(t-\tau)-h_2 P_n(t)} \quad , \end{cases}$$

setting

(17)
$$P_n(t) = \int_\tau^t p_n(s)\,ds \ .$$

The sum of equations (16) gives a differential equation in $P_n(t)$:

(17)
$$P_n' = p_{n1}(\tau)e^{-m_1(t-\tau)-h_1 P_n} + P_{n2}(\tau)e^{-m_2(t-\tau)-h_2 P_n} \ .$$

This equation can be integrated in three special cases:

    1) when the limiting coefficients are equal:

(18)
$$h_1 = h_2 = h \ ,$$

    2) when the coefficients of adult mortality are equal:

(19)
$$m_1 = m_2 = h \ ,$$

    3) when the following proportionality holds:

(20)
$$\frac{m_1}{h_1} = \frac{m_2}{h_2} = V \ .$$

We are going to see that these three hypotheses, although different, give qualitatively similar results. From this one can conclude that this would also apply to the general case. Since we are not interested in direct numerical applications, it is enough to consider the special cases.

4. **Equal limiting coefficients.** Setting $h_1 = h_2 = h$ in (17), one has

(21)
$$e^{hP_n} P'_n = p_{n1}(\tau)e^{-m_1(t-\tau)} + p_{n2}(\tau)e^{-m_2(t-\tau)} ,$$

which, on integration, gives

(22)
$$e^{hP_n} = 1 + \frac{hp_{n1}(\tau)}{m_1}\left[1 - e^{-m_1(t-\tau)}\right] + \frac{hp_{n2}(\tau)}{m_2}\left[1 - e^{-m_2(t-\tau)}\right] .$$

On the other hand, by integrating equations (14) one gets

(23)
$$\begin{cases} q_{n1}(t) = q_{n1}(\sigma_1)e^{-\mu_1(t-\sigma_1)} , \\[2mm] q_{n2}(t) = q_{n2}(\sigma_2)e^{-\mu_2(t-\sigma_2)} . \end{cases}$$

One can now consider the chain of relations

(24)
$$\begin{cases} p_{n+1,1}(\tau) = q_{n1}(1+\tau) = q_{n1}(\sigma_1)e^{-\mu_1(1+\tau-\sigma_1)} , \\[2mm] p_{n+1,2}(\tau) = q_{n2}(1+\tau) = q_{n2}(\sigma_2)e^{-\mu_2(1+\tau-\sigma_2)} \end{cases}$$

(25)
$$q_{n1}(\sigma_1) = e^{\rho_1}p_{n1}(\sigma_1), \; q_{n2}(\sigma_2) = e^{\rho_2}p_{n2}(\sigma_2) ,$$

(26)
$$\begin{cases} p_{n1}(\sigma_1) = p_{n1}(\tau)e^{-m_1(\sigma_1-\tau)-hP_n(\sigma_1)} , \\[2mm] p_{n2}(\sigma_2) = p_{n2}(\tau)e^{-m_2(\sigma_2-\tau)-hP_n(\sigma_2)} , \end{cases}$$

from which one derives the required relationship between the initial sizes of the two consecutive generations:

(27)
$$
\begin{cases}
\dfrac{P_{(n+1),1}(\tau)}{P_{n1}(\tau)} = e^{E_1 - hP_n(\sigma_1)} \\[4ex]
\dfrac{P_{(n+1),2}(\tau)}{P_{n2}(\tau)} = e^{E_2 - hP_n(\sigma_2)} .
\end{cases}
$$

As in the case of a single group, the expressions E_1 and E_2 play the roles of coefficients of increase. If one or the other of these expressions is negative, the corresponding variety will tend to zero. Therefore, if one is looking for an equilibrium state, both E_1 and E_2 must be positive, and λ_1, λ_2 must be much larger than unity. Assume, then, that

(28)
$$
E_1 > 0 \quad , \quad E_2 > 0
$$

or, which amounts to the same

(29)
$$
\lambda_1 > e^{m_1(\sigma_1 - \tau) + \mu_1(1 + \tau - \sigma_1)} \gg 1 ,
$$
$$
\lambda_2 > e^{m_2(\sigma_2 - \tau) + \mu_2(1 + \tau - \sigma_2)} \gg 1 .
$$

Under these conditions the equations (27) admit three limit points:

(30)
$$
\lim p_{n1}(\tau) = 0; \quad \lim p_{n2}(\tau) = \frac{e^{E_2} - 1}{1 - e^{-(\sigma_2 - \tau)m_2}} \cdot \frac{m_2}{h} ,
$$

(31)
$$
\lim p_{n1}(\tau) = \frac{e^{E_1} - 1}{1 - e^{-\mu_1(\sigma_1 - \tau)}} \cdot \frac{\mu_1}{h} ; \quad \lim p_{n2}(\tau) = 0 ,
$$

and the third point which is defined by the linear equations

$$(32) \quad \begin{cases} u_1\left[1-e^{-m_1(\sigma_1-\tau)}\right] + u_2\left[1-e^{-m_2(\sigma_1-\tau)}\right] = e^{E_1} - 1 , \\\\ u_1\left[1-e^{-m_1(\sigma_2+\tau)}\right] + u_2\left[1-e^{-m_2(\sigma_2-\tau)}\right] = e^{E_2} - 1 , \end{cases}$$

having set

$$(33) \quad u_1 = \frac{h}{m_1} \lim p_{n1}(\tau) \quad , \quad u_2 = \frac{h}{m_2} \lim p_{n2}(\tau) \quad .$$

The fixed point (32) is admissable in two cases:

Either

$$(34) \quad \begin{cases} \dfrac{e^{E_1} - 1}{1-e^{-m_1(\sigma_1-\tau)}} > \dfrac{e^{E_2} - 1}{1-e^{-m_1(\sigma_2-\tau)}} \quad , \\\\ \dfrac{e^{E_1} - 1}{1-e^{-m_2(\sigma_1-\tau)}} < \dfrac{e^{E_2} - 1}{1-e^{-m_2(\sigma_2-\tau)}} \quad , \end{cases}$$

or

$$(35) \quad \begin{cases} \dfrac{e^{E_1} - 1}{1-e^{-m_1(\sigma_1-\tau)}} < \dfrac{e^{E_2} - 1}{1-e^{-m_1(\sigma_2-\tau)}} \quad , \\\\ \dfrac{e^{E_1} - 1}{1-e^{-m_2(\sigma_1-\tau)}} > \dfrac{e^{E_2} - 1}{1-e^{-m_2(\sigma_2-\tau)}} \quad . \end{cases}$$

It is easy to see that this fixed point is only stable under conditions (34). In case (35), at the limit, either the fixed point (30) or the fixed point (31) is attained, that is to say that one or the other of the two varieties will finally disappear. Whether the process will tend toward one or the other of these two limit states depends on the initial conditions.

It would be interesting to examine the probability of the limit states (32), namely of the stable coexistence of the two varieties. Assuming that all the values of the coefficients are equally probable (which is certainly wrong) the probability in question would be 1/4. This probability must in fact be smaller, and, therefore, the local splitting of a species into two coexisting groups must be rather rare.

5. Equal coefficients of adult mortality. Now let $h_1 \neq h_2$, $m_1 = m_2 = m$. Equation (17) becomes

(36)
$$\frac{P_n'}{p_{n1}(\tau)e^{-h_1 P_n} + p_{n2}(\tau)e^{-h_2 P_n}} = e^{-m(t-\tau)} \quad ,$$

and its integral is

(37)
$$1 - e^{-m(t-\tau)} = m \int_0^{P_n(t)} \frac{ds}{p_{n1}(\tau)e^{-h_1 s} + p_{n2}(\tau)e^{-h_2 s}} \quad .$$

The inversion of this definite integral makes it possible to derive P_n as a function of t . On the other hand, the equations (16), (23-26) still hold, and one derives from them relationships between the initial adult populations in the two consecutive generations:

(38)
$$\frac{p_{n+1,1}(\tau)}{p_{n1}(\tau)} = e^{E_1 - h_1 P_n(\sigma_1)} \quad , \qquad \frac{p_{n+1,2}(\tau)}{p_{n2}(\tau)} = e^{E_2 - h_2 P_n(\sigma_2)} \quad ,$$

having set, in this case,

$$(39) \quad \begin{cases} E_1 = \rho_1 - m(\sigma_1 - \tau) - \mu_1(1 + \tau - \sigma_1) \, , \\[2em] E_2 = \rho_2 - m(\sigma_2 - \tau) - \mu_2(1 + \tau - \sigma_2) \, . \end{cases}$$

Suppose that these expressions are positive. In the case of coexistence the limit values

$$v_1 = \lim_{n \to \infty} p_{n1}(\tau) \, , \quad v_2 = \lim_{n \to \infty} p_{n2}(\tau)$$

satisfy the equations

$$E_1 = h_1 \lim_{n \to \infty} p_n(\sigma_1) \, , \qquad E_2 = h_2 \lim_{n \to \infty} p_n(\sigma_2) \, ,$$

which can be easily reduced to the following relations

$$(40) \quad \begin{cases} F_1(v_1, v_2) = m \int_0^{\frac{E_1}{h_1}} \frac{ds}{v_1 e^{-h_1 s} + v_2 e^{-h_2 s}} - \left[1 - e^{-m(\sigma_1 - \tau)} \right] = 0 \, , \\[3em] F_2(v_1, v_2) = m \int_0^{\frac{E_2}{h_2}} \frac{ds}{v_1 e^{-h_1 s} + v_2 e^{-h_2 s}} - \left[1 - e^{-m(\sigma_2 - \tau)} \right] = 0 \, . \end{cases}$$

F_1 and F_2 are decreasing functions of v_1 and v_2. For very small values of these variables F_1 and F_2 are very large and positive. For very large values of v_1 and v_2 the functions F_1 and F_2 are negative, and they tend to finite limits. One can easily show that there is a positive point of intersection if either

$$(41) \quad \begin{cases} \dfrac{e^{E_1}-1}{1-e^{-m(\sigma_1-\tau)}} > \dfrac{e^{\frac{h_1 E_2}{h_2}}-1}{1-e^{-m(\sigma_2-\tau)}} \,, \\[3em] \dfrac{e^{\frac{h_2 E_1}{h_1}}-1}{1-e^{-m(\sigma_1-\tau)}} < \dfrac{e^{E_2}-1}{1-e^{-m(\sigma_2-\tau)}} \,, \end{cases}$$

or the reverse inequalities are satisfied. A stable coexistence is only possible in the case (41). In the other case either of the two varieties will eventually disappear. The tendency toward one or the other of these limit states depends on the initial condition.

6. Proportional coefficients. When the adult mortalities and the limiting coefficients satisfy the proportionality (20), equation (17) becomes

$$(42) \qquad x_n' = p_{n1}(\tau)e^{-h_1 x_n} + p_{n2}(\tau)e^{-h_2 x_n} + v \,,$$

setting

$$(43) \qquad x_n = v(t-\tau) + \int_\tau^t p_n \, ds \,.$$

By integration one finds

$$(44) \qquad t - \tau = \int_0^{x_n(t)} \frac{ds}{p_{n1}(\tau)e^{-h_1 s} + p_{n2}(\tau)e^{-h_2 s} + v} \,,$$

and, as in the previous case, $x_n(t)$ is computed by inverting this integral.

The relationship between the initial adult sizes of two consecutive generations takes the form

$$(45) \quad \begin{cases} \dfrac{P_{n+1,1}(\tau)}{P_{n1}(\tau)} = e^{\rho_1-\mu_1(1+\tau-\sigma_1)-h_1 x(\sigma_1)} \; , \\[4mm] \dfrac{P_{n+1,2}(\tau)}{P_{n2}(\tau)} = e^{\rho_2-\mu_2(1+\tau-\sigma_2)-h_2 x(\sigma_2)} \end{cases}$$

and the limit point is at the intersection of the two curves

$$(46) \quad \begin{cases} \rho_1-\mu_1(1+\tau-\sigma_1) - h_1 x(\sigma_1) = 0 \; , \\[4mm] \rho_2-\mu_2(1+\tau-\sigma_2) - h_2 x(\sigma_2) = 0 \; . \end{cases}$$

One can give to these equations the form

$$(47) \quad \begin{cases} F_1(v_1,v_2) = \displaystyle\int_0^{\frac{\rho_1-\mu_1(1+\tau-\sigma_1)}{h_1}} \dfrac{ds}{v+v_1 e^{-h_1 s} + v_2 e^{-h_2 s}} - (\sigma_1-\tau) = 0 \; , \\[8mm] F_2(v_1,v_2) = \displaystyle\int_0^{\frac{\rho_2-\mu_2(1+\tau-\sigma_2)}{h_2}} \dfrac{ds}{v+v_1 e^{-h_1 s} + v_2 e^{-h_2 s}} - (\sigma_2-\tau) = 0 \; . \end{cases}$$

As in the preceding case, the study of these functions shows that the two varieties can coexist if certain inequalities are satisfied. If not, one or the other of the two varieties will disappear.

These examples show that the physiological segregation can produce a stable state of mutual isolation between two very close varieties of the same species. One could as well show that the same phenomenon may occur in the case of other forms of reproductive isolation. Genetic variability will then transform such two close varieties into two well differentiated species.

References

1. Kostitzin, V. A. Sur la loi logistique et ses généralistions. Acta Biotheoretica 5, 155 (1940).

Comments on the note by Mr. Régnier and Miss Lambin:

Study of a case of microbial competition (Bacillus coli-Staphylococcus aureus)[1]

V. Volterra

If we consider the case of a single bacterial species, we see that its growth does not follow Pearl's law of the logistic curve. When each species is alone the number of individuals reaches a maximum at a certain time, and then it declines. The logistic curve corresponds to a population in which the resource available to each individual decreases as their numbers increase. We might assume that the form of growth observed by the authors depends on the accumulation of toxic substances in the culture medium, or on analogous factors resulting from their metabolism. I have tried to work out a calculation based upon this hypothesis.

It can be supposed that the intoxication of the cultural medium at time t, owing to the presence of $N(\tau)$ bacteria during the time interval $(\tau, \tau + d\tau)$, is given by

$$N(\tau)f(t-\tau)d\tau$$

In this way the growth rate of the species at time t is decreased by the amount

$$\int_0^t N(\tau)f(t-\tau)d\tau$$

owing to the presence of bacteria in the medium during the time interval $(0, t)$.

One would then have the integro-differential equation

$$\frac{dN(t)}{dt} = \left[\varepsilon - hN(t) - \int_0^t N(\tau)f(t-\tau)d\tau\right] N(t)$$

in which ε is the gross rate of growth, and $-hN(t)$ a term corresponding to the Pearl effect.

The preceding equation can be simplified by neglecting this last term, and by assuming that $f = c$ is constant. Setting $\int_0^t N(\tau)d\tau = n(t)$, the integro-differential equation is reduced to the differential equation

$$\frac{d^2 n(t)}{dt^2} = (\varepsilon - cn) \frac{dn}{dt} .$$

By integration this gives

$$N = \frac{N_0 (\alpha + \beta)^2 e^{\lambda t}}{(\beta e^{\lambda t} + \alpha)^2}$$

in which N_0 is the initial number of bacteria, and α , $-\beta$ are the roots of the equations

$$\frac{1}{2} cx^2 - \varepsilon x - N_0 = 0 , \quad \lambda = \frac{1}{2} c (\alpha + \beta) .$$

The maximum, $\dfrac{c}{8(\alpha + \beta)^2}$, is attained at the time $\frac{1}{\lambda} \ell n (\alpha / \beta)$.

In the case of two coexisting species one obtains the system of integro-differential equations

$$\frac{dN_1(t)}{dt} = \left\{ \varepsilon_1 - \left[h_{11}N_1(t) + h_{12}N_2(t) \right] - \int_0^t \left[N_1(\tau)f_{11}(t-\tau) + N_2(\tau)f_{12}(t-\tau) \right] dt \right\} N_1(t)$$

$$\frac{dN_2(t)}{dt} = \left\{ \varepsilon_2 - \left[h_{21}N_1(t) + h_{22}N_2(t) \right] - \int_0^t \left[N_1(\tau)f_{21}(t-\tau) + N_2(\tau)f_{22}(t-\tau) \right] dt \right\} N_2(t)$$

under the same assumptions as in the previous case. This reduces to the differential equations

$$\frac{d^2 n_1}{dt^2} = \left[\varepsilon_1 - c_{11} n_1(t) - c_{12} n_2(t) \right] \frac{dn_1}{dt}$$

$$\frac{d^2 n_2}{dt^2} = \left[\varepsilon_2 - c_{21} n_1(t) - c_{22} n_2(t) \right] \frac{dn_2}{dt}$$

whose integral is

$$c_{21} \frac{dn_1}{dt} + c_{12} \frac{dn_2}{dt} = c_{21} \varepsilon_1 n_1 + c_{12} \varepsilon_2 n_2 - c_{11} c_{21} \frac{n_1^2}{2}$$

$$- c_{12} c_{21} n_1 n_2 - c_{12} c_{22} \frac{n_2^2}{2} + K$$

in which K is a constant, and n_1 and n_2 are positive increasing quantities.

N_1 and N_2 increase toward maxima which they respectively reach when $\varepsilon_1 = c_{11} n_1 + c_{12} n_2$ and $\varepsilon_2 = c_{21} n_1 + c_{22} n_2$. Thereafter they decline. c_{12} represents the toxic action of the second species on the first, and c_{21} that of the first on the second. These coefficients can be approximately evaluated from data on the growth curves of the two species.

References

1. Régnier, J., and S. Lambin. Etude d'un cas d'antagonisme microbien (Bacillus coli - Staphylococcus auereus). Comptes rendus de l'Ac. des Sciences 199, 1682 (1934).

The integro-differential equations for the toxic contamination of a medium

V. A. Kostitzin

1. **General equations.** Volterra's integro-differential equation

$$(1) \qquad p' = \varepsilon p - hp^2 - cp \int_0^t p(\tau)k(t-\tau)dt$$

represents the growth of a population which is intoxicated by its own catabolic products. It differs from the logistic by the presence of the integral term. By introducing this term one implicitly assumes that the increase of catabolic products in the medium is, at any one time t , proportional to the population size p(t). This hypothesis is a legitimate one as long as one deals with the catabolic products produced by living individuals. Besides this, however, one must also take into account the permanence of dead individuals in the medium. Now the number of individuals which die in the time interval $(\tau,\tau+d\tau)$ is not proportional to $p(\tau)$. Non-linear factors such as, for instance, competition affect both the birth and the death rates.[1]

Let a(t)dt be the increase of the population due to reproduction, and b(t)dt the decline due to mortality.   One easily finds the following expressions for these functions:

$$(2) \qquad \begin{cases} a(t) = p\left[n-\nu p-\gamma \int_0^t k(t-u)p(u)du - \lambda \int_0^t g(t-u)b(u)du\right] , \\[2em] b(t) = p\left[m+\mu p+\delta \int_0^t k(t-u)p(u)du + \eta \int_0^t g(t-u)b(u)du\right] . \end{cases}$$

In these formulas the first integral term expresses the toxic action of catabolic products, and the second corresponds to the toxic action of dead individuals.   One obtains $p'(t)$ by subtracting the second equation from the first, that is,

$$(3) \qquad p' = p\left[\varepsilon - hp - c \int_0^t k(t-\tau)p(u)du - \lambda \int_0^t g(t-\tau)b(u)du\right]$$

in which $\varepsilon = n - m$, $h = \nu + \mu$, $c = \gamma + \delta$, $\lambda = \xi + \eta$. Equations (2_2) and (3) are the system which allows us to calculate the functions $b(t)$ and $p(t)$.

2. <u>Integrable case</u>. Suppose that the two residual actions are constant, $g(t) \equiv 1$, $k(t) \equiv 1$, and set

$$P = \int_0^t p(s)ds \quad , \qquad B = \int_0^t b(s)ds \quad .$$

The equations (2) and (3) become

$$(4) \qquad a(t) = p(n-\nu p-\gamma P-\xi B)$$

$$(5) \qquad \left\{ \begin{array}{l} b(t) = B'(t) = p(m+\mu p+\delta P+\eta B) \ , \\[2em] p' = p(\varepsilon-hp-cP-\lambda B) \ . \end{array} \right.$$

Let us introduce a new independent variable $z = P(t)$; equations (5) then become linear:

$$(6) \qquad \frac{dp}{dz} = \varepsilon-cz-hp-\lambda B \quad ; \qquad \frac{dB}{dz} = m+\delta z+\mu p+\eta B \ .$$

Denoting by ρ_1 , ρ_2 the roots of the characteristic equation,

$$(7) \qquad \rho^2 + \rho(h-\eta) + \lambda\mu - \eta h = 0 \quad ,$$

the solutions of system (6) can be expressed in the form

$$
(8) \quad
\left\{
\begin{array}{l}
p = F(z) = C_0 + C_1\, e^{\rho_1 z} + C_2\, e^{\rho_2 z} + C'z \, , \\[2em]
B = \Phi(z) = D_0 + D_1\, e^{\rho_1 z} + D_2\, e^{\rho_2 z} + D'z \, .
\end{array}
\right.
$$

The initial conditions and the substitution allows to calculate their coefficients. The time t is determined by integration as a function of P :

$$
(9) \qquad t = \int_0^P \frac{dz}{F(z)} \, ,
$$

which completely solves the problem.

3. <u>Direct enumerations</u>. In a direct assessment of the number of individuals in a bacterial population it is practically impossible to distinguish, in the total number, dead from living individuals. Let $q(t)$ be this number. Suppose that of the $b(\tau)d\tau$ individuals which died in the time interval $(\tau,\tau+d\tau)$, $b(\tau)\, \ell(t-\tau)$ are still distinguishable as individual units at the time t . Under these conditions the toal number, $q(t)$, is given by

$$
(10) \qquad q(t) = p(t) + \int_0^t \ell(t-\tau)b(\tau)d\tau \, .
$$

Suppose that $p(t)$ tends to zero. If $p(t)$ and $q(t)$ pass through some maxima, the first maximum of $p(t)$ will precede the first maximum of $q(t)$. This theoretical prediction is fully confirmed by observations.[2]

Another interesting question is posed by the limit of $q(t)$, for which there arise two cases. It can happen that the limit of $\ell(t)$ is finite so that $q(t)$ will also tend toward a finite limit. It can also happen, and this is much more likely, that the limit of $\ell(t)$ is zero. In this case the limit of

q(t) is also zero while the ratio q(t)/p(t) tends to a limit which is larger than
1 . In particular, when p and B are given by the formulas (8), one finds for
the limit

$$(11) \qquad \lim_{t \to \infty} \frac{q(t)}{p(t)} = 1 + \Phi'(P_\infty) \int_0^\infty \ell(\tau) \, e^{-\tau F'(P_\infty)} \, dt$$

in which P_∞ denotes the limit of P(t) , and it is naturally assumed that the
integral is finite. The observations fully confirm the theoretical predictions.

References

1. Kostitzin, V. A. Biologie mathématique. Paris: A. Colin, 1937.

2. See page 100 in the Thesis of A. Kaplan-Brille. Influence du nombre des
    microbes etc., Paris 1931.

Comments on the toxic action of the medium relative to the note by Mr. Régnier and Miss Lambin[1]

V. Volterra and V. A. Kostitzin

In a note on microbial competition[2] Regnier and Lambin have studied the development of *Staphylococcus aureus* and *Bacillus coli* in a liquid medium (meat broth with peptone and salt), both in pure and mixed cultures. In the first case the population follows a pattern of growth completely different from that of Verhulst-Pearl, passing through a maximum, and then declining very rapidly. In a comment to this note, and in a volume in collaboration with D'Ancona[3], Volterra has attempted to explain this phenomenon through the action of toxic substances accumulating in the medium from the metabolism of the bacterial population.

From the mathematical point of view the problem is reduced to an integro-differential equation which, in the case of constant intoxication, is reduced to a differential equation. Volterra has given an approximate solution which qualitatively follows the experimental curve rather well. Kostitzin[4] has further developed the problem through an exact solution which is also in good agreement with observation. He has also solved a more general problem.

In a new note[1] Regnier and Lambin studied the development of a *Bacillus coli* population in a saline medium containing, in different experiments, different proportions of peptone. When there is no peptone at all one observes an extinction of the population which is very rapid, *especially at the onset*. A minimal concentration of nutrients is, however, enough to promote population growth with a very characteristic maximum, followed by a decline. When the position of these maxima is plotted as a function of the initial concentration of peptone one gets a curve very similar to the logistic.

Volterra and Kostitzin have tried to get an idea of the form of this curve, starting with Kostitzin's formulas. These formulas predict a maximum, p_m , as a function of the vital coefficients: ε (fission), h (mutual actions, competition), c (toxic action). One can suppose *a priori* that fission should increase with the concentration of the nutrient, q , and that, on the contrary, it should decrease from competition and toxic action. In the following table we summarize the numerical calculations by Kostitzin:

q	0.001	0.005	0.020	0.050	0.100	0.25	0.5	1.0	2.5	5.0	10	20
ε	0.33	0.37	0.46	0.63	0.93	0.99	1.07	1.13	1.18	1.23	1.24	1.24
$10^3 h$ ...	144	93	39	30	24	12.3	7.3	5.0	3.9	3.4	3.3	3.1
$10^5 c$ ...	430	370	78	93	60	31	11.7	6.7	4.6	4.1	3.3	3.2
$10^{-6} p_m$ ..	1.8	3.0	16.7	17.2	35.8	70.0	134.0	214.0	286.0	348.0	363.5	380.6

They show that the behaviors of the coefficients correspond to the theoretical predictions. One sees, on the other hand, that for high concentrations of peptone, the coefficients ε, h, c have a clearly asymptotic behavior. This is very well explained by the existence of an upper physiological limit for the assimilation of the nutrient, as well as for the speed of growth and cellular division. Therefore, assuming that all the coefficients are functions of the concentration of the nutrient, q , one would have the following approximate limits:

$$\lim_{q\to\infty} \varepsilon = 1.2 \quad , \qquad \lim_{q\to\infty} h = 0.003 \quad , \qquad \lim_{q\to\infty} c = 0.00003 \; .$$

This hypothesis is also relevant to the figures in the first note by Regnier and Lambin[2]. Taking into account the different census techniques in the two series of experiments, one finds the following asymptotic values of the coefficients:

$$\varepsilon \; = \; 1.1 \; , \qquad h = 0.003 \; , \qquad c = 0.00003 \; .$$

Now, the concentration of nutrients in the first series of experiments by Regnier and Lambin was certainly higher than 20. The agreement is perfect.

Consider now the formula

$$p_m = \frac{\varepsilon}{h} - \frac{c}{h^2} \ln \left(1 + \frac{\varepsilon h}{c} - p_0 \frac{h^2}{c} \right)$$

which expresses the maximum, p_m , as a function of ε, h, c. It is clear that p_m , considered as a function of q , necessarily tends toward a finite limit, in very good agreement with the observations. Volterra's hypothesis of cumulative toxic action accounts very well for the new experiments by Régnier and Lambin.

References

1. Régnier, J., and S. Lambin. Étude sur le croit microbien en fonction de la quantité de substance nutritive des milieux de culture. Comptes rendus d l'Ac. des Sciences 207, 1263 (1938).

2. Régnier, J., and S. Lambin. Étude d'un cas d'antagonisme microbien (Bacillus coli - Staphylococcus aureus). Comptes rendus de l'Ac. des Sciences 199, 1682 (1934).

3. Volterra, V., and U. D'Anconna. Les associations biologiques au point de vue mathématique. (Actualités Scientifiques et Industrielles, 243). Paris: Hermann et Cie, 1934.

4. Kostitzin, V. A. Sur l'intoxication d'un milieu par les produits cataboliques d'une population. Comptes rendus de l'Ac. des Sciences 201, 516-518 (1935).

   Kostitzin, V. A. Sur une généralisation des équations biologiques dans le cas d'une population intoxiquée par les produits de son activité chimique. Comptes rendus de l'Ac. des Sciences 204, 1683 (1937).

   Kostitzin, V. A. Biologie mathématique. Paris: A. Colin, 1937.

PART II

COMPETITION AND PREDATION

INTRODUCTION

Volterra wrote several monographs on the interactions within and among species. Most of his many papers fall into two broad categories--either they are mathematically technical treatments or they are popular accounts in the context of empiracal observations. Among the latter only "Variazioni e fluttuazioni del numero d'individui in specie animali conviventi" (1926), a very elementary review, has been translated into English (Chapman, 1931). Perhaps because this paper is so widely cited, it has contributed to a false impression that Volterra had produced only minor extensions and mathematical refinements of Lotka'a treatment of interacting populations. The more elementary papers may also convey the impression that mathematical considerations were of only marginal value. It is for these reasons that we have not chosen to include such excellent popular reviews as "Applications des mathématiques à la biologie" (1937), or "Les associations biologiques au point de vue mathématique" (1935) written with D'Ancona.

Most of Volterra's original research papers are very technical and dry, especially for biologists. Much the same applies to the notes of his "Lecons sur la théorie mathématique de la lutte pour la vie" (1931b) compiled by Brelot.

The major monograph which we have translated, "Variazioni e fluttuazioni del numero d'individui in specie animali conviventi" (1927), has the same title as Volterra's elementary 1926 review. However, the later work contains a number of most interesting and novel approaches which did not appear in the earlier review. Among them are n-species models in which the vital coefficients are either functions of the numbers or periodic functions of time, and integro-differential predator-prey models with delayed effects of the interactions.

This monograph is, in our opinion, his only general work which strikes a happy compromise between the two extremes discussed above. It is also didactically useful since it starts by treating only two species cases through quadratic models applied to competition and predation. In this form of competition for a single limiting resource, the species which utilizes it with higher efficiency is the only one which eventually survives. The predation model is the same as that developed independently by Lotka, but Volterra's treatment is more thorough. Volterra worked out the

behavior of oscillations of any amplitude, and proved that the averages of the
numbers over one period are the same as the equilibrium values. One of the most
interesting consequences of this model is the "principle of the perturbation of the
averages." If both the predator and the prey are removed at a constant rate in
proportion to their numbers, the average number of the prey species will tend, up
to a point, to increase, and that of the predator will tend to decrease.

Part I of Volterra's monograph continues by examining the ten distinct cases
which arise according to the possible signs of the constant parameters in the more
general quadratic model of the interactions between two species. Part I concludes
by studying the upper bound below which an increase in exploitation of a predator-
prey system will increase the population of the prey. This effect has important
implications for biological control, and its robustness is verified by later, more
general, treatments.

Part II considers more general models in the case of any number of species.
In the case of pure competition for a single limiting resource the result is much
the same as for two species: that species with the highest efficiency in utilizing
the resource will be the only one to survive. In this model the effect of
competition is taken into account by a very general function of the numbers.

The case of pure predation is an extension to n species of Lotka's equations,
and leads to results which are similar to those obtained for only two species. The
major difference is that stable coexistence is possible only for an even number of
species, and that the undamped oscillations are not periodic. Volterra emphasized
that much the same treatment also holds when the "vital coefficients" are general
functions of the numbers, rather than being constants. The pecularities of such
systems, called <u>conservative</u>, led Volterra to consider them to be of very limited
practical value.

Within the very restrictive domain of quadratic, differential equation models,
those which include competition as well as predation should be somewhat more
"realistic." Such models were termed <u>dissipative</u> by Volterra. The conditions for
the coexistence of the n species are essentially determined by a set of n algebraic
inequalities of degree n, n - 1, ...1 involving the coefficients of interaction.

This shows that coexistence is a precarious affair, which becomes more so the higher the number of species. When n is high any perturbation resulting in a change of the coefficients is likely to upset the balance, and lead to the extinction of several species. This may perhaps "explain" why, in the more stable climate of the tropics, ecosystems contain very high numbers of species. On the other hand, in the more variable temperate climates each single ecosystem tends to be much less diverse, and a major component of diversity is that among ecosystems.

Part III extends the treatment of conservative systems to the case of periodic variations in the vital coefficients, as in the cycle of the seasons. The problem of extinction and the consequences of the introduction of a new species are also considered. The case of an association of three species is worked out in detail in an appendix.

Part IV begins with a discussion of the inadequacies of differential equation models for biological associations. By definition such models cannot account for delayed effects, satiation, and a number of other biological realities. To account for the delayed effects of feeding on reproduction in a predator-prey association Volterra introduced a system of two integro-differential equations under the admittedly unsatisfactory restriction of a constant age distribution. The system leads to a stable, aperiodic oscillation within precise bounds. In this case too a uniform removal of individuals of both species has the general effect of increasing the average population of the prey, and of decreasing that of the predator.

A number of special considerations not included in the first edition of this monograph are reported in an appendix added to a 1962 re-edition. Particularly interesting among them is a further peculiarity of conservative systems. They enjoy a "time reversal" property by which a system with reversed signs for the growth and interaction constants behaves as a mirror image of the original system with respect to time.

Among the many papers by Volterra which refine the study of conservative and dissipative associations, we choose to follow with the second part of his "Principes de biologie mathématique" (1937b). The first part of this paper overlaps broadly with his earlier monograph (1927), and only certain major points of reference have

been quoted from it. This paper is particularly interesting since it contrasts
more clearly than the 1927 monograph the limited biological significance of
conservative models with the wealth of general indications they can still provide.

The conservative property implies even more severe restrictions on the
interaction parameters than those discussed in the 1927 monograph. If one considers
the total quantity of life in a conservative ecosystem this is affected only in an
illusory way by the interactions among species. An appropriate change in the scale
by which the different species are "weighted" makes the contribution of the
interactions to the total quantity of life disappear. This paper also deals with a
refinement of the concept of predator and prey in the case of more than two species.
It distinguishes among species which are only preyed upon, those which only prey
on others, and those which both prey on some species and are preyed upon by others.
This distinction makes possible a more general statement of the principle of the
perturbation of the average than was presented in Volterra (1927).

Most of the second half of this paper deals with the mathematical properties
which underlie conservative systems. As in the first part of the present volume,
in the much simpler case of the Verhulst-Pearl equation, these systems can also
be derived from variational principles. Their laws of motion correspond in fact to
a minimum of a quantity Volterra called "vital action." They can be handled with
much the same canonical treatment as in the mechanics of material systems.

We conclude Volterra's contribution to this volume with a short paper originally
in English on models for the delayed effects of interactions: "The general
equations of biological strife in the case of historical actions" (1939a). It
summarizes, in simple form, the major results obtained for this case in a number
of previous papers.

This collection of papers leaves out several original contributions by Volterra
to mathematical ecology. Among them, particularly worthy of mention is the
treatment of age specific competition in species with two or more life stages,
including the case of cannibalism. These treatments may be found, among other
places, in the 1935 monograph jointly written with D'Ancona, and in D'Ancona's
review (1954, in English). The reader is also referred to Brelot's (1931) paper on

integro-differential models of predator-prey systems for additional developments.

The fourth paper, by Lotka (1932), extends the quadratic model of competition between two species to account for more than one limiting resource. By introducing two different, linear functions of exploitation of the continuously renewed resources, the coexistence of the two species becomes possible. This qualitative result is more frequently attributed to Gause on the basis of his 1934 book. These works were seminal to the controversial concepts of "competitive exclusion," "ecological displacement," and "niche width."

The fifth paper, originally in Italian, by A. Kolmogoroff, has also a great seminal importance. Kolmogoroff uses Lotka's original point of departure, namely the general differential equations of the form (1936, cf. also 1972):

$$\frac{dN_1}{dt} = K_1(N_1, N_2) \cdot N_1$$

$$\frac{dN_2}{dt} = K_2(N_1, N_2) \cdot N_2 .$$

Instead of dealing with simple analytical forms of the functions, as Lotka liked to do, Kolmogoroff specified for them only a few qualitative conditions such as the signs of the slopes and the values at some limit or critical point, which describe a predator-prey interaction. These conditions take into account a number of "realistic" effects, such as satiation. By direct, qualitative methods Kolmogoroff showed that, besides a node or a focus, limit cycles are also among the most likely behaviors. This possibility was already implicit in the general theory of differential equations, but this is the first paper in which the existence of limit cycles is explicitly pointed out in an ecological context. An excellent abridged version of this paper, in English, has already been provided in Rescigno and Richardson (1967) in discussing analogous approaches to competition.

The last paper of this section, by Kostitzin, again has limit cycles as a central theme. He shows that they can easily occur in a simple cubic model for two interacting species. This leads Kostitzin to a general discussion of the importance of periodic and quasi-periodic behaviors as the only tool which is likely to make some sense out of the apparent "chaos" of population numbers.

In his "Biologie mathématique" (1937a) Kostitzin dealt with competition and predation in a variety of more specific situations. Particularly interesting among them is the case in which critical values in the densities of prey species determine whether or not they will be preyed upon by other species. This situation is described by different systems of equations, each holding in a distinct range of the variables. The structure and the results of this model are wholly analogous to his treatment of glaciations (1935, cf., Part V). It is also analogous to the ecological application of "relaxation oscillations" by Gause and Witt (1934). Other interesting treatments by Kostitzin (1937a) include the extension of competition to the case in which the different toxic metabolites of two species accumulate in the medium, the case of two predators competing for the same prey, and treatments of life stage- or size-specific interactions between predators and their prey.

It is interesting to compare the different "styles" of the works in this section. For example, a comparison between Lotka's paper, "The Growth of Mixed populations: ...", and Volterra's "The Variations ..." suggests very large differences in "philosophy of science". Lotka emerges largely as an "empiricist", a "curve fitter", primarily interested in quantitative predictions of special phenomena (cf. also the papers by Lotka in Part III of this collection). Volterra emerges as an "hypothetico-deductivist", primarily interested in providing general qualitative explanations in terms of direct causation. In this characterization, Kostitzin falls somewhere in between Lotka and Volterra; Kolmogoroff falls squarely in line with Volterra.

Volterra and Kostitzin, nevertheless, eagerly tested their theories on empirical data, and they even employed quantitative "fitting", as in their joint paper, "Comments on the Toxic Action . . ." (cf. Part I of this collection), or in Kostitzin and Kostitzin (1931; cf. also, Kostitzin's monograph in Part III of this collection). Both men, however, tended to keep their quantitative testing somehow "separate" from their general theoretical treatments. The more empirical aspects of their results were only alluded to in discussions of their general theoretical treatments, and explicit tests of their theories were considered in separate works.

The paper by Lotka on competition is a simple, explicit, local, and numerical treatment of a very special case of Volterra's general class of **dissipative systems**: namely, the case of just two species sharing a "mixed diet." Volterra had previously considered competition _per se_ only in the case of a "single resource", having been influenced possibly by the ideas of Liebig and Elton on the overriding importance of a single "limiting factor" in determining the distribution and abundance of any one species. The "single limiting factor" case also deserves treatment since its global behavior is more readily understood than for most other cases. It has also important implications for directional natural selection at a genotypic level (cf. Part V).

Notice how Lotka in his discussion of competition places great emphasis on demography and economics proper, much as he had placed a great emphasis on physics proper, in terms of flows of mass and energy, in his "Physical biology" (1924). Volterra on the other hand approaches the "struggle for existence" in terms of analogies with physics, but downplaying or disregarding the direct role of gross physical quantities, such as mass and energy. Also, Lotka deals directly with economics but without reference to general theories on it. Volterra, on the other hand, hardly at all mentions economics, but he approaches animal ecology through a general notion of "value", bearing some analogies to theories of value in economics.

Notice also how part of the "philosophical" differences which set Lotka apart from the other authors, and are so blatant in the works considered here (cf. also Part III), almost vanish in Lotka's later and more comprehensive "Théorie Analytique ...", especially Vol. 1 (1934). These and other "late", general purpose monographs in this tradition (Kostitzin, 1937, Volterra and D'Ancona, 1935 and D'Ancona, 1942, 1954) then reach a "near consensus" which sets them clearly apart from the prevailing "philosophical mood" of their times.

VARIATIONS AND FLUCTUATIONS IN THE NUMBERS OF

COEXISTING ANIMAL SPECIES

VITO VOLTERRA

CONTENTS

Preliminary Considerations

1. I would like to present a few studies on the coexistence of species in the same habitat. Ordinarily, they either compete for the same nutritional resources, or they feed on each other. However, they may also help one another.[1]

In order to approach the question mathematically it is convenient to start with hypotheses which, although departing from reality, give an approximate image of it. The representation will be a gross one, at least to begin with, but at the same time it will be simple. One can deal with it analytically, and verify quantitatively, or even just qualitatively, whether the results agree with observation. One can thereby test the correctness of the initial hypothesis and lay the ground for new results.

To facilitate the analysis it is convenient to represent the phenomenon schematically, by isolating those factors one wishes to examine, assuming they act alone, and by neglecting the others. Thus, for instance, the fluctuations in the numbers of fishes living in a certain habitat may depend on the state of the fauna, and on other environmental conditions. I have started by studying *only the intrinsic phenomena* due to the voracity and feritility of the coexisting species. I therefore place myself in the ideal condition in which only intrinsic causes operate, and all the others can be neglected.

Later (Part II, §8) I have also considered the superposition of these factors and periodical environmental variations.

----

[1]Dr. Umberto D'Ancona had mentioned several times to me his collections of data on fishing during the periods encompassing the war. He had asked me if it was possible to explain mathematically the results he was getting on the temporal variations in the composition of species. This question led me to set up the problem as in the following pages, to work it out, and to establish a number of principles stated at the end of Part I, §2, and Part II, §5. The two of us had been working independently, and we were very pleased with the agreement of our results, obtained respectively by analysis and by observation. By fishing man perturbs the natural state of variation of prey and predator species, increasing the quantity of the prey species, and decreasing that of the predator.

This might serve as justification for taking the liberty to publish these investigations which are analytically simple, but which, neverthless, seemed novel to me.

2. Before moving to general cases, I preferred to consider a few special ones. This allowed me to become oriented in a field which was completely new to me.[2] In Part I, I concentrated on two special cases. One deals with two species, isolated in the same habitat, which compete for the same nutritional resources (§1). The other is a case of two species, one feeding on the other. The prey is only limited by the predator, and it would otherwise increase without bounds (§2).

I then considered in Part I, §4 all the cases arising from the coexistence of two species, according to whether their interactions are favourable or unfavourable.

By suitably approaching the two cases considered in §§1 and 2, one derives differential equations representing them. Their integrals can be obtained, and they display the patterns by which the two species increase or decrease. In the second case the calculations point to the existence of periodic fluctuations of the two species, and their period can also be determined. Data on fishing seem to show that such fluctuations effectively occur in nature.

These fluctuations and their periods are regulated by three general principles. The third principle also predicts the perturbations in the mean quantities of the two species produced by external actions which destroy the individuals of both species. It predicts an average increase of the prey and an average decrease of the predator. This result seems also to agree with fishing data, the perturbing action being the artificial one of fishing by man.

Following this result it was obvious to investigate levels of destruction of both species within which the prey is favoured. It is intuitive that with

---

[2] My first publication in this area appeared in the "Memorie della R. Accademia Nazionale dei Lincei", classe di Scienze fisico-matematiche e naturali, Serie VI, vol. II, fasc. III, 1926. After publishing this paper I became aware of the analytical treatment by Ross of malaria epidemics. I also became aware that Dr. Lotka in his Elements of Physical Biology, New York, 1925, had considered the case of two species which I treat in Part I, §2. He had arrived by other means at the integral, his phase-space representation, and the period of small oscillations. Nevertheless, the general principles which I derive in the same section, and the different cases treated in other sections of Part I, are novel and original results. Also original are the applications of these principles, and the treatment of the coexistence of n species in conservative and dissipative associations. I regret I was not able to refer in the paper mentioned above to this interesting work by Dr. Lotka, which contains a variety of mathematical treatments of chemical and biological problems.

increasing destruction of the two species a limit must be reached beyond which both
are depleted. Therefore, Part I, §5 deals in detail with this *optimal limit*. It
has the character of an upper limit rather than a maximum. As one approaches it,
the average quantity of the prey keeps increasing. Once this limit is reached, however, the predator is depleted, and the prey species tends toward a value which is
lower than the previous averages.

3. The cases treated in the first two sections display qualitatively
different behaviors. That of the first is asymptotic, and that of the second is
periodic. They can, thus, be considered as two different typical situations. The
second case, which leads to a stable condition, has to be compared with those of
unstable nature treated in §4.

New principles introduced in the second part of this work allow us to deal
with more general cases. General equations representing an association of any number of species feeding on each other are derived. Fluctuations also occur in this
case, and the three general principles, derived in the case of two species, also
apply. I will then distinguish between *conservative* and *dissipative* biological
associations. Later I will study the superposition of the free fluctuations due to
the fertility and voracity of the different species, and the forced fluctuations due
to periodical environmental factors. I will finally examine the perturbations
produced in an association by adding a new species.

This is followed by an appendix in which the general theory is applied to
a limited habitat in which three species coexist. The first species feeds on the
second, and this on the third, while this last species derives its nourishment from
the habitat. It is as if a species had a parasite and this parasite itself had a
parasite. At the end I will take into account age distribution and other delayed
effects, leading to integro-differential equations.

4. One's first inclination would be to approach the problem in terms of
stochastic models, but these are not suitable to our goals. In approaching such
problems one tries at first to roughly describe the phenomenon in words. Then these

words are translated into mathematical terms, which lead to differential equations. The methods of mathematical analysis then lead us much farther than one could go with ordinary language and reasoning. Precise mathematical principles can be derived and they do not seem to contradict the results of observations. Rather, the most important of these principles seems to be in perfect agreement with the data.[3]

Analytically, the study of fluctuations or oscillations in the numbers of coexisting species is rather different from the ordinary study of oscillations. The classical theory of oscillations only applies to linear equations, while the present study must deal, in general terms, with non-linear equations. In this case one has in fact to deal with fluctuations which are generally not small. Only under the hypothesis that fluctuations are small, can one deal with approximations, neglecting second order terms, such that the general theory of linear differential or integro-differential equations is again relevant.

5. Following the ideas discussed above, the treatment is simplified by assuming that species increase in a continuous manner. That is to say, we assume that the quantity of living beings in a species, instead of being an integer number, is measured by a real and positive number which varies continuously. In nature breeding generally takes place during different specific seasons. We will ignore this aspect, and assume, instead, that births take place continuously at any one time. It is also assumed that the number of births is proportional to the number of individuals in a species at any one time. The same also applies to deaths. According to whether births outnumber deaths, or vice versa, the number of individuals increases or decreases. This implies that the individuals of each species are considered to be homogeneous, or that variations in age and size among individuals are neglected.

Assume that a species lives in isolation, or that it is not affected by other species. As long as the birth and death rates do not change, the number of

---

[3]For this data see the Memoria CXXVI in this series by Dr. Umberto D'Ancona, Dell' influenza della stasi peschereccia nel periodo 1914-18 sul patrimonio ittico dell'Alto Adriatico, in which the theoretical and practical consequences of the above results are discussed (Cf. §2, Number 9).

individuals, N, will change according to

$$\frac{dN}{dt} = nN - mN = (n-m)N$$

where t denotes the time, n and m denote respectively the birth and death rates, both constants. Setting n - m = ε , one has

$$(I) \quad \frac{dN}{dt} = \varepsilon N \,, \qquad\qquad (II) \quad N = N_0 e^{\varepsilon t}$$

where N_0 is the number of individuals at time zero. The constant ε is called the growth rate of the species. If it is positive the species will increase, while the species will be depleted if it is negative. If the birth and death rates change, ε will also change, either with time, with N , or with other factors. In this case (I) still holds true, but, clearly, (II) does not.

PART ONE

BIOLOGICAL ASSOCIATIONS OF TWO SPECIES

§1. Two species competing for the same resource.

1. Assume there are two species living in the same habitat, and denote the numbers of their respective individuals by N_1 and N_2. Let ε_1 and ε_2 be the values which their growth rates would have if the common resource were always available in sufficient quantity to fully satisfy their needs. We would then have

$$\frac{dN_1}{dt} = \varepsilon_1 N_1, \quad \frac{dN_2}{dt} = \varepsilon_2 N_2 \qquad (\varepsilon_1 > 0, \ \varepsilon_2 > 0).$$

But, as the numbers of individuals in the two species continue to increase, the amount of this resource available to each individual will decrease. Let us assume, therefore, that the resource is decreased by the amount $h_1 N_1$ through the action of the first species, and, similarly, by the amount $h_2 N_2$ through the action of the second. On the whole, then, the available resource is decreased by the amount $h_1 N_1 + h_2 N_2$. This decrease effects each species differently, so that their growth rates are reduced to

(1) $\qquad \varepsilon_1 - \gamma_1(h_1 N_1 + h_2 N_2) \qquad\qquad \varepsilon_2 - \gamma_2(h_1 N_1 + h_2 N_2)$.

We have, then, the differential equations

(2_1) $$\frac{dN_1}{dt} = (\varepsilon_1 - \gamma_1(h_1 N_1 + h_2 N_2))N_1$$

(2_2) $$\frac{dN_2}{dt} = (c_2 - \gamma_2(h_1 N_1 + h_2 N_2))N_2$$

in which all the constants, ε_1, ε_2, h_1, h_2, γ_1, γ_2, are positive.

2. From the preceding equations it follows that

(3_1)
$$\frac{d \ln N_1}{dt} = \varepsilon_1 - \gamma_1(h_1 N_1 + h_2 N_2)$$

(3_2)
$$\frac{d \ln N_2}{dt} = \varepsilon_2 - \gamma_2(h_1 N_1 + h_2 N_2)$$

and, therefore

(4)
$$\gamma_2 \frac{d \ln N_1}{dt} - \gamma_1 \frac{d \ln N_2}{dt} = \varepsilon_1 \gamma_2 - \varepsilon_2 \gamma_1$$

which is the same as

(5)
$$\frac{d \ln \frac{N_1^{\gamma_2}}{N_2^{\gamma_1}}}{dt} = \varepsilon_1 \gamma_2 - \varepsilon_2 \gamma_1 \quad .$$

Integrating, and passing from logarithms to numbers, we finally obtain

(6)
$$\frac{N_1^{\gamma_2}}{N_2^{\gamma_1}} = C \, e^{(\varepsilon_1 \gamma_2 - \varepsilon_2 \gamma_1)t}$$

where C is a constant.

3. Let us suppose that the constants in the above equation satisfy the special relationship

$$\varepsilon_1 \gamma_2 - \varepsilon_2 \gamma_1 = 0$$

so that

$$\frac{\varepsilon_1}{\gamma_1} = \frac{\varepsilon_2}{\gamma_2} = K$$

and

(7)
$$\frac{N_1^{\gamma_2}}{N_2^{\gamma_1}} = C \quad .$$

From this one has

$$N_2 = \frac{1}{C^{\frac{1}{\gamma_1}}} \, N_1^{\frac{\gamma_2}{\gamma_1}}$$

which, substituted in (2_1), gives

$$\frac{dN_1}{N_1 \left\{ \varepsilon_1 - \gamma_1 \left(h_1 N_1 + \frac{h_2}{C^{\frac{1}{\gamma_1}}} N_1^{\frac{\gamma_2}{\gamma_1}} \right) \right\}} = dt \quad .$$

The variables are then separated, and we have

$$t - t_0 = \int_{N_1^0}^{N_1} \frac{dN_1}{N_1 \left\{ \varepsilon_1 - \gamma_1 \left(h_1 N_1 + \frac{h_2}{C^{\frac{1}{\gamma_1}}} N_1^{\frac{\gamma_2}{\gamma_1}} \right) \right\}}$$

where N_1^0 is the number of individuals of the first species at the initial time, t_0 .

Three cases are possible according to the initial numbers of the two species, N_1^0 and N_2^0 :

1st case
$$K > h_1 N_1^0 + h_2 N_2^0 \quad ;$$

N_1 and N_2 will increase, preserving the relationship (7), that is,

$$\frac{N_1^{\gamma_2}}{N_2^{\gamma_1}} = \frac{N_1^0{}^{\gamma_2}}{N_2^0{}^{\gamma_1}} \quad .$$

They will tend asymptotically to values such that

(9) $$h_1 N_1 + h_2 N_2 = K$$

2nd case $$K < h_1 N_1^0 + h_2 N_2^0 \quad ;$$

N_1 and N_2 will decrease, again preserving the relationship (7), and tending asymptotically to values satisfying (9).

3nd case $$h_1 N_1^0 + h_2 N_2^0 = K$$

in which case N_1 and N_2 remain constant.

It is evident, however, that the probability that (7) will actually be met in nature is vanishingly small.

4. If the binomial $\varepsilon_1 \gamma_2 - \varepsilon_2 \gamma_1$ is non-zero, we can as well suppose it is positive. If not, it can be made positive by exchanging species 1 with species 2. In this case

$$\lim_{t=\infty} \frac{N_1^{\gamma_2}}{N_2^{\gamma_1}} = \infty \quad .$$

When N_1 is larger than or equal to $\dfrac{\varepsilon_1}{\gamma_1 h_1}$, it follows from (2_1) that $\dfrac{dN_1}{dt}$ is negative, and, therefore, N_1 cannot increase beyond some limit. Then, necessarily, N_2 must tend to zero, and it becomes easy to compute the asymptotic expression for N_1 .

When N_2 becomes small enough to be neglected, equation (2_1) becomes

$$\frac{dN_1}{dt} = (\varepsilon_1 - \gamma_1 h_1 N_1) N_1$$

or, separating the variables

$$dt = \frac{dN_1}{N_1(\varepsilon_1 - \gamma_1 h_1 N_1)} .$$

Integrating, and passing from logarithms to numbers, we obtain

$$\frac{N_1}{\varepsilon_1 - \gamma_1 h_1 N_1} = C_0 e^{\varepsilon_1 t}$$

where C_0 is a constant. One has, therefore,

$$N_1 = \frac{C_0 \varepsilon_1 e^{\varepsilon_1 t}}{1 + \gamma_1 h_1 C_0 e^{\varepsilon_1 t}} = \frac{C_0 \varepsilon_1}{e^{-\varepsilon_1 t} + \gamma_1 h_1 C_0} .$$

That is to say that N_1 tends asymptotically to the value $\frac{\varepsilon_1}{\gamma_1 h_1}$, increasing or decreasing according to whether C_0 is positive or negative. We may summarize these results in the following statement: *If* $\frac{\varepsilon_1}{\gamma_1} > \frac{\varepsilon_2}{\gamma_2}$ *the second species tends to extinction, and the first species tends to the number of individuals* $\frac{\varepsilon_1}{\gamma_1 h_1}$.

5. Generally these equations cannot be integrated, but there is a special case in which integration is easily accomplished. If we assume that, approximately, $\gamma_1 = \gamma_2 = \gamma$ one can set $c^{-\frac{1}{\gamma}} = c$. We then have

$$N_2 = c N_1 e^{(\varepsilon_2 - \varepsilon_1)t}$$

or

$$N_2 e^{-\varepsilon_2 t} = c N_1 e^{-\varepsilon_1 t} .$$

Setting

$$N_1 e^{-\varepsilon_1 t} = M_1 \qquad\qquad N_2 e^{-\varepsilon_2 t} = M_2$$

we obtain

$$M_2 = c M_1$$

and Equation (3) becomes

$$\frac{d \ln M_1}{dt} = - \gamma(h_1 N_1 + h_2 N_2) = - \gamma(h_1 e^{\varepsilon_1 t} M_1 + h_2 e^{\varepsilon_2 t} M_2)$$

$$= -\gamma M_1 (h_1 e^{\varepsilon_1 t} + h_2 c \, e^{\varepsilon_2 t}) \ .$$

That is to say,

$$\frac{d M_1}{M_1^2} = - \gamma(h_1 e^{\varepsilon_1 t} + h_2 c \, e^{\varepsilon_2 t}) dt$$

or, integrating,

$$\frac{1}{M_1} = \gamma\left(\frac{h_1}{\varepsilon_1} e^{\varepsilon_1 t} + \frac{h_2 c}{\varepsilon_2} e^{\varepsilon_2 t}\right) + C'$$

where C' is a constant. Then

$$M_1 = \frac{1}{\gamma\left(\dfrac{h_1}{\varepsilon_1} e^{\varepsilon_1 t} + \dfrac{h_2 c}{\varepsilon_2} e^{\varepsilon_2 t}\right) + C'} \ , \qquad M_2 = \frac{c}{\gamma\left(\dfrac{h_1}{\varepsilon_1} e^{\varepsilon_1 t} + \dfrac{h_2 c}{\varepsilon_2} e^{\varepsilon_2 t}\right) + C'}$$

from which

$$N_1 = \frac{e^{\varepsilon_1 t}}{\gamma\left(\dfrac{h_1}{\varepsilon_1} e^{\varepsilon_1 t} + \dfrac{h_2 c}{\varepsilon_2} e^{\varepsilon_2 t}\right) + C'} \ , \qquad N_2 = \frac{c \, e^{\varepsilon_2 t}}{\gamma\left(\dfrac{h_1}{\varepsilon_1} e^{\varepsilon_1 t} + \dfrac{h_2 c}{\varepsilon_2} e^{\varepsilon_2 t}\right) + C'} \ .$$

It is easy to verify the general statement which was proposed in the previous section for the special case in which integration could be carried out. If $\varepsilon_1 > \varepsilon_2$, one has in fact

$$\lim_{t=\infty} N_1 = \frac{\varepsilon_1}{\gamma_1 h_1} \ , \qquad \lim_{t=\infty} N_2 = 0$$

6. We have assumed that the numbers of individuals in the two species

affect their growth rates, ε_1 and ε_2, according to (1), in which N_1 and N_2 appear as linear terms. With even greater generality one can assume that the growth rates are given, instead, by

$$\varepsilon_1 - \gamma_1 F(N_1, N_2) \qquad \varepsilon_2 - \gamma_2 F(N_1, N_2)$$

Here $F(N_1, N_2)$ is a continuous and positive function which is zero only when $N_1 = N_2 = 0$. It is an increasing function of both N_1 and N_2, and it tends to infinity with either one of its arguments. Except for these conditions, the function $F(N_1, N_2)$ is an arbitrary one.

In this case Equations (2_1) and (2_2) are replaced by

$$\frac{d N_1}{dt} = (\varepsilon_1 - \gamma_1 F(N_1, N_2))N_1 , \qquad \frac{d N_2}{dt} = (\varepsilon_2 - \gamma_2 F(N_1, N_2))N_2 .$$

Equations (4), (5), (6) and their consequences continue to hold. In particular, if $\varepsilon_1 \gamma_2 - \varepsilon_2 \gamma_1 > 0$, N_2 tends asymptotically to zero, that is species 2 tends to extinction.

The asymptotic behavior of species 1 is then represented by

$$dt = \frac{dN_1}{N_1 (\varepsilon_1 - \gamma_1 F(N_1, 0))} .$$

Let us begin in this formula with the initial values N_1^0, 0, and suppose that

$$\varepsilon_1 - \gamma_1 F(N_1^0, 0) > 0$$

Since $\varepsilon_1 - \gamma_1 F(\infty, 0) < 0$ the equation

(10) $$\varepsilon_1 - \gamma_1 F(N_1, 0) = 0$$

must have one or more roots which are larger than N_1^0. Then N_1 will increase, tending asymptotically toward the smallest of the roots. If, on the other hand, $\varepsilon_1 - \gamma_1 F(N_1^0, 0) < 0$, there must exist roots of (10) which are smaller than N_1^0 since $\varepsilon_1 - \gamma_1 F(0,0) > 0$. In this case N_1 will decrease and tend asymptotically toward the largest of the roots.

§2. Two species, one feeding on the other.

1. Let N_1 and N_2 be the numbers of individuals in the two species. If the second species were not present, the first would grow at a rate $\varepsilon_1 > 0$. Let us also suppose that the second species would die out for lack of food if the first were not present. It would have, in this case, a negative growth rate, $-\varepsilon_2$. If each of the two species were alone we would have

$$(11_1) \qquad \frac{d\,N_1}{dt} = \varepsilon_1 N_1 \; , \qquad\qquad (11_2) \qquad \frac{d\,N_2}{dt} = -\varepsilon_2 N_2 \; .$$

But, if they are together and species 2 feeds on species 1, ε_1 will decrease and $-\varepsilon_2$ will increase. The more abundant is species 2, the more ε_1 is decreased, and the more abundant is species 1, the more $-\varepsilon_2$ is increased. In order to give the simplest representation of this relationship we shall suppose that ε_1 decreases in proportion to N_2 , by the amount $\gamma_1 N_2$, and that $-\varepsilon_2$ increases in proportion to N_1 , by the amount $\gamma_2 N_1$. We then have the system of differential equations

$$(A_1) \qquad \frac{d\,N_1}{dt} = (\varepsilon_1 - \gamma_1 N_2) N_1 \; , \qquad\qquad (A_2) \qquad \frac{d\,N_2}{dt} = (-\varepsilon_2 + \gamma_2 N_1) N_2 \quad .$$

It might seem very naive to assume that the coefficients of growth and decline are linear, in N_2 and N_1 respectively. However, as we shall see in §5, this can be justified in terms of the expected number of encounters between individuals of the two species. It is also worth noting that the same procedure of integration employed in this section applies as well to coefficients which are arbitrary functions of N_2 and N_1 respectively.

2. The constant parameters ε_1 and ε_2 represent the birth and death rates of the two species, while γ_1 measures the susceptibility of species 1 to predation and γ_2 measures the predatory ability of species 2. Both γ_1 and γ_2 are increased by an increase in the predatory ability of species 2, and they are likewise decreased by an improvement in the protection of species 1. The dimensions

of ε_1 and ε_2 are derived by integrating (11_1) and (11_2). If the two species were not to interact, their numbers would change according to

$$N_1 = C_1 e^{\varepsilon_1 t} \qquad\qquad N_2 = C_2 e^{-\varepsilon_2 t}$$

where C_1 and C_2 are, respectively, the values of N_1 and N_2 at $t = 0$. Set, for instance, $N_1 = 2C_1$, $N_2 = \frac{1}{2} C_2$, and denote by t_1 the time required for species 2 to decrease to one half. Such quantities are related to the coefficients of growth by

$$\varepsilon_1 = \frac{\ln 2}{t_1} = \frac{0.693}{t_1} \quad , \qquad \varepsilon_2 = \frac{\ln 2}{t_2} = \frac{0.693}{t_2}$$

so that the dimension of ε_1 and ε_2 is t^{-1} . It is always possible to choose the time units so that $\varepsilon_1 = 1$. Indeed, by appropriately selecting the time unit, the time required for species 1 to increase by a factor of $e = 2.728$ will be $e = e^{\varepsilon_1}$, and, therefore, $\varepsilon_1 = 1$. Analogous reasoning applies to ε_2 .

Setting

(12)
$$\frac{\varepsilon_2}{\gamma_2} = K_1 \ , \qquad \frac{\varepsilon_1}{\gamma_1} = K_2 \ ,$$

Equations (A_1) and (A_2) will be at the steady state, namely $\dfrac{d\,N_1}{dt} = \dfrac{d\,N_2}{dt} = 0$, when $N_1 = K_1$ and $N_2 = K_2$. One has, then

$$\gamma_1 = \frac{\varepsilon_1}{K_2} \ , \qquad \gamma_2 = \frac{\varepsilon_2}{K_1} \ .$$

Equations (A_1) and (A_2) are integrated by observing that

(13)
$$\frac{d\,\dfrac{N_1}{K_1}}{dt} = \varepsilon_1 \left(1 - \frac{N_2}{K_2}\right)\frac{N_1}{K_1} \ , \qquad \frac{d\,\dfrac{N_2}{K_2}}{dt} = -\varepsilon_2\left(1 - \frac{N_1}{K_1}\right)\frac{N_2}{K_2} \ .$$

Setting

(14)
$$N_1 = K_1 n_1 , \quad N_2 = K_2 n_2$$

the preceding equations reduce to

$$(A_1') \quad \frac{dn_1}{dt} = \varepsilon_1 (1-n_2)n_1 , \qquad (A_2') \quad \frac{dn_2}{dt} = -\varepsilon_2(1-n_1)n_2$$

Multiplying these equations by ε_2 and ε_1 , respectively, and summing, we have

(15)
$$\frac{d}{dt} (\varepsilon_2 n_1 + \varepsilon_1 n_2) = \varepsilon_1 \varepsilon_2 (n_1 - n_2) \quad .$$

Multiplying by $\dfrac{\varepsilon_2}{n_1}$ and $\dfrac{\varepsilon_1}{n_2}$, respectively, and summing, we have

$$\frac{\varepsilon_2}{n_1} \frac{dn_1}{dt} + \frac{\varepsilon_1}{n_2} \frac{dn_2}{dt} = \varepsilon_1 \varepsilon_2 (n_1 - n_2)$$

that is,

(16)
$$\frac{d}{dt} (\ln n_1^{\varepsilon_2} + \ln n_2^{\varepsilon_1}) = \varepsilon_1 \varepsilon_2 (n_1 - n_2) \quad .$$

Equating the first two members of Equations (15) and (16) we obtain

$$\frac{d}{dt} (\varepsilon_2 n_1 + \varepsilon_1 n_2) = \frac{d}{dt} (\ln n_1^{\varepsilon_2} + \ln n_2^{\varepsilon_1}) \quad .$$

Integrating, and passing from logarithms to numbers, one then has

$$n_1^{\varepsilon_2} n_2^{\varepsilon_1} = C \, e^{\varepsilon_2 n_1 + \varepsilon_1 n_2}$$

where C is a positive constant. From this

(17)
$$\left(\frac{n_1}{e^{n_1}} \right)^{\varepsilon_2} = C \left(\frac{n_2}{e^{n_2}} \right)^{-\varepsilon_1}$$

and from (A_1') and (A_2') it follows that

$$dt = \frac{dn_1}{\varepsilon_1(1-n_2)n_1} = \frac{dn_2}{-\varepsilon_2(1-n_1)n_2} \; .$$

The integration is then performed by expressing n_2 as a function of n_1, or vice versa, through the integral (17).

4. It is more convenient, however, to discuss the solution directly, in particular integral (17). To this end set

(18)
$$x = \left(\frac{n_1}{e^{n_1}} \right)^{\varepsilon_2} = C \left(\frac{n_2}{e^{n_2}} \right)^{-\varepsilon_1}$$

and consider the curves Γ_1 and Γ_2, having x as their ordinate and respectively n_1 and n_2 as abscissas (Figure 1).

The quantity

(19)
$$\frac{d}{dn_1} \cdot \frac{n_1}{e^{n_1}} = e^{-n_1}(1-n_1)$$

will be positive when $n_1 < 1$, and negative when $n_1 > 1$. Then, as n_1 varies

between 0 and ∞, x increases from 0 to the maximum $\left(\frac{1}{e} \right)^{\varepsilon_2}$ when $n_1 = 1$, after which it decreases to 0. On the other hand, as n_2 varies from 0 to ∞, x decreases from ∞ to the minimum $C e^{\varepsilon_1}$ when $n_2 = 1$, after which it continues to increase, tending to ∞ with n_2. The shapes of Γ_1 and Γ_2 are then as in Figure 1.

The constant C is determined from (17) as a function of the initial values of n_1 and n_2, and it satisfies the condition

$$C \leq e^{-(\varepsilon_1 + \varepsilon_2)} \; .$$

If $C < e^{-(\varepsilon_1+\varepsilon_2)}$, that is $e^{-\varepsilon_2} > C\, e^{\varepsilon_1}$, for each value of x between Ce^{ε_1} and $e^{-\varepsilon_2}$ there will correspond two values of n_1 and n_2 , except at the maxima and minima of the two curves. Having set the two curves side by side as in Figure 1, let us draw two straight horizontal lines through their vertexes, C_1 and C_2 . Consider the portions of the two curves, $A_1C_1B_1$ and $A_2C_2B_2$, between the horizontal lines. Let $a_1 < 1$ and $b_1 > 1$ be the abscissas of A_1 and B_1 , and let $a_2 < 1$ and $b_2 > 1$ be the abscissas of A_2 and B_2 .

We will then try to construct the curve λ which has n_1 on the abscissa and n_2 on the ordinate. Suppose first that the point A_2 corresponds to the point C_1 . Then move the point G_1 along the segment C_1B_1 of the curve Γ_1 . Correspondingly, the point G_2 will move on curve Γ_2 along the segment A_2C_2 , so that G_1 and G_2 are on the same horizontal line as x . Then, the values $n_2 = g_2$ will correspond to $n_1 = g_1$, where g_1 and g_2 are, respectively, the abscissas of G_1 and G_2 . Then, as n_1 increases from 1 to b_1 , n_2 increases from a_2 to 1 . That is to say, one moves in Figure 2 from the point R_2 of coordinates $(1, a_2)$, to the point S_1 of coordinates $(b_1, 1)$. Continuing, as n_1 decreases from b_1 to 1 , n_2 increases from 1 to b_2 . That is to say, one moves in Figure 2 from the point S_1 of coordinates $(b_1, 1)$, to the point S_2 of coordinates $(1, b_2)$. Continuing again, both n_1 and n_2 decrease, moving from the point S_2 to the point R_1 in Figure 2. Finally, n_1 increases and n_2 decreases moving, again in Figure 2, from the point R_1 to the point R_2.

Having returned to the starting point, one runs periodically through the closed cycle in Figure 2. It follows from (18), and also from Figure 1, that when n_1 and n_2 again take their previous values, x will also resume its previous value.

5. From (18) it follows that $\ln x = \varepsilon_2(\ln n_1 - n_1)$, so that, deriving with respect to time and taking into account (A_1'),

$$\frac{1}{x}\frac{dx}{dt} = \varepsilon_2\left(\frac{1}{n_1} - 1\right)\frac{dn_1}{dt} = \varepsilon_1\varepsilon_2(1-n_1)(1-n_2) ,$$

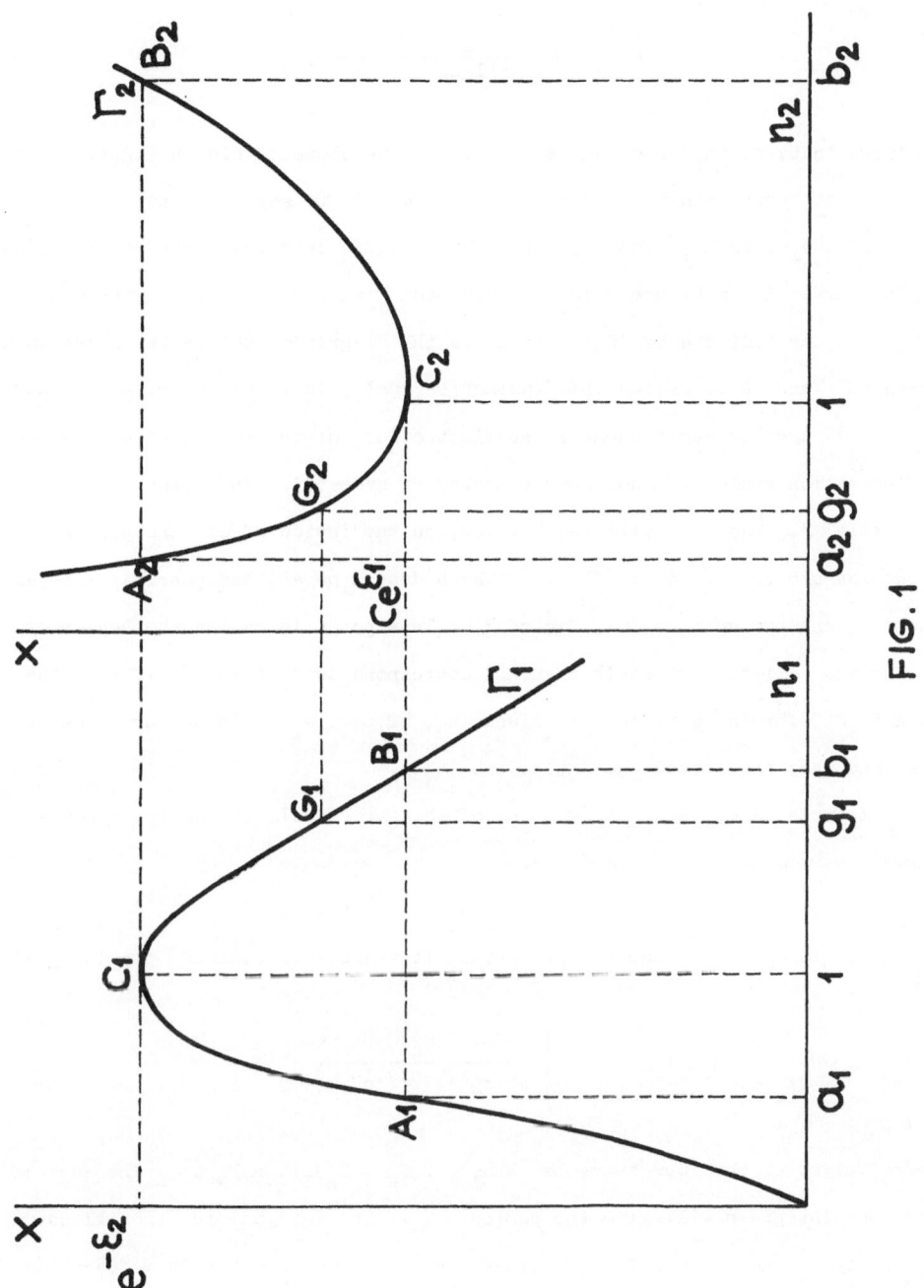

FIG. 1

that is,

$$dt \ = \ \frac{dx}{\varepsilon_1 \varepsilon_2 x (1-n_1)(1-n_2)} \ .$$

It follows that, as n_1 and n_2 move through the closed cycle in Figure 2, t will increase by the constant quantity T . That is to say, n_1 and n_2 and, by virtue of (14), also N_1 and N_2 will be periodic functions of time of period T . The curve Λ in Figure 3 is obtained from Figure 2 multiplying the abscissa by K_1 and the ordinate by K_2 . It gives the diagram of the cyclic relationships between N_1 and N_2 , called the *fluctuation cycle*, in which $K_1(b_1 - a_1)$ and $K_2(b_2 - a_2)$ are the amplitudes of the fluctuations of the two species. In general, the fluctuation cycle will not have a center of symmetry. The point Ω of coordinates K_1 and K_2 will be, however, on the inside of all the possible fluctuation cycles Φ, Λ, Ψ, X, ... which depend on all the possible initial numbers of the two species (Cf. Figure 3). This point is called the *center of fluctuation.* All the curves in Figure 3 correspond to different values of the constant C , for the same set of values ε_1, ε_2, γ_1, γ_2 . These curves never cross, and they are nested one inside the other.

In this case, then, *the numbers of the individuals in the two species fluctuate periodically with period T.*

6. In order to compute the period, T, we have to evaluate the integral

$$\int \frac{dx}{\varepsilon_1 \varepsilon_2 x (1-n_1)(1-n_2)}$$

successively over the four segments $R_2 S_1$, $S_1 S_2$, $S_2 R_1$, $R_1 R_2$. The sum of these four integrals will give the period, T . The integral function blows up to infinity at each of the four vertexes, R_1, R_2, S_1, S_2, but it is easy to recognize that the order of infinity is such that the integrals converge. The above integral shows that the period, T, depends only on ε_1, ε_2, and C . In the

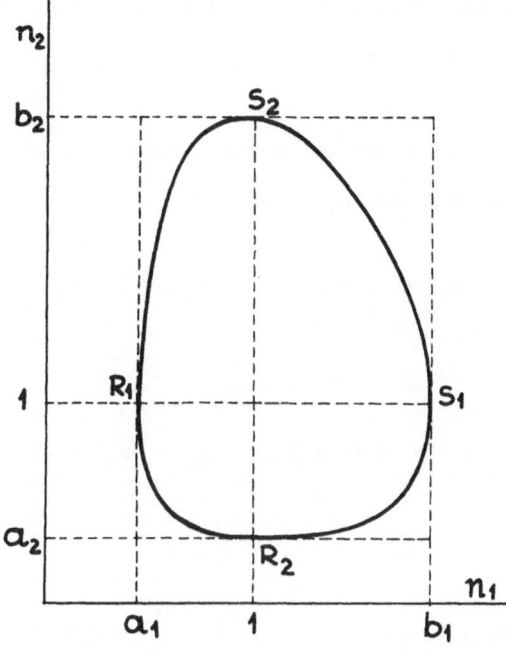

FIG. 2

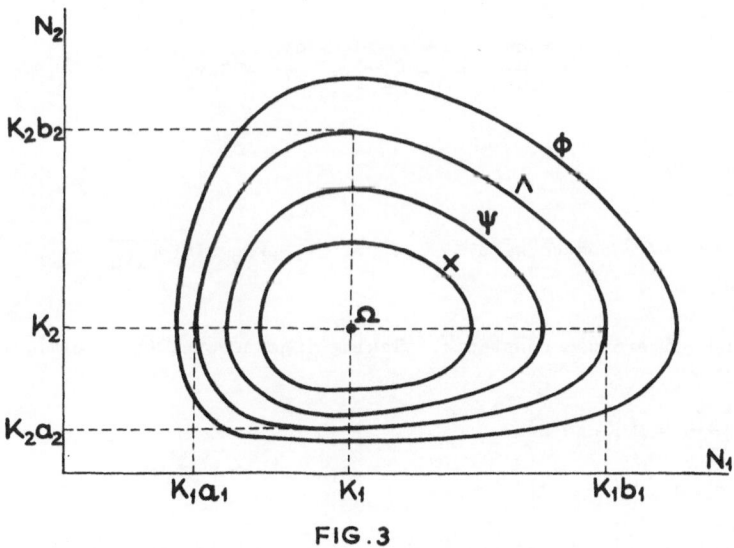

FIG. 3

following section we shall compute the approximate period in the case of small oscillations.[4]

7. It is easy to consider the approximation for small fluctuations, starting from Equations (A_1) and (A_2). Setting

$$(20) \qquad n_1 = 1 + v_1 \qquad n_2 = 1 + v_2$$

we have

$$(21) \qquad N_1 = K_1(1+v_1) , \qquad N_2 = K_2(1+v_2) .$$

Equations (A_1') and (A_2') become

$$(A_1'') \quad \frac{dv_1}{dt} = -\varepsilon_1 v_2 - \varepsilon_1 v_1 v_2 , \qquad (A_2'') \quad \frac{dv_2}{dt} = \varepsilon_2 v_1 + \varepsilon_2 v_1 v_2 .$$

If the fluctuations are small, v_1 and v_2 can be considered to be small quantities of first order, so that the second order terms in the equations above can be neglected. Then

$$\frac{dv_1}{dt} = -\varepsilon_1 v_2 , \quad \frac{dv_2}{dt} = \varepsilon_2 v_1$$

are integrated by

$$v_1 = L\sqrt{\varepsilon_1} \cos(\sqrt{\varepsilon_1 \varepsilon_2}\, t + \alpha) , \qquad v_2 = L\sqrt{\varepsilon_2} \sin(\sqrt{\varepsilon_1 \varepsilon_2}\, t + \alpha)$$

where L and α are two constants. Taking into account (12), (14), (20), and setting $L \dfrac{\varepsilon_1 \varepsilon_2}{\gamma_1 \gamma_2} = E$, one has

---

[4] For the exact computation of the period one can see the previously cited paper in the Proceedings of the R. Accademia dei Lincei, Part I, §3, Number 5 (Editors' note: see also Appendix 1).

$$N_1 = \frac{\varepsilon_2}{\gamma_2} + \frac{\gamma_1}{\sqrt{\varepsilon_1}} E \cos(\sqrt{\varepsilon_1 \varepsilon_2}\ t + \alpha)$$

(22)

$$N_2 = \frac{\varepsilon_1}{\gamma_1} + \frac{\gamma_2}{\sqrt{\varepsilon_2}} E \sin(\sqrt{\varepsilon_1 \varepsilon_2}\ t + \alpha)$$

so that their common period is $\dfrac{2\ \pi}{\sqrt{\varepsilon_1 \varepsilon_2}}$.

We would have arrived at the same value by directly computing the integral T in the previous section, and by neglecting higher order terms. We can then assume that the period of the fluctuation cycle is approximately given by

$$T = \frac{2\ \pi}{\sqrt{\varepsilon_1 \varepsilon_2}}\ .$$

Denoting by t_1 and t_2 , respectively, the times in which species 1 doubles, and species 2 halves, as in 2 above, we have

$$T = \frac{2\ \pi\ \sqrt{t_1 t_2}}{0.693} = 9.06\sqrt{t_1 t_2}\ \ .$$

The fluctuation cycle becomes an ellipse having as center the fluctuation center, and as half axes

$$E\ \frac{\gamma_1}{\sqrt{\varepsilon_1}}\ ,\quad E\ \frac{\gamma_2}{\sqrt{\varepsilon_2}}\ .$$

The amplitudes of the fluctuations are then

$$f_1 = 2\ E\ \frac{\gamma_1}{\sqrt{\varepsilon_1}}\ ,\quad f_2 = 2\ E\ \frac{\gamma_2}{\sqrt{\varepsilon_2}}\ ,$$

and their ratio

$$\frac{f_1}{f_2} = \frac{\gamma_1}{\gamma_2}\ \sqrt{\frac{\varepsilon_2}{\varepsilon_1}}\ \ .$$

The family of the fluctuation cycles comprise, then, a set of homothetic ellipses, centered on the fluctuation center (See Figure 4).

8. Let us now deal with the mean numbers of individuals of the two species during a cycle. We start again from Equations (A_1') and (A_2'), and divide both members by n_1 and n_2, getting

$$\frac{d \ln n_1}{dt} = \varepsilon_1(1-n_2) , \qquad \frac{d \ln n_2}{dt} = - \varepsilon_2(1-n_1) .$$

Integrating between two times t' and t'', in which n_1 and n_2 assume, respectively, the values n_1', n_1'' and n_2', n_2'',

$$\ln \frac{n_1''}{n_1'} = \varepsilon_1 \left[(t''-t') - \int_{t'}^{t''} n_2 dt \right] , \quad \ln \frac{n_2''}{n_2'} = -\varepsilon_2 \left[(t''-t') - \int_{t'}^{t''} n_1 dt \right]$$

By extending the range of integration to one period, T, the left sides go to zero, and

$$T = \int_0^T n_1 dt = \int_0^T n_2 dt ,$$

that is to say,

$$\frac{1}{T} \int_0^T n_1 dt = \frac{1}{T} \int_0^T n_2 dt = 1 .$$

The average values of n_1 and n_2 in a period are, then, both equal to 1, and from (14)

$$\frac{1}{T} \int_0^T N_1 dt = K_1 = \frac{\varepsilon_2}{\gamma_2} , \quad \frac{1}{T} \int_0^T N_2 dt = K = \frac{\varepsilon_1}{\gamma_1} .$$

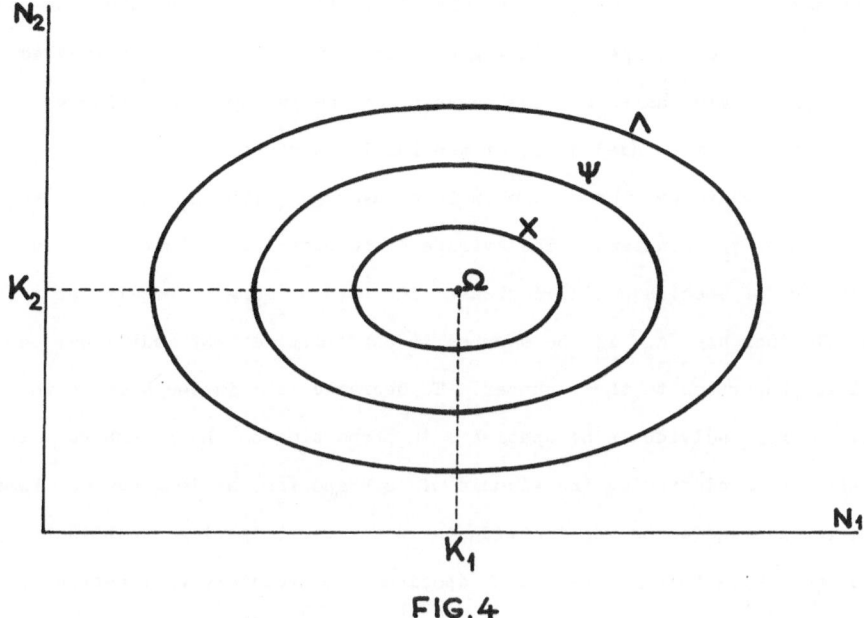

FIG. 4

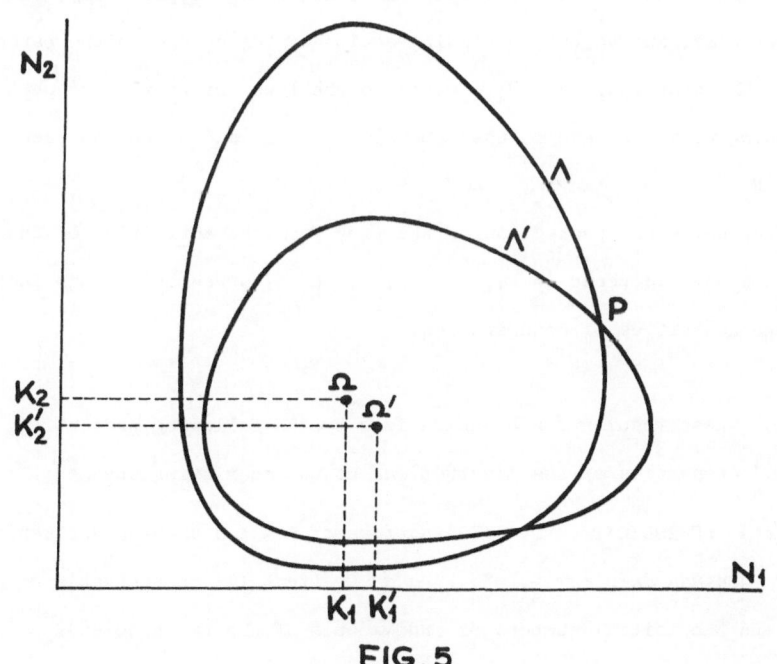

FIG. 5

That is to say,   the coordinates of the fluctuation center are the mean values of the numbers of the two species during a cycle. It follows that as long as ε_1, ε_2, γ_1, γ_2 remain unaltered, the averages of the two species during a cycle will always be the same, independent of the initial numbers.

Let us now discuss the change in these averages with ε_1 and ε_2 while keeping γ_1 and γ_2 constant. The average of species 1 will increase with ε_2 and the average of species 2 will decrease with ε_1 as long as the latter remains positive. To increase ε_2 is the same as to uniformly destroy individuals of species 2 in proportion to their number. To decrease ε_1 is the same as to uniformly destroy individuals of species 1 in proportion to their numbers. It follows that,by so destroying individuals of both species, as long as ε_1 remains positive, the average number of individuals in the first species, that is the prey, will increase, while that of the second species, the predator, will decrease. Figure 5 represents the switch from a cycle Λ , corresponding to ε_1, ε_2 to a cycle Λ' of parameters $\varepsilon_1' < \varepsilon_1$ and $\varepsilon_2' > \varepsilon_2$ (γ_1 and γ_2 remain unchanged, and $\varepsilon_1' > 0$). One can imagine that this switch might take place at a time corresponding to the point P in which the two cycles meet. At this time, the numbers of individuals in the two species do not change appreciably, but they will change over time, due to the continuing action of the change in the parameters ε_1 and ε_2 . The center of Λ', Ω', moves to the lower right with respect to Ω , corresponding to a decrease in the mean value of N_2 and to an increase in the mean value of N_1 .

Increasing the protection of the prey from the activities of the predator is reflected by a decrease in γ_1 and γ_2 , which corresponds to an increase in the average quantities of both species.

9. These results can be summarized by the following properties, called the *fundamental properties of the fluctuations* of two coexisting species:

1st) PERIODICITY. *The fluctuations of the two species are periodic, and the period depends only on* ε_1, ε_2 *, and* C (that is, on the rates of growth and decline, and the initial numbers of individuals of the two species).

2nd) CONSERVATION OF THE AVERAGES. *The averages of the numbers of individuals in the two species do not depend of the initial numbers, and they remain constant as long as the coefficients of growth, decline, defense, and offense of the two species,* ε_1, ε_2, γ_1, γ_2, *also remain constant.*

3rd) PERTURBATION OF THE AVERAGES. *If one uniformly destroys the individuals of the two species in proportion to their numbers, the average number of the prey increases, and that of the predator decreases. On the other hand an increase in the protection of the prey increases the averages of both species.*[5]

When the fluctuations are small one has the following approximate properties:

1st) *The small fluctuations are isochronic, that is their period is affected only negligibly by the initial numbers and by the conditions of defense and offense.*

2nd) *The period of the fluctuation is proportional to the square root of the product of the time in which the prey, if alone, would double, and of that in which the predator, if alone, would halve* ($T = 9.06 \sqrt{t_1 t_2}$).

3rd) *The constant destruction of a given proportion of the predator accelerates the fluctuations while similar destruction of the prey slows them down. By simultaneously destroying in the same way individuals of both species, the ratio of the amplitude of the fluctuation of the prey to that of the predator increases.*

The verification of these properties in nature seems to be easier in the case of fish communities in which some species are carnivorous. A continuous process of fishing can then provide a uniform destruction of individuals of the various species.

The virtual cessation of fishing during World War I, and its subsequent resumption, are changes comparable to those considered above from one cycle to

---

[5]This law only holds within certain limits described in Paragraph 8, namely, as long as the growth rate of the prey remains positive. In §5 the limits within which the destruction of both species favors the prey will be studied in greater detail. (Editors' note: cf. also Appendix 2.)

another. The relative abundance of species in the catch provides a measure of the actual abundance of the various species. In this way the data on fishing provide data on fluctuations, and they appear to agree with the mathematical predictions derived above.[6]

§3. Fluctuation Diagrams.

1. To diagram the variation of n_1 and n_2 as a function of time it is convenient to start from Equation (A_1')

$$\frac{dn_1}{dt} = \varepsilon_1(1-n_2)n_1 \quad ,$$

and assume, for simplicity, that $\varepsilon_1 = 1$.

Figure 6-A reproduces Figure 1. From it one can derive $1 - n_2$ as a function of n_1 , and then graphically construct the product of $(1-n_2)$ and n_1 . In Figure 6-A, subdivide in six parts the portions of the curves representing the excursions of the two species. The same subdivision, 0, 1, 2, 3, 4, 5, 6, 7, is also employed on the abscissa of Figure 6-B on which the arbitrary segment OP = p is also marked. On the ordinate axis we also mark the positive segments Oa, Ob, Oc, and the negative segments Od, Oe, Of, derived from Figure 6-A. In Figure 6-B we project the points a, b, c, d, e, f, from the point P, and draw through the points 2 and 6 the lines parallel to Pa and Pd respectively. Similarly, we draw

---

[6]See the note in §1 regarding Dr. D'Ancona's data. The property of the perturbation of the averages was anticipated by Charles Darwin. While discussing the struggle for existence he wrote:

"The amount of food for each species of course gives the extreme limit to which each can increase; but very frequently it is not the obtaining food, but the serving as prey to other animals which determines the average numbers of a species. Thus, there seems to be little doubt that the stock of partridges, grouse and hares on any large estate depends chiefly on the destruction of vermin. If not one head of game were shot during the next twenty years in England, and, at the same time, if no vernim were destroyed, there would, in all probability, be less game than at present, although hundreds of thousands of game animals are now annually shot." Ch. Darwin, The origin of species by means of natural selection, or the preservation of favoured races in the struggle for life. Sixth edition, with corrections to 1871. London, John Murray, 1882 (pages 53-54).

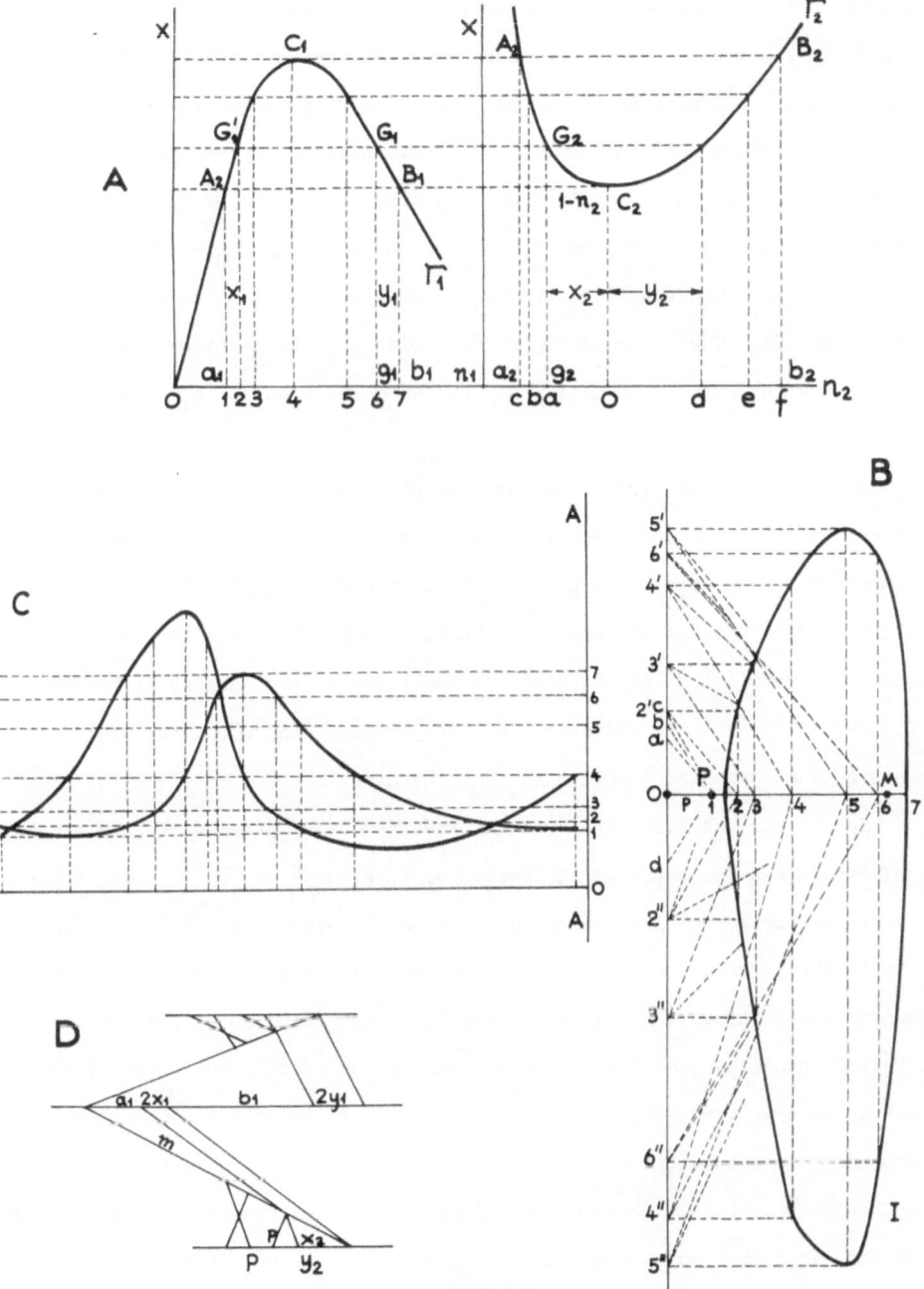

FIG. 6

through the points 3 and 5 the lines parallel to Pb and Pe , and through point 4 we draw the lines parallel to Pc and Pf . We then determine their intersections with the ordinate axis. In this way we obtain the points $2'$, $3'$, $4'$, $5'$, $6'$, and $2''$, $3''$, $4''$, $5''$, $6''$ in Figure 6-B. Their distances from the origin, 0 , taken with their sign and multiplied by p , give the values of $\frac{dn_1}{dt}$ corresponding to the values $02, 03, 04, 05, 06$ for n_1 . The quantity $\frac{dn_1}{dt}$ will be positive when n_1 increases and negative when n_1 decreases. It will be zero when n_1 equals 01 and 07. The curve in Figure 6-B with n_1 as abscissa and $\frac{dn_1}{dt}$ as ordinate gives the rate of variation of n_1 as a function of n_1 itself.

2. We shall now derive the curve which gives n_1 on the ordinate as a function of time, t, on the abscissa. Take any point M on the abscissa $(OM = m)$ in Figure 6-B. From this point project the points $2'$, $3'$, $4'$, $5'$, $6'$, and $2''$, $3''$, $4''$, $5''$, $6''$. By a convenient choice of the time unit, this straight line can be assumed to be parallel to the tangents of the curve we are seeking at the points $02, 03, 04, 05, 06$ on the ordinate. At the points 01 and 07 on the ordinate the tangents to the curve are parallel to the time axis.

Let us now subdivide the straight line AA drawn parallel to the ordinate axis of Figure 6-B according to the scale, 0, 1, 2, 3, 4, 5, 6, 7. Let us then draw a set of horizontal lines perpendicular to AA through the dividing points, and denote them by the same numbers. Since AA is the ordinate in Figure 6-C, the horizontal lines denote the heights of points on the curve we wish to derive. We arbitrarily choose on line 2 the corresponding point of this curve (See Figure 6-C). Through it we draw the parallel to the line $M2'$ of Figure 6-B until it meets line 3. We take this intersection as the point of the curve we are seeking which lies on the horizontal line 2. Passing from line 2 to line 3 the ordinate of the curve, n_1, is increased by $(\Delta n)_{23} = 23$. Denoting by $(\overline{\Delta t})_{23}$ the increase in the abscissa, from the method of construction we have

(23)
$$\frac{(\Delta n)_{23}}{(\overline{\Delta t})_{23}} = \frac{02'}{m} \ .$$

Denoting by $(\Delta t)_{23}$ the increase in time, from what was said in Number 1 we have, approximately,

$$\frac{(\Delta n)_{23}}{(\Delta t)_{23}} = 02'.p$$

and, therefore,

$$\frac{(\overline{\Delta t})_{23}}{(\Delta t)_{23}} = m \cdot p \quad.$$

This means that by $(\overline{\Delta t})_{23}$ we denote the measure of $(\Delta t)_{23}$ in the scale of Figure 6-B.

From the point obtained on line 3 we draw the parallel to M3' until it meets line 4. The point thus derived is the approximate position of the curve which lies on line 4.

In fact n_1 has increased by the amount $(\Delta n)_{34} = 34$, and, since the abscissa has increased by $(\overline{\Delta t})_{34}$, by the method of construction one has

$$\frac{(\Delta n)_{34}}{(\overline{\Delta t})_{34}} = \frac{03'}{m} \quad.$$

At the same time, since $(\Delta t)_{34}$ is the corresponding time increment, we must have (See Number 1)

$$\frac{(\Delta n)_{34}}{(\Delta t)_{34}} = 03' \cdot p \quad.$$

From this

$$\frac{(\overline{\Delta t})_{34}}{(\Delta t)_{34}} = m \cdot p$$

so the time scale has remained unchanged.

Continuing in this way we find the next approximate points of the required curve which lie on the straight lines 5 and 6. This procedure, however, cannot be applied to the points which lie on the lines 1 and 7 since the tangents at those points are parallel to the abscissa. We will need special procedures to estimate their abscissas.

3. The portion C_2G_2 in Figure 6-A can be conveniently approximated by a parabola, so that it can be described by the equation

$$1 - n_2 = \sqrt{2q\zeta} \; ,$$

in which q is the parameter of the parabola, and ζ is the difference between the ordinate of the generic point and that of the vertex, C_2. Because of (A_1'), and since $\varepsilon_1 = 1$,

$$\frac{dn_1}{dt} = n_1 \sqrt{2q\zeta} \quad .$$

Assuming that the segment A_1G_1' is a straight line, we have

(24) $$\zeta = \alpha(n_1 - a_1)$$

where $0a_1 = a_1$, and α is constant.

Let us set

(25) $$n_1 = a_1 + \xi^2 \; .$$

Through (24) and (25) we have

$$\frac{dn_1}{dt} = 2\xi \frac{d\xi}{dt} = (a_1 + \xi^2) \sqrt{2q\alpha\xi^2} \; ,$$

that is,

$$2 \frac{d\xi}{dt} = \sqrt{2q\alpha}(a_1 + \xi^2) \; .$$

From this we derive

$$dt = \frac{2d\xi}{\sqrt{2q\alpha}(a_1 + \xi^2)} \; ,$$

and then

$$t - t_1 = \frac{2}{\sqrt{2q\alpha}} \; \sqrt{a_1} \; \text{arc tan} \; \frac{\xi}{\sqrt{a_1}} \quad ,$$

where t_1 denotes the time corresponding to the ordinate 01 in Figure 6-A.

Since t_2 is the time corresponding to the ordinate 02 of the same figure, we have

$$t_2 - t_1 = \frac{2}{\sqrt{2qa}} \; \frac{1}{\sqrt{a_1}} \; \text{arc tan} \; -\frac{\xi_{12}}{\sqrt{a_1}} \quad ,$$

having set

$$\xi_{12} = \sqrt{x_1}$$

where x_1 is the distance 12 in Figure 6-A. Setting now $1 - a = x_2$, we obtain $x_2^2 = 2q \, \zeta_1$ since ζ_1 is the value of ζ corresponding to the point G_2 . Then, because of (24) and (25),

$$\sqrt{2q\alpha} = x_2 \; \sqrt{\frac{\alpha}{\zeta_1}} = \frac{x_2}{\sqrt{x_1}}$$

and, therefore,

$$t_2 - t_1 = \frac{2}{\left(\frac{x_2}{\sqrt{x_1}}\right)\sqrt{a_1}} \; \text{arc tan} \; \sqrt{\frac{x_1}{a_1}} = \frac{2}{x_2} \sqrt{\frac{x_1}{a_1}} \; \text{arc tan} \; \sqrt{\frac{x_1}{a_1}} \quad .$$

Since $\sqrt{\frac{x_1}{a_1}}$ is small we can take, approximately,

$$\text{arc tan} \; \sqrt{\frac{x_1}{a_1}} = \sqrt{\frac{x_1}{a_1}}$$

and, therefore,

$$t_2 - t_1 = \frac{2x_1}{x_2 a_1} \quad .$$

Let us denote by $\overline{t_2 - t_1}$ the interval $t_2 - t_1$ measured in the time
scale of Figure 6-C. Taking into account the results of Number 2, we have

$$\frac{\overline{t_2 - t_1}}{t_2 - t_1} = m.p \quad .$$

Therefore

$$\overline{t_2 - t_1} = \frac{2x_1}{x_2 a_1} \; mp \; = \; 2x_1 \frac{m}{a_1} \cdot \frac{p}{x_2}$$

The interval $\overline{t_2 - t_1}$ is the difference of the abscissas of the points
on the desired curve which lie respectively on the straight lines 1 and 2 of
Figure 6-C.

The graphic construction of $\overline{t_2 - t_1}$ is then performed, through the
preceding formula, in Figure 6-D. It is convenient to construct in a similar way
also the other three segments which locate the vertices of the curve with respect
to their neighboring points. This is also carried out in Figure 6-D. The
remaining points are obtained by the procedure in Number 2. Because of periodicity
we can prolong the curve giving n_1 as a function of t .

From Figure 6-A we then obtain n_2 as a function of n_1 , and immediately
build a curve having t as the abscissa and n_2 as the ordinate. The two diagrams
of n_1 and n_2 as a function of t are given in Figure 7 on a different scale,
and reproduced for three periods.

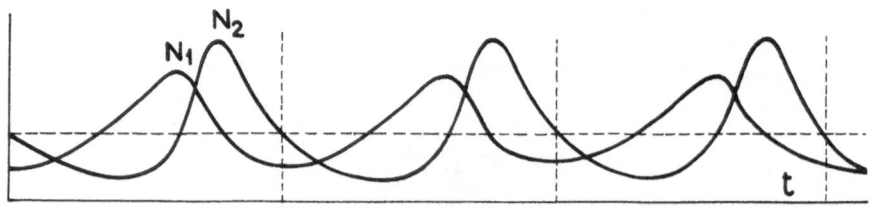

FIG. 7

§4. The effects of different interactions between two coexisting species

1. Consider two coexisting species, and let N_1 and N_2 be their numbers. The number of encounters between individuals of the two species during one time unit will be proportional to $N_1 N_2$. Assume it is given by $\alpha N_1 N_2$, where α is a constant. Let λ_1 and λ_2 be the coefficients of increase or decline of the two species when they are in isolation. In the case treated above λ_1 was positive and λ_2 was negative, and the encounters were unfavourable to the first species, the prey, and favourable to the second, the predator. Denote by β_1 the increase of individuals in the first species, and by β_2 that in the second species, which result from a given number of encounters, n . In the previous case we would have chosen β_1 negative and β_2 positive.

In a time interval dt the changes in the two species will be respectively

$$dN_1 \;=\; \lambda_1 N_1 dt + \frac{\beta_1}{n} \alpha \, N_2 N_1 dt$$

$$dN_2 \;=\; \lambda_2 N_2 dt + \frac{\beta_2}{n} \alpha \, N_1 N_2 dt \; .$$

Setting

$$\frac{\beta_1}{n} \alpha \;=\; \mu_1 \,, \quad \frac{\beta_2}{n} \alpha \;=\; \mu_2$$

the previous equations become

(26)
$$\begin{cases} \dfrac{dN_1}{dt} \;=\; N_1(\lambda_1 + \mu_1 N_2) \\[2mm] \dfrac{dN_2}{dt} \;=\; N_2(\lambda_2 + \mu_2 N_1) \end{cases} .$$

By making explicit the signs of the constants for the case above, that is by setting

$$\begin{aligned} \lambda_1 &= \varepsilon_1 & \lambda_2 &= -\varepsilon_2 \\ \mu_1 &= -\gamma_1 & \mu_2 &= \gamma_2 \end{aligned}$$

we regain Equations (A_1) and (A_2) of §2, that is,

$$\frac{dN_1}{dt} = N_1(\epsilon_1 - \gamma_1 N_2)$$

$$\frac{dN_2}{dt} = N_2(-\epsilon_2 + \gamma_2 N_1) \ .$$

In this manner we justify the statement in §2, Number 1, about choosing coefficients of increase which are linear in N_1 and N_2 .

2. Let us now consider the Equations (26) without specifying the signs of the coefficients, which can be either positive or negative. We assume they represent the growth of two coexisting species in which λ_1 and λ_2 are the *coefficients of growth*, while μ_1 and μ_2 are the *incremental coefficients of encounter*. The signs of λ_1 and λ_2 indicate whether the species will increase or decrease when alone. The signs of μ_1 and μ_2 indicate whether the encounters will be favourable or unfavourable to the two species. If, for instance, λ_1 and λ_2 are positive and μ_1 , μ_2 negative, both species will increase when alone and their encounters will be unfavourable to both. We can then consider all the possible cases.[7]

3. From Equations (26) it follows that

$$\mu_2 \frac{dN_1}{dt} - \mu_1 \frac{dN_2}{dt} = \lambda_1 \mu_2 N_1 - \lambda_2 \mu_1 N_2$$

$$\lambda_2 \frac{d \ln N_1}{dt} - \lambda_1 \frac{d \ln N_2}{dt} = \lambda_2 \mu_1 N_2 - \lambda_1 \mu_2 N_1$$

---

[7] It is clear that many phenomena of interest to medicine can be considered as dependent on the encounters and interactions among different species (the human species and pathogens, a parasite and its host). Therefore, the fluctuations of epidemics are related to the theories developed here. Cf. Sir Ronald Ross, The prevention of Malaria, Second edition 1911; Martini, Berechnungen und Beobachtungen zur Epidemiologie der Malaria (Gente, Hamburg, 1921); A. J. Lotka and F. R. Sharpe, Contribution to the Analysis of Malaria Epidemiology (American Journ. of Hygiene. Vol. III); Lotka, Op. cit. A few of these case are discussed in Numbers 8 and 9 of this Section.

and summing side by side

$$\mu_2 \; \frac{dN_1}{dt} + \lambda_2 \; \frac{d \ln N_1}{dt} \; = \; \mu_1 \; \frac{dN_2}{dt} + \lambda_1 \; \frac{d \ln N_2}{dt} \; .$$

Integrating and passing from logarithms to numbers, one has

(27)
$$N_1^{\lambda_2} \, e^{\mu_2 N_1} \; = \; C N_2^{\lambda_1} \, e^{\mu_1 N_2}$$

where C is a positive constant.

To study the behavior of the process, that is the curve represented by (27), it is convenient to construct the two curves

$$x \; = \; C' \, N_1^{\lambda_2} \, e^{\mu_2 N_1} \; , \qquad x \; = \; C'' \, N_2^{\lambda_1} \, e^{\mu_1 N_2}$$

where C' and C" are two constants such that $\frac{C''}{C'} = C$. We can then jointly consider the two curves which have respectively N_1 and N_2 as abscissas, and x as ordinate. We draw them side by side so that their abscissas are on the same straight line, as in Figure 1. Proceeding then as in §2, we can construct the curve with N_1 as abscissa and N_2 as ordinate.

4. There are four types of curves corresponding to the equation

$$x \; = \; C \, N^{\lambda} \, e^{\mu N} \; ,$$

according to the signs of λ and μ (C is a positive constant). They are represented in Figures 8, 9, 10, and 11. An horizontal line would intersect curves I and II at only one point, and it would either intersect the curves III and IV at two real points, or not at all.

All the possible cases are derived by jointly considering a curve of one type with a second curve of the same or different type. We then have 10 typical cases, as many as the combinations with repetition among 4 objects taken two at a

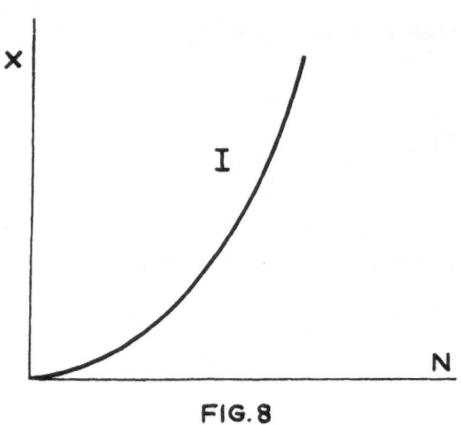

FIG. 8

$\lambda > 0, \mu > 0$

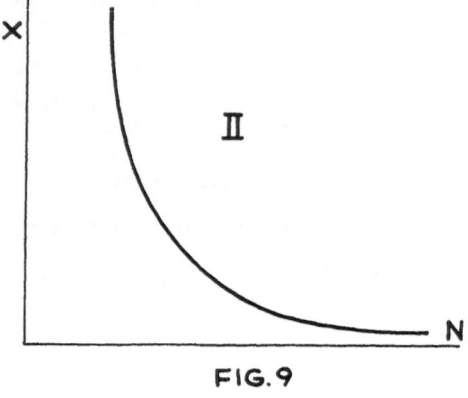

FIG. 9

$\lambda < 0, \mu < 0$

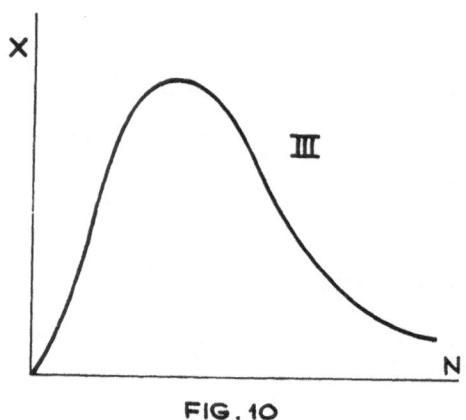

FIG. 10

$\lambda > 0, \mu < 0$

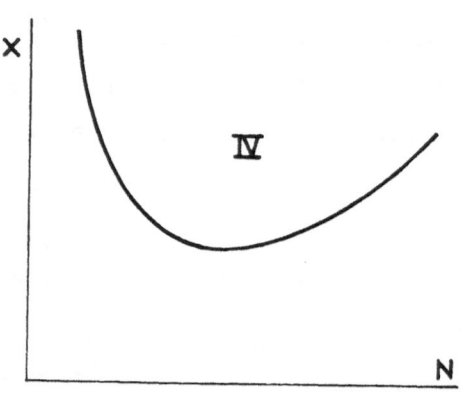

FIG. 11

$\lambda < 0, \mu > 0$

time. They can be represented by the symbols

$$(I, I), (I, II), (I, III), (I, IV)$$
$$(II, II), (II, III), (II, IV)$$
$$(III, III), (III, IV)$$
$$(IV, IV)$$

By (II, III), for instance, we refer to the curve obtained by combining

$$x = C' \, N_1^{\lambda'} \, e^{\mu' N_1}$$

in which $\lambda' < 0$, $\mu' < 0$, with the curve

$$x = C'' \, N_2^{\lambda''} \, e^{\mu'' N_2}$$

in which $\lambda'' > 0$, $\mu'' < 0$, and in which C' and C'' are two positive constants. The equation of this curve is, therefore

$$N_1^{\lambda'} \, e^{\mu' N_1} = C N_2^{\lambda''} \, e^{\mu'' N_2} \quad (\lambda' < 0, \, \mu' < 0, \, \lambda'' > 0, \, \mu'' < 0)$$

in which C is a constant.

5. Notice that case (II, II) is obtained from case (I, I) simply by changing in Equation (27) the signs of the four coefficients, $\lambda_1, \lambda_2, \mu_1, \mu_2$. This is equivalent to changing in this formula C by $\frac{1}{C}$. Nothing changes, then, from the point of view of the curve connecting N_1 with N_2. The same applies to the following combination of cases

(II, III) and (I, IV); (III, III) and (IV, IV);
(I, III) and (II, IV).

The only cases which give rise to different curves are then

(I, I) (I, II) (I, III) (II, III) (III, III) (III, IV) .

We have already considered in detail the last of these cases in §2. The others

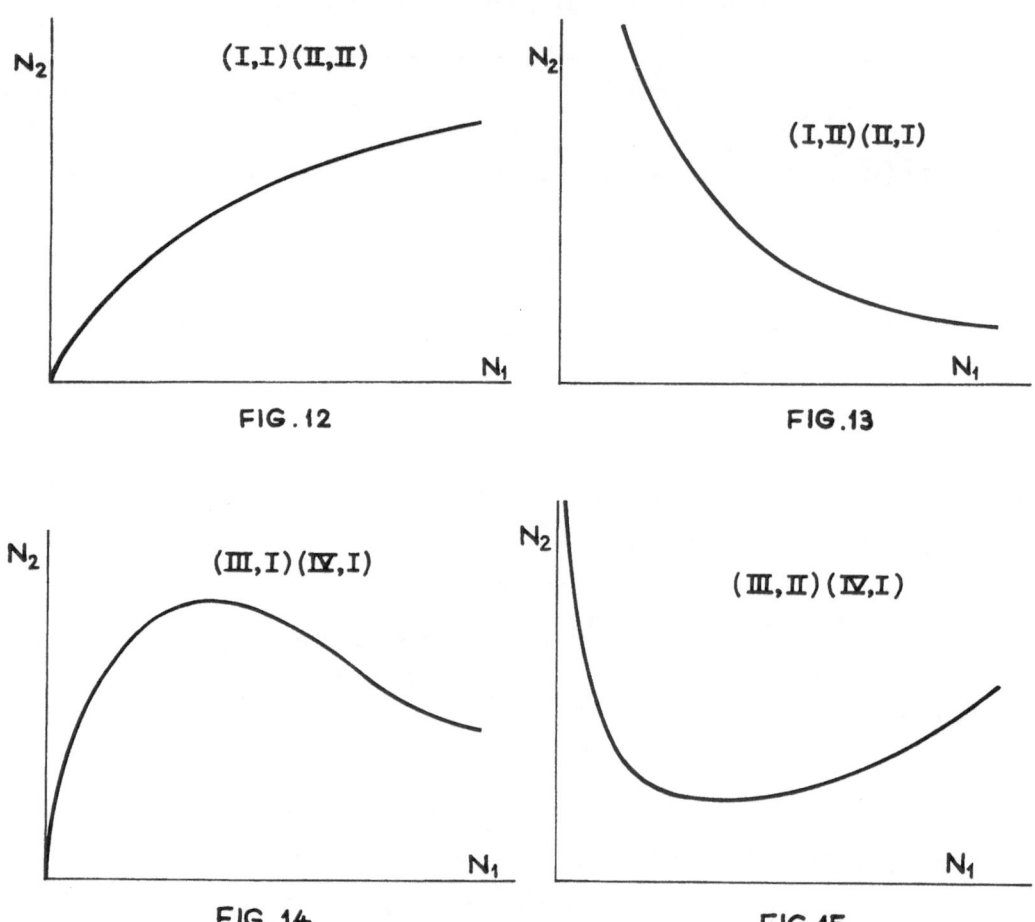

FIG.12

FIG.13

FIG. 14

FIG.15

are represented below in Figures 12, 13, 14, and 15 respectively, while Figures 16, 16', and 16" correspond to case (III, III).

6. When maximum x of the curve (N_1, x) is smaller than maximum x of the curve (N_2, x), case (III, III) will be represented by a curve with two branches as in Figure 16. When the reverse occurs, the two branches are as in Figure 16', which is the same as exchanging N_1 with N_2 in the preceding case. If the two maxima coincide the two branches intersect at an angle, as in Figure 16". To compute this angle let us divide side by side the Equations (26), obtaining

$$\frac{dN_2}{dN_1} = \frac{\mu_2}{\mu_1} \frac{N_2}{N_1} \left[\frac{\frac{\lambda_2}{\mu_2} + N_1}{\frac{\lambda_1}{\mu_1} + N_2} \right] .$$

The intersection is reached when

$$N_1 = -\frac{\lambda_2}{\mu_2}, \qquad N_2 = -\frac{\lambda_1}{\mu_1}$$

and, therefore, when

$$\lim \frac{dN_2}{dN_1} = \frac{\mu_2}{\mu_1} \frac{\lambda_1 \mu_2}{\lambda_2 \mu_1} \lim \frac{\frac{\lambda_2}{\mu_2} + N_1}{\frac{\lambda_1}{\mu_1} + N_2} = \frac{\mu_2}{\mu_1} \frac{\lambda_1 \mu_2}{\lambda_2 \mu_1} \lim \frac{dN_1}{dN_2} .$$

That is to say,

$$\left(\lim \frac{dN_2}{dN_1} \right)^2 = \frac{\mu_2^2}{\mu_1^2} \frac{\lambda_1}{\lambda_2}$$

or

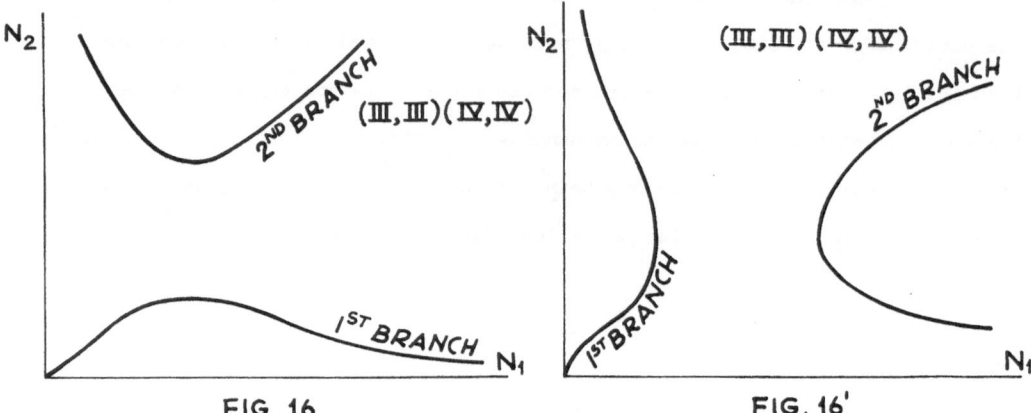

FIG. 16   FIG. 16'

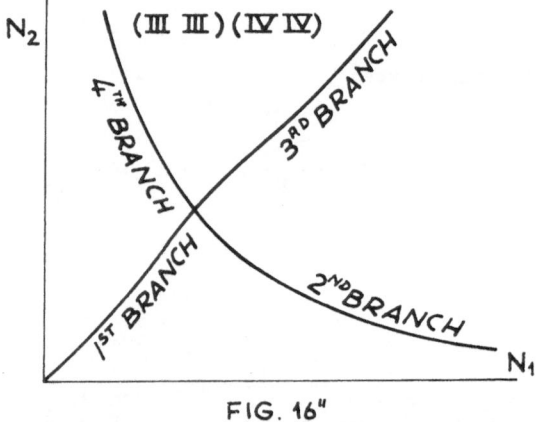

FIG. 16"

$$\lim \frac{dN_2}{dN_1} = \pm \frac{\mu_2}{\mu_1} \sqrt{\frac{\lambda_1}{\lambda_2}} \; .$$

Therefore, the trigonometric tangent of the angle formed by the two tangents at this double point is

$$\frac{2\mu_1\mu_2\sqrt{\lambda_1\lambda_2}}{\mu_1^2\lambda_2 - \mu_2^2\lambda_1} \; .$$

7. We will use the preceding figures to examine all the possible cases, except those studied in §2. Notice that to change the signs of the four coefficients in Equations (26) is the same as changing t to $-t$. It will be enough, then, to change the direction in which the curve is traced through time.

We can summarize the results in the table at the top of the next page. In this table l.r. indicates that the curve is to be traced from left to right as time progresses; r.l., from right to left; d., downward; u., upward.

8. The only cases in which coexistence can be achieved are (III, IV), (III, III), and (IV, IV). In the last two cases coexistence is only possible when the two maxima or the two minima of the curves of type III or of type IV are the same, an event of vanishingly small probability. In the former case the equilibrium is stable. If the system is displaced from equilibrium it will fluctuate in the vicinity of the equilibrium, and such fluctuations can be as small as one pleases (see §2). In cases (III, III) or (IV, IV) the equilibrium could potentially be reached through any one of the four brances which converge to the double point (Cf. Figure 16"). It is easy to see, however, that this would take an infinitely large amount of time. Suppose, on the other hand, that the system is already at equilibrium, which can only occur with vanishingly small probability. Any small perturbation would displace the system farther and farther from this equilibrium state.

Figure 12 $\begin{cases}(\text{I}, \quad \text{I}) \quad \lambda_1 > 0, \ \mu_1 > 0, \ \lambda_2 > 0, \ \mu_2 > 0 \quad \text{l.r.} \\ (\text{II}, \quad \text{II}) \quad \lambda_1 < 0, \ \mu_1 < 0, \ \lambda_2 < 0, \ \mu_2 < 0 \quad \text{r.l.}\end{cases}$

Figure 13 $\begin{cases}(\text{I}, \quad \text{II}) \quad \lambda_1 < 0, \ \mu_1 < 0, \ \lambda_2 > 0, \ \mu_2 > 0 \quad \text{r.l} \\ (\text{II}, \quad \text{I}) \quad \lambda_1 > 0, \ \mu_1 > 0, \ \lambda_2 < 0, \ \mu_2 < 0 \quad \text{l.r.}\end{cases}$

Figure 14 $\begin{cases}(\text{III}, \ \text{I}) \quad \lambda_1 > 0, \ \mu_1 > 0, \ \lambda_2 > 0, \ \mu_2 < 0 \quad \text{l.r.} \\ (\text{IV}, \ \text{II}) \quad \lambda_1 < 0, \ \mu_1 < 0, \ \lambda_2 < 0, \ \mu_2 > 0 \quad \text{r.l.}\end{cases}$

Figure 15 $\begin{cases}(\text{III}, \ \text{II}) \quad \lambda_1 < 0, \ \mu_1 < 0 \ \ \lambda_2 > 0, \ \mu_2 < 0 \quad \text{r.l.} \\ (\text{IV}, \ \text{I}) \quad \lambda_1 > 0, \ \mu_1 > 0, \ \lambda_2 < 0, \ \mu_2 > 0 \quad \text{l.r.}\end{cases}$

Figure 16 $\begin{cases}(\text{III},\text{III}) \quad \lambda_1 > 0, \ \mu_1 < 0, \ \lambda_2 > 0, \ \mu_2 < 0 \ \begin{cases}\text{l.r., 1st branch} \\ \text{r.l., 2nd branch}\end{cases} \\ (\text{IV}, \ \text{IV}) \quad \lambda_1 < 0, \ \mu_1 > 0, \ \lambda_2 < 0, \ \mu_2 > 0 \ \begin{cases}\text{r.l., 1st branch} \\ \text{l.r., 2nd branch}\end{cases}\end{cases}$

Figure 16' $\begin{cases}(\text{III},\text{III}) \quad \lambda_1 > 0, \ \mu_1 < 0, \ \lambda_2 > 0, \ \mu_2 < 0 \ \begin{cases}\text{u., 1st branch} \\ \text{d., 2nd branch}\end{cases} \\ (\text{IV}, \ \text{IV}) \quad \lambda_1 < 0, \ \mu_1 > 0, \ \lambda_2 < 0, \ \mu_2 > 0 \ \begin{cases}\text{d., 1st branch} \\ \text{u., 2nd branch}\end{cases}\end{cases}$

Figure 16" $\begin{cases}(\text{III},\text{III}) \quad \lambda_1 > 0, \ \mu_1 < 0, \ \lambda_2 > 0, \ \mu_2 < 0 \ \begin{cases}\text{l.r., 1st \& 2nd branches} \\ \text{r.l., 3rd \& 4th branches}\end{cases} \\ (\text{IV}, \ \text{IV}) \quad \lambda_1 < 0, \ \mu_1 > 0, \ \lambda_2 < 0, \ \mu_2 > 0 \ \begin{cases}\text{r.l., 1st \& 2nd branches} \\ \text{l.r., 3rd \& 4th branches}\end{cases}\end{cases}$

In all the reamining cases either both species tend to extinction, or the numbers in one or both species tend to infinity. They are, then, also unstable. In this way we have considered all the possible combinations of signs for the coefficients λ_1, λ_2, μ_1, μ_2 .

It would also be useful to consider the cases in which some of these coefficients are zero. Although such cases have vanishingly small probabilities of occurance they are of interest as transitional from one case to the other of the cases considered above. Among them, we will only consider, in the next section, the case in which $\lambda_1 = 0$ and $\lambda_2 < 0$, $\mu_1 < 0$, $\mu_2 > 0$. This case is particularly interesting, since it is the transition stage between type (III, IV) and type (IV,II). It also indicates how the other cases may be treated.

The results of all the cases considered so far, with the exception of those having vanishingly small probability, can be summarized as follows: *Two species can coexist in a stable and permanent manner only in the case, (III,IV), considered in §2.* In all the other cases either both species go extinct, or one species goes extinct and the other increases without bounds. Clearly, this last possibility is just a theoretical one since in practice no species can grow indefinitely. A deeper discussion of this last condition will be presented in Part II, §6.

9. Up to now we have dealt with the numerical dependence between N_1 and N_2, neglecting time dependence. The bionomial $\lambda_1 + \mu_1 N_2 = 0$ corresponds either to a maximum or to a minimum of N_1, and $\lambda_2 + \mu_2 N_1 = 0$ corresponds either to a maximum or to a minimum of N_2. Denoting by M_1 and M_2 these extremes, the binomials are infinitesimal of degree ½ with respect to $N_1 - M_1$ and $N_2 - M_2$. It follows that

$$\int \frac{dN_1}{N_1(\lambda_1 + \mu_1 N_2)} \quad , \quad \int \frac{dN_2}{N_2(\lambda_2 + \mu_2 N_1)}$$

do not diverge at the points of maximum or minimum mentioned above.

Let us examine what happens when N_1 and N_2 diverge. From the preceding results (see Figures 12-16") we see that when one of the quantities goes to ∞, the other goes either to 0 or to ∞. In the former case t goes to ∞, in the latter N_1 and N_2 take the same order of infinity. In fact, from Equations (26), as N_1 and N_2 both tend to ∞, one has

$$\lim \frac{dN_1}{dN_2} = \lim \frac{\dfrac{\lambda_1}{N_2} + \mu_1}{\dfrac{\lambda_2}{N_1} + \mu_2} = \frac{\mu_1}{\mu_2} \; .$$

In this case, then, t remains finite, as also revealed by the two integrals above. It follows that the numbers of the two species, N_1 and N_2, diverge to infinity in a finite time.

In all the other cases considered in this section the curves are traced in an infinite time. Each extreme corresponds, in fact, to at least one of the quantities N_1 or N_2 going to zero. Going back to the table in Number 7, and comparing it with Figures 12 through 16", one sees that N_1 and N_2 diverge simultaneously in the cases

Figure 12 (I,I); Figure 15 (IV,I); Figure 16 (IV,IV) 2nd branch;
Figure 16' (IV,IV) 2nd branch; Figure 16" (IV,IV) 3rd branch.

In all such cases the encounters are favourable to both species, corresponding to a positive value for both μ_1 and μ_2. These cases cannot be realized as such, since both N_1 and N_2 increase without bounds.

§5. <u>Limits within which the destruction of two species favours the prey species.</u>

1. We have shown in §2, Number 8 that by decreasing the rate of growth of the prey, ε_1, and increasing the rate of decline of the predator, ε_2, the average number of the prey increases, and that of the predator decreases. This property was also expressed in Number 9 as follows: *If one destroys the individuals of the two species in proportion to their numbers, the average number of the prey increases, and that of the predator decreases.* We have also pointed out that this property holds as long as ε_1 remains positive.

2. We will now study the details of this phenomenon. Let us denote by $\alpha\lambda$ the ratio of the number of individuals, n_1, of species 1,which are removed from the association per unit time,to the total number of individuals of species 1. $\beta\lambda$ will denote the analogous ratio for species 2. During a time interval dt,

$$n_1 dt \;=\; \alpha\lambda N_1 dt \qquad \text{and} \qquad n_2 dt \;=\; \beta\lambda N_2 dt$$

individuals, respectively of the two species,will be removed from the association. Therefore, Equations (A_1) and (A_2) are modified by substituting ε_1 and ε_2 respectively with

$$\varepsilon_1 - \alpha\lambda , \qquad \varepsilon_2 + \beta\lambda .$$

One might suppose that the ratio

$$(n_2 : N_2) : (n_1 : N_1) = \frac{\beta}{\alpha} = \vartheta$$

between the relative rates of removal or destruction of the two species depends *only on the means by which this removal or destruction takes place.* On the other hand, *the intensity of this removal or destruction* can be considered to depend on λ . An increase in λ while α and β are held constant means that the removal is intensified, but carried out in the same way. On the other hand, a change in the ratio $\vartheta = \frac{\beta}{\alpha}$ would signify a change in the means by which the removal is carried out.

Consider as a concrete example two coexisting species of fish, the second feeding on the first. To increase λ without changing either α or β means that fishing is intensified, but that the fishing procedures remain unchanged. A change in the latter, on the other hand, would be reflected in a change of $\vartheta = \frac{\beta}{\alpha}$.

3. The Equations (A_1) and (A_2) then become

$$(28) \qquad \frac{dN_1}{dt} = (\varepsilon_1 - \alpha\lambda - \gamma_1 N_2)N_1$$

$$(28') \qquad \frac{dN_2}{dt} = (- \varepsilon_2 - \beta\lambda + \gamma_2 N_1)N_2 .$$

If $\varepsilon_1' = \varepsilon_1 - \alpha\lambda > 0$ there will be a fluctuation of period T (§2, Number 5). The number of individuals of the first species removed in a time dt is

$$\alpha\lambda N_1 dt ,$$

so that the total number removed during a period T is

$$\int_0^T \alpha \lambda N_1 dt \ .$$

From this the mean number of individuals removed from species 1 per unit time is

$$P \ = \ \frac{1}{T} \int_0^T \alpha \lambda N_1 dt \ = \ \frac{\alpha \lambda}{T} \int_0^T N_1 dt \ .$$

From §2, Number 8, one has

$$\frac{1}{T} \int_0^T N_1 dt \ = \ \frac{\varepsilon_2 + \beta \lambda}{\gamma_2} \ = \ \frac{\varepsilon_2}{\gamma_2} \ ,$$

and, therefore,

$$P \ = \ \frac{\alpha \lambda (\varepsilon_2 + \beta \lambda)}{\gamma_2} \ .$$

Since $\varepsilon_1 - \alpha \lambda > 0$, the upper limit of λ is $\frac{\varepsilon_1}{\alpha}$, and the upper limit of P is, therefore,

$$P_m \ = \ \frac{\varepsilon_1 (\varepsilon_2 + \vartheta \varepsilon_1)}{\gamma_2} \ .$$

With reference to the specific example of fishing one can infer that, as long as the same method of fishing is continued, the catch during a cycle will indefinitely approach P_m , but will not become larger. This upper limit is an increasing function of the ratio ϑ .

4. If λ becomes larger than $\dfrac{\varepsilon_1}{\alpha}$, so that

$$\varepsilon_1 - \alpha\lambda < 0 \; ,$$

the fluctuation will cease, and both species will tend toward extinction (see §4 Number 5). We will be in the case denoted in §4 as type (IV,II), in which the curve in Figure 14 is traced from right to left.

It is of interest to examine the limit case in which λ reaches the value $\dfrac{\varepsilon_1}{\alpha}$. This case corresponds to the transition from type (III,IV) to type (IV,II), and, as mentioned in §4, it is the only transition case we will examine. This case is helpful in the present context, and it is also a guideline for the treatment of the other transitional cases.

When $\lambda = \dfrac{\varepsilon_1}{\alpha}$, Equations (28) and (28') become

$$(29) \qquad \frac{dN_1}{dt} \;=\; - \, \gamma_1 N_1 N_2 \; ,$$

$$(29') \qquad \frac{dN_2}{dt} \;=\; (- \, \varepsilon_2'' + \gamma_2 N_1) N_2$$

where

$$\varepsilon_2'' \;=\; \varepsilon_2 + \varepsilon_1 \, \vartheta \; .$$

Their integral (Cf. §4, Number 2) is

$$(30) \qquad N_1^{\varepsilon_2''} \, e^{-\gamma_2 N_1} \;=\; C \, e^{\gamma_1 N_2}$$

in which C is a positive constant.

Setting

$$x \;=\; N_1^{\varepsilon_2''} \, e^{-\gamma_2 N_1} \;=\; C \, e^{\gamma_1 N_2} \; ,$$

we obtain the curves Γ_1 and Γ_2 represented in Figure 17, having (N_1,x) and (N_2,x) respectively as coordinate axes. By applying the same procedure employed in §2, Number 4 (Cf. Figures 1 and 2) we are able to draw the integral curve of

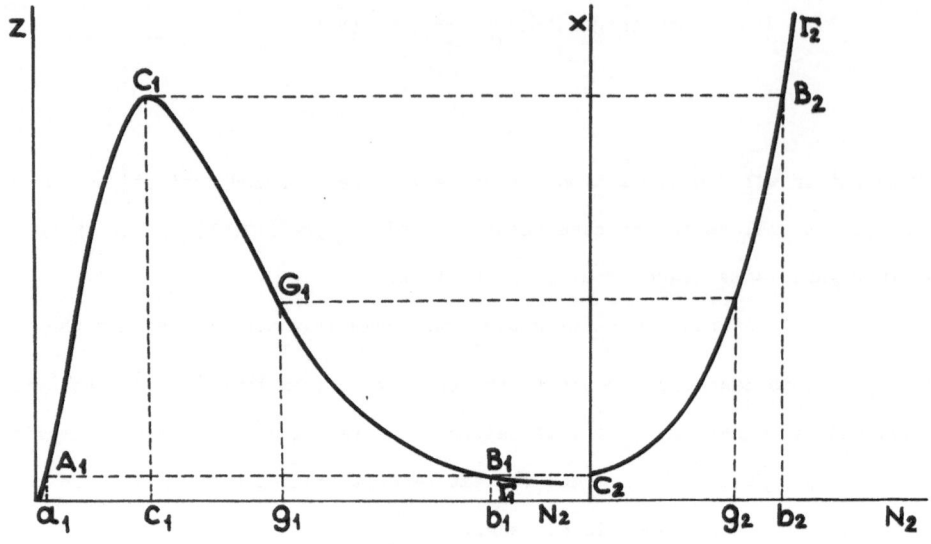

FIG . 17

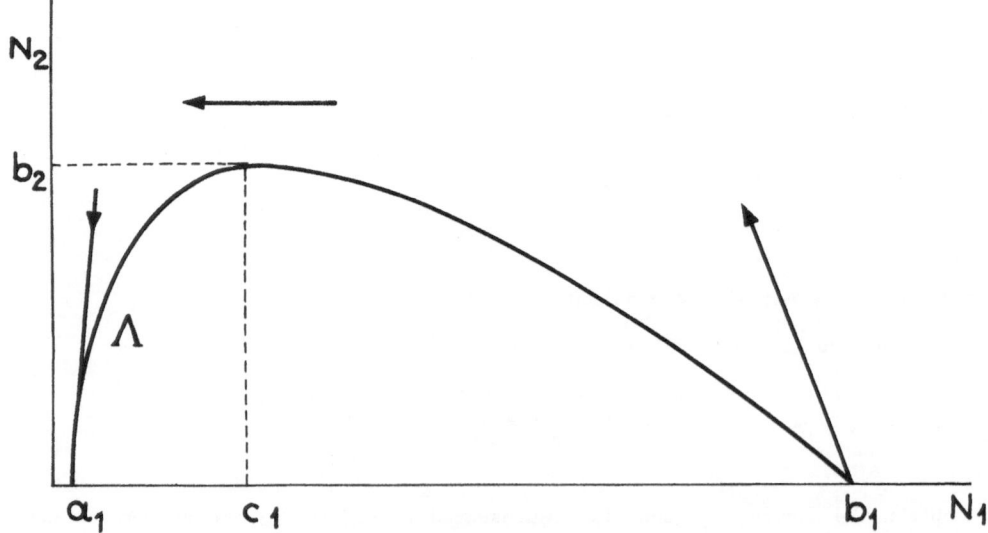

FIG . 18

Equation (30) (See Figure 18).

The minimum and maximum values, a_1 and b_1, of N_1 are the real roots of the equation

(30')
$$N_1^{\varepsilon_2''} e^{-\gamma_2 N_1} = C .$$

The maximum of N_2, b_2, corresponds to the value c_1 of N_1. Since

$$c_1 = \frac{\varepsilon_2''}{\gamma_2}$$

we have

$$b_2 = \frac{\varepsilon_2''(\ln c_1 - 1) - \ln C}{\gamma_1} .$$

On the other hand, we have

$$\ln C = \varepsilon_2'' \left(\ln a_1 - \frac{a_1}{c_1} \right)$$

so that

$$b_2 = \frac{-\varepsilon_2'' \left(\ln \dfrac{a_1}{c_1} + \dfrac{c_1 - a_1}{c_1} \right)}{\gamma_1} =$$

$$= \frac{\varepsilon_2''}{\gamma_1} \left(\frac{1}{2} \left(\frac{c_1 - a_1}{c_1} \right)^2 + \frac{1}{3} \left(\frac{c_1 - a_1}{c_1} \right)^3 \cdots \cdots \right) .$$

From Equations (29) and (29') it follows that

$$\tan \phi = \frac{dN_2}{dN_1} = \frac{\gamma_2 N_1 - \varepsilon_2''}{\gamma_1 N_1} = \frac{\gamma_2}{\gamma_1} \frac{N_1 - c_1}{N_1}$$

in which ϕ denotes the angle between a generic tangent to the curve and the N_1

axis. Let us denote by ϕ_{a_1} and ϕ_{b_1} the values of ϕ at the extremes a_1 and b_1 . We have

$$\tan \phi_{b_1} = \frac{\gamma_2}{\gamma_1} \frac{b_1-c_1}{b_1} \, , \quad \tan \phi_{a_1} = \frac{\gamma_2}{\gamma_1} \frac{a_1-c_1}{a_1} \, .$$

As t increases the curve is traced from right to left, namely, N_1 continues to decrease. Let us compute now the length of time it takes for the number of individuals in the first species, N_1, to be reduced from N_1^0 to N_1' , where

$$a_1 < N_1^0 < b_1 \, , \quad a_1 < N_1' < N_1^0 \, .$$

From (29) this time is seen to be

$$t = \int_{N_1'}^{N_1^0} \frac{dN_1}{\gamma_1 N_1 N_2} = \int_{N_1^0}^{N_1'} \frac{-dN_1}{\gamma_1 N_1 N_2} \, .$$

From (30) we obtain

$$\gamma_1 N_2 = \varepsilon_2'' \left(\ln N_1 - \frac{N_1}{c_1} \right) - \ln C = \varepsilon_2'' \left(\ln \frac{N_1}{a_1} - \frac{N_1-a_1}{c_1} \right)$$

so that, as N_1 tends to a_1 , $\gamma_1 N_2$ tends to zero by the same order as $N_1 - a_1$. We then have

$$\int_{a_1}^{N_1^0} \frac{dN_1}{\gamma_1 N_1 N_2} = \infty \, .$$

This shows that, starting from any point on the curve Λ in Figure 18, we will tend to approach the point a_1 asymptotically, without ever reaching it. In other terms, from any initial starting point, the number of individuals in the first species will tend asymptotically to

$$a_1 < c_1 = \frac{\epsilon_2''}{\gamma_2} ,$$

while the second species will tend towards extinction.

When $\epsilon_1' = \epsilon_1 - \alpha\lambda$ is zero, the cyclic curves in Figure 3 take, at the limit, the form of the curve in Figure 18. The lower portion of the previous curve approaches the straight line segment $a_1 b_1$ in Figure 18. However, while the curves in Figure 3 can be traced periodically, the segment $a_1 b_1$ can never be traced since it would take an infinite amount of time to reach the point a_1. For all practical purposes, at any point on the segment $a_1 b_1$, $N_2 = 0$ and N_1 appears to be constant.

The three curves, I, II, and III, which are represented in Figure 19, all originate at point P. Their equations are respectively:

$$\text{I.} \quad N_1^{\epsilon_2'} \; e^{-\gamma_2 N_1} = C' \; N_2^{-\epsilon_1'} \; e^{\gamma_1 N_2}$$

$$\text{II.} \quad N_1^{\epsilon_2''} \; e^{-\gamma_2 N_1} = C'' \; e^{\gamma_1 N_2}$$

$$\text{III.} \quad N_1^{\epsilon_2'''} \; e^{-\gamma_2 N_1} = C''' \; N_2^{h} \; e^{\gamma_1 N_2} .$$

The first is a cyclic curve corresponding to $\epsilon_1' > 0$, and, therefore, of type (III,IV). The second curve is a transitional one from type (III,IV) to type (IV,II) as seen in Figure 18. The third is a curve of type (IV,II), corresponding to $\epsilon_1 - \alpha\lambda$ negative and equal to $-h$. One has, furthermore,

$$0 < \epsilon_2' < \epsilon_2'' < \epsilon_2'''$$

Such curves represent, respectively, the cases in which the intensity of removal is less than, equal to, and greater than the limit $\dfrac{\epsilon_1}{\alpha}$ (Cf. §4, Number 5).

The fluctuations which correspond to the curves of type I give a typical example of an upper limit which is not a maximum. In fact, as the intensity of

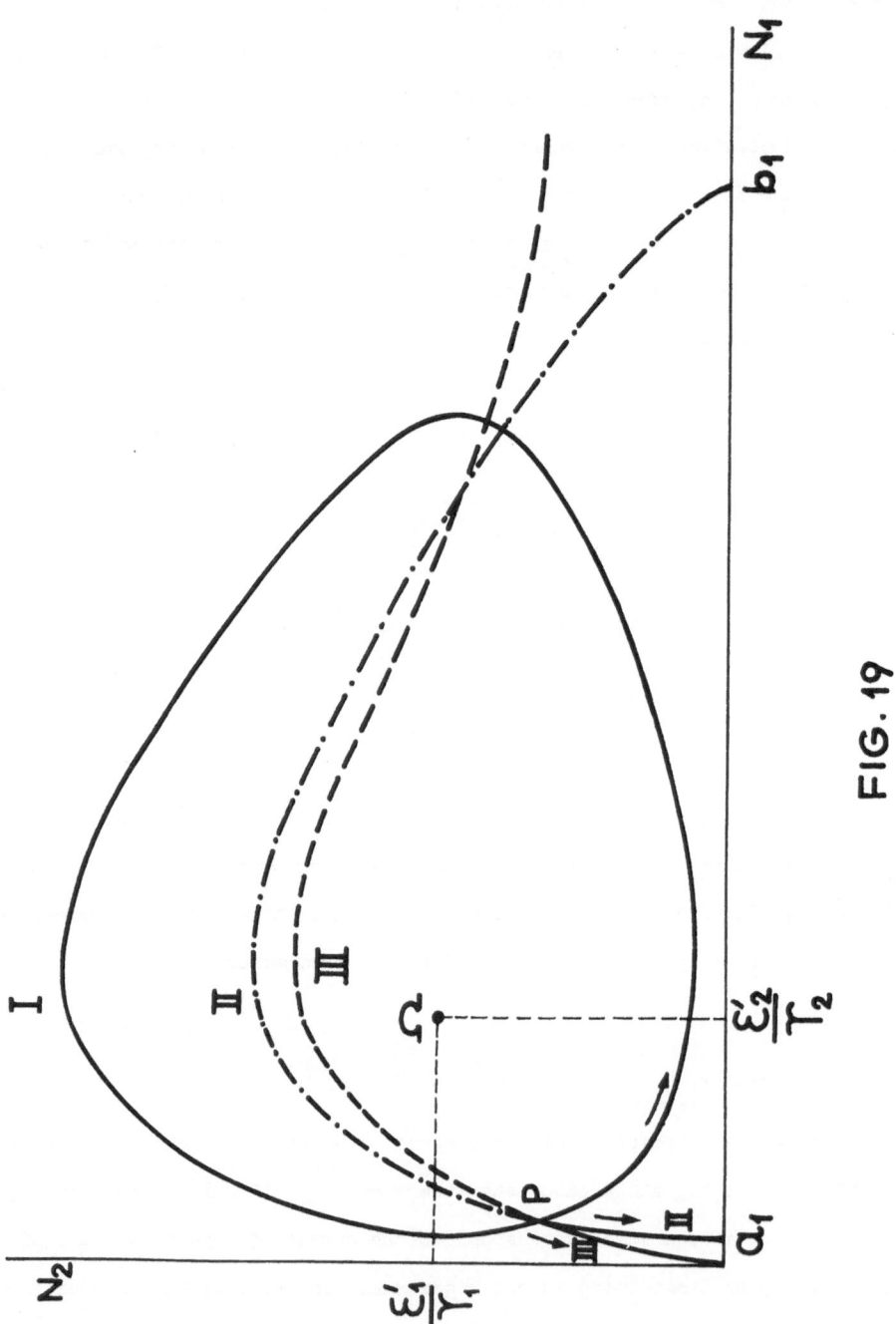

FIG. 19

removal, λ , approaches the limit $\dfrac{\varepsilon_1}{\alpha}$, the average number of individuals in the

first species, $\dfrac{\varepsilon_2'}{\gamma_2}$, increases toward $C_1 = \dfrac{\varepsilon_2''}{\gamma_2}$. However, it cannot reach this

value since, as λ reaches the limit $\dfrac{\varepsilon_1}{\alpha}$, the fluctuations cease. The number of

individuals in the first species then tends toward a_1 , below the previous

averages, while the second species tends toward extinction. When the intensity

of removal exceeds this limit, both species will tend toward extinction. The solid

portions of the curves, and the arrows connected to them, indicate how such changes

are produced.

To conclude let us remark that, as the intensity of removal increases to

$\dfrac{\varepsilon_1}{\alpha}$, the average number of individuals in the first species increases. At the

same time, however, the period of the fluctuation over which this average must

be computed continues to increase without bounds.

PART TWO

BIOLOGICAL ASSOCIATIONS OF SEVERAL SPECIES

§1. The case of any number of species competing for the same resource.

1. The treatment of competition between two species for a single resource is easily extended to any number of species. Let there be n species, which, in isolation, would have growth rates $\varepsilon_1, \varepsilon_2, \ldots, \varepsilon_n$. Denote by $F(N_1, N_2, \ldots, N_n)dt$ the decrease in the amount of common resource in an interval dt , where $N_1, N_2, \ldots, N_n$ are the numbers of individuals in each species. This function is zero only when $N_1 = N_2 = \ldots N_n = 0$. Otherwise the function is positive, and it increases indefinitely with each of the N_r . For simplicity we might as well take F to be linear, that is

$$F(N_1, N_2 \ldots N_n) = \alpha_1 N_1 + \alpha_2 N_2 + \ldots + \alpha_n N_n$$

where the coefficients α_r are positive. But we shall leave F in its general form.

The presence of N_1 individuals of the first species, N_2 of the second species, etc. will affect their coefficients of growth. The ε_r's reduce to $\varepsilon_r - \gamma_r F(N_1, \ldots N_n)$, in which the positive coefficient γ_r measures the effect of the reduced availability of resource on the growth rates of the species.

We will then have the differential equations

$$(31) \qquad \frac{dN_r}{dt} = N_r(\varepsilon_r - \gamma_r F(N_1 \ldots N_n)) , \quad (r = 1, 2 \ldots n)$$

from which it follows that

$$\frac{1}{\gamma_r N_r} \frac{dN_r}{dt} - \frac{1}{\gamma_s N_s} \frac{dN_s}{dt} = \frac{\varepsilon_r}{\gamma_r} - \frac{\varepsilon_s}{\gamma_s} \quad .$$

Integrating and passing from logarithms to numbers, one has

$$\frac{N_r^{\frac{1}{\gamma_r}}}{N_s^{\frac{1}{\gamma_s}}} = C e^{\left(\frac{\varepsilon_r}{\gamma_r} - \frac{\varepsilon_s}{\gamma_s}\right) t}$$

where C is a positive constant.

2. Let us order the ratios $\frac{\varepsilon_r}{\gamma_r}$ according to their magnitudes,[8] namely we suppose that

$$\frac{\varepsilon_1}{\gamma_1} > \frac{\varepsilon_2}{\gamma_2} > \frac{\varepsilon_3}{\gamma_3} \ldots > \frac{\varepsilon_n}{\gamma_n} .$$

Then, if r < s , we have

$$\lim_{t=\infty} \frac{N_r^{\frac{1}{\gamma_r}}}{N_s^{\frac{1}{\gamma_s}}} = \infty .$$

This implies that, as time increases, N_r can either increase without bounds, or that

$$\lim_{t=\infty} N_s = 0 .$$

However, the former possibility must be excluded since F increases indefinitely with N_r , so that the right-hand side of (31) becomes negative when N_r exceeds a given limit. It follows that the upper limit of N_r must be finite and, therefore, that all of the species except the first will tend to disappear. The asymptotic behavior of N_1 is then the same as that for the case of just two species.

---

[8] We do not consider the cases of equality since their probability is vanishingly small.

§2. The case of any number of predators and prey.

1. Let us consider the case of n species, and suppose that encounters between two individuals of different species are always either favorable to one species and unfavorable to the other, or of no consequence to either species. If N_r is the number of individuals of species r , and N_s the number of species s , the probability of an encounter between two individuals of the different species will be proportional to $N_r N_s$. We may then assume that the number of encounters per unit time is given by $m_{rs} N_r N_s$. Let p_{rs} be the probability that an encounter between an individual of species r and an individual of species s results in the destruction of the r individual. Then, in a unit time, $m_{rs} p_{rs} N_r N_s$ individuals of this species will be destroyed. Let us now evaluate the effect of this destruction on the number of individuals in the other species.

A rough evaluation of this effect can be made as follows. Denote by β_1, β_2, ... β_n the average weights of the individuals in each of the n species, and by P_1, P_2, ... P_n, respectively, the total weights of their individuals. The numbers of individuals in each species is then

$$N_1 = \frac{P_1}{\beta_1} , \ \ldots \ N_r = \frac{P_r}{\beta_r} , \ \ldots \ N_n = \frac{P_n}{\beta_n} \ .$$

When an individual of species r is eaten by an individual of species s the weight P_r decreases to $P_r - \beta_r$, while the weight P_s increases to $P_s + \beta_r$. The numbers of individuals in each of the two species then change, respectively, to

$$\frac{P_r - \beta_r}{\beta_r} \ = \ N_r - 1, \quad \frac{P_s + \beta_r}{\beta_s} \ = \ N_s + \frac{\beta_r}{\beta_s}$$

In this very rough evaluation the encounters between individuals of species r and species s decrease the number of individuals in species r per unit time by the amount

$$m_{rs} p_{rs} \ N_r N_s \quad .$$

Similarly, the number of individuals in species s will increase per unit time by the amount

$$m_{rs}p_{rs} N_r N_s \frac{\beta_r}{\beta_s} .$$

Setting $m_{rs}p_{rs}\beta_r = a_{rs}$, species r will decrease by $\frac{1}{\beta_r} a_{rs} N_r N_s$ and species s will increase by $\frac{1}{\beta_s} a_{rs} N_r N_s$. Alternatively, supposing that a_{sr} is negative, and setting $a_{rs} = -a_{sr}$, we can say that, because of their encounters, the respective numbers in species r and in species s will change per unit time by

$$\frac{1}{\beta_r} a_{sr} N_r N_s , \quad \frac{1}{\beta_s} a_{rs} N_r N_s .$$

The analogous increments in the time interval dt would then be

$$\frac{1}{\beta_r} a_{sr} N_r N_s dt , \quad \frac{1}{\beta_s} a_{rs} N_r N_s dt .$$

The same reasoning applies for any other pair of species. In other words, the numbers $\frac{1}{\beta_1}$, $\frac{1}{\beta_2}$, $\dots$ $\frac{1}{\beta_n}$ represent the *equivalent numbers* of individuals in the different species. To say that $\frac{1}{\beta_r}$ individuals of species r are transformed into $\frac{1}{\beta_s}$ individuals of species s , is in fact the same as saying that $\frac{1}{\beta_r}$ individuals of species r are *equivalent* to $\frac{1}{\beta_s}$ individuals of species s . Above, as a very rough approximation, we assumed that such *equivalent numbers* are the reciprocals of the average weights. Generally, this will not be the case. The above reasoning, however, still applies since it is enough that such *equivalent numbers* can be defined, no matter what they are.

2. Denote by ε_r the growth rate that species r would have it alone. Now, because of its coexistence with the other species, the variation in the number of its individuals in a time interval dt will be

$$dN_r = \varepsilon_r N_r dt + \frac{1}{\beta_r} \sum_{1}^{n} a_{sr} N_r N_s dt \quad .$$

From this we have the system of differential equations

$$\frac{dN_r}{dt} = \left(\varepsilon_r + \frac{1}{\beta_r} \sum_{1}^{n} a_{sr} N_s \right) N_r \quad , \quad (r = 1,2,\ldots\ n)$$

which is the same as

$$(B) \qquad \beta_r \frac{dN_r}{dt} = \left(\varepsilon_r \beta_r + \sum_{1}^{n} a_{sr} N_s \right) N_r \quad , \quad (r = 1,2,\ldots\ n)$$

in which

$$a_{rs} = - a_{sr} \ , \quad a_{rr} = 0 \ , \quad \beta_1 \ , \ \beta_2 \ \ldots\ \beta_n > 0 \ .$$

In the case of just one predator and one prey, we considered the equations (Part I, §2, Number 1)

$$\frac{dN_1}{dt} = (\varepsilon_1 - \gamma_1 N_2) N_1 \ , \quad \frac{dN_2}{dt} = (-\varepsilon_2 + \gamma_2 N_1) N_2 \ .$$

These assume the same form as (B) by setting

$$\gamma_1 = \frac{a_{12}}{\beta_1} \ , \qquad \gamma_2 = \frac{a_{21}}{\beta_2}$$

and by writing ε_2 instead of $-\varepsilon_2$, supposing that ε_2 is negative. We see , then, that in this case there is no need to explicitly introduce a set of *equivalent numbers*.

As another example, suppose we have an association of n species so that the first feeds on the second, the second on the third, and so on up to the nth species. We will then have the n equations

$$\frac{dN_1}{dt} = (\varepsilon_1 + \gamma_1' N_2)N_1 \quad , \qquad \frac{dN_2}{dt} = (\varepsilon_2 - \gamma_2 N_1 + \gamma_2' N_3)N_2$$

$$\frac{dN_3}{dt} = (\varepsilon_3 - \gamma_3 N_2 + \gamma_3' N_4)N_3 \;,\cdots \frac{dN_n}{dt} = (\varepsilon_n - \gamma_n N_{n-1})N_n$$

in which the γ_2, γ_3, $\cdots \gamma_n$ and the γ_1', γ_2', $\cdots \gamma_{n-1}'$ are positive. We can then choose the numbers a_{21}, a_{32}, a_{43}, $\cdots a_{n,n-1}$, and β_1, β_2, $\cdots \beta_n$ such that

$$\gamma_1' = \frac{a_{21}}{\beta_1} \;, \quad \gamma_2 = \frac{a_{21}}{\beta_2} \;, \quad \gamma_2' = \frac{a_{32}}{\beta_2} \;, \quad \gamma_3 = \frac{a_{32}}{\beta_3} \;, \quad \gamma_3' = \frac{a_{43}}{\beta_3} \;, \; \cdots \gamma_n = \frac{a_{n,n-1}}{\beta_n}$$

while all the other a_{sr}'s are zero. In this way the above equations assume the form (B).

As a further example consider four species, and suppose that both the first and the third species feed upon species two, while only species three feeds upon species four. The corresponding equations would be

$$\frac{dN_1}{dt} = (\varepsilon_1 + \gamma_1 N_2)N_1 \;, \quad \frac{dN_2}{dt} = (\varepsilon_2 - \gamma_2 N_1 - \gamma_2' N_3)N_2$$

$$\frac{dN_3}{dt} = (\varepsilon_3 + \gamma_3 N_2 + \gamma_3' N_4)N_3 \;, \quad \frac{dN_4}{dt} = (\varepsilon_4 - \gamma_4 N_3)N_4$$

in which γ_1, γ_2, γ_3, γ_4, and γ_2', γ_3' are positive numbers. They can be re-written in the form (B),

$$\frac{dN_1}{dt} = (\varepsilon_1 + \frac{a_{21}}{\beta_1} N_2)N_1 \;, \quad \frac{dN_2}{dt} = (\varepsilon_2 + \frac{a_{12}}{\beta_2} N_1 + \frac{a_{32}}{\beta_2} N_3)N_2$$

$$\frac{dN_3}{dt} = (\varepsilon_3 + \frac{a_{23}}{\beta_3} N_2 + \frac{a_{43}}{\beta_3} N_4)N_3 \;, \quad \frac{dN_4}{dt} = (\varepsilon_4 + \frac{a_{34}}{\beta_4} N_3)N_4 \;,$$

by conveniently choosing β_1, β_2, β_3, β_4, a_{12}, a_{23}, a_{34}, and by taking all the

the other a_{sr}'s to be zero.

In all of these cases, as in many others one might conceive, the numbers β_1, β_2, ... β_n are obtained without any special hypothesis as to their meaning.

3. A number of general theorems directly follow from Equations (B):

It is impossible for all of the species to go extinct if at least one species maintains a positive rate of growth.

Suppose, in fact, that $\varepsilon_r > 0$ and

$$N_1, N_2, ... N_n < \eta , \quad \frac{1}{\beta_r} \sum_1^n |a_{sr}| = p_r ,$$

so that we have

$$\frac{1}{N_r} \frac{dN_r}{dt} > \varepsilon_r - p_r\eta .$$

If all the numbers N_1, N_2, ... N_n were to tend to zero, we could find a value of time, t_0, such that, when $t \geq t_0$ we would have

$$\eta < \frac{\varepsilon_r}{p_r}$$

and, therefore,

$$\varepsilon_r - p_r\eta = \ell > 0 .$$

From this one would have

$$N_r > N_r^0 e^{\ell(t-t_0)}$$

where N_r^0 is the value of N_r at $t = t_0$. It follows that, as t increases indefinitely, N_r will also indefinitely increase, in contrast with the assumption $N_r < \eta$.

If the growth rates are all negative, the species will all go extinct, while, if they are all positive, the numbers in each species will increase without bounds.

Suppose, in fact, that in the Equations (B)

$$\varepsilon_r < -\varepsilon \ , \qquad (r = 1, 2, \ldots n)$$

where ε is a positive quantity. Summing these equations, side by side, we have

$$\sum_1^n \beta_r \frac{dN_r}{dt} < - \varepsilon \sum_1^n \beta_r N_r$$

that is,

$$\frac{d}{dt} \ln \sum_1^n \beta_r N_r < - \varepsilon$$

Integrating, and passing from logarithms to numbers, we get

$$\sum_1^n \beta_r N_r < \sum_1^n \beta_r N_r^0 e^{-\varepsilon t}$$

in which N_r^0 denotes the initial value of N_r , showing that all the N_r's tend to zero.

Much in the same way, if we had

$$\varepsilon_r > \varepsilon \ , \quad (r = 1, 2, \ldots n)$$

in which ε was positive, we would obtain

$$\sum_1^n \beta_r N_r > \sum_1^n \beta_r N_r^0 e^{\varepsilon t}$$

so that all the N_r's could not be bounded above by a finite number. Combining the results of these two theorems leads to this proposition: *All the species will go to extinction if and only if all their rates of growth are negative.*

4. In order that N_1, N_2, ... N_n remain constant, they must satisfy the system of equations

$$\frac{dN_1}{dt} = \frac{dN_2}{dt} = \ldots = \frac{dN_n}{dt} = 0$$

which, from (B), give

(B')
$$\varepsilon_r \beta_r + \sum_1^n a_{sr} N_s = 0 .$$

We shall call these equations the *steady state equations*. Their determinant,

(C)
$$\begin{vmatrix} 0 & , a_{21}, & a_{31}, & \ldots & a_{n1} \\ a_{12}, & 0 & , a_{32}, & \ldots & a_{n2} \\ a_{13}, & a_{23}, & 0 & , \ldots & a_{n3} \\ \ldots\ldots\ldots\ldots\ldots\ldots\ldots\ldots \\ a_{1n}, & a_{2n}, & a_{3n}, & \ldots & 0 \end{vmatrix} ,$$

which we shall call the *fundamental determinant*, is *anti*-symmetric. It will therefore be positive if n is even, and zero if n is odd. It is then convenient to treat the cases in which the number of species is *even* separately from the cases in which the number is *odd*.

§3. <u>An even number of coexisting species.</u>

1. At the end of the previous section we distinguished between the cases of even and odd numbers of species. Let us start with the former case.

Since $a_{sr} = -a_{rs}$, we have

(32)
$$\sum_1^n \beta_r \frac{dN_r}{dt} = \sum_1^n \varepsilon_r \beta_r N_r$$

(33)
$$\beta_r \frac{d \ln N_r}{dt} - \varepsilon_r \beta_r = \sum_s a_{sr} N_s , \quad (r = 1, 2 \ldots n) .$$

The fundamental determinant

$$
\begin{vmatrix}
0 & , a_{21}, & a_{34}, & \cdots\cdots & a_{n1} \\
a_{12}, & 0 & , a_{32}, & \cdots\cdots & a_{n2} \\
a_{13}, & a_{23}, & 0 & , \cdots\cdots & a_{n3} \\
\cdots & \cdots & \cdots & \cdots\cdots & \cdots \\
a_{1n}, & a_{2n}, & a_{3n}, & \cdots\cdots & 0
\end{vmatrix}
$$

is of even order. It is antisymmetrical, and generally non-zero. We will assume that it is always *positive* since it is a square, and the probability that it is zero is vanishingly small.

Denote by A_{sr} the conjugate of a_{sr}, so that

$$
\sum_{1}^{n} {}_r A_{hr} a_{sr}
\begin{cases}
= 0 , h \gtrless s \\[2em]
= 1 , h = s .
\end{cases}
$$

From (33) we get

$$
N_h = \sum_{1}^{n} {}_r A_{hr} \left(\beta_r \frac{d \ln N_r}{dt} - \varepsilon_r \beta_r \right)
$$

from which it follows that

$$
\sum_{1}^{n} {}_h \varepsilon_h \beta_h N_h = \sum_{1}^{n} {}_h \varepsilon_h \beta_h \sum_{1}^{n} {}_r A_{hr} \left(\beta_r \frac{d \ln N_r}{dt} - \varepsilon_r \beta_r \right) .
$$

But $A_{hr} = - A_{rh}$, and, therefore,

$$
\sum_{1}^{n} {}_h \varepsilon_h \beta_h \sum_{1}^{n} {}_r A_{hr} \varepsilon_r \beta_r = 0 ,
$$

by which the preceding equation becomes

$$
\sum_{1}^{n} {}_h \varepsilon_h \beta_h N_h = \sum_{1}^{n} {}_h \sum_{1}^{n} {}_r A_{hr} \beta_h \beta_r \varepsilon_h \frac{d \ln N_r}{dt} = \sum_{1}^{n} {}_r q_r \beta_r \frac{d \ln N_r}{dt}
$$

in which

(34)
$$q_r = \sum_{1}^{n} A_{hr} \beta_h \epsilon_h \ .$$

Because of (32) we have

$$\sum_{1}^{n} \beta_r \frac{dN_r}{dt} = \sum_{1}^{n} q_r \beta_r \frac{d \ln N_r}{dt} \ ,$$

that is,

$$\frac{d}{dt} \sum_{1}^{n} \beta_r (N_r - q_r \ln N_r) = 0 \ .$$

Integrating, and passing from logarithms to numbers, we have the following integral

$$\left(\frac{e^{N_1}}{N_1^{q_1}} \right)^{\beta_1} \left(\frac{e^{N_2}}{N_2^{q_2}} \right)^{\beta_2} \cdots \cdots \left(\frac{e^{N_n}}{N_n^{q_n}} \right)^{\beta_n} = C \ .$$

in which C is a positive constant.

Assuming that all the q_r's are non-zero, this equation can also be written as

$$\left(\frac{e^{\frac{N_1}{q_1}}}{\frac{N_1}{q_1}} \right)^{q_1 \beta_1} \left(\frac{e^{\frac{N_2}{q_2}}}{\frac{N_2}{q_2}} \right)^{q_2 \beta_2} \cdots \cdots \left(\frac{e^{\frac{N_n}{q_2}}}{\frac{N_n}{q_n}} \right)^{q_n \beta_n} = C'$$

in which $C' = C q_1^{q_1 \beta_1} \ q_2^{q_2 \beta_2} \ \cdots \ q_n^{q_n \beta_n}$, is a new positive constant. Setting

$$n_r = \frac{N_r}{q_r}$$

we have

$$\left(\frac{e^{n_1}}{n_1}\right)^{q_1\beta_1} \left(\frac{e^{n_2}}{n_2}\right)^{q_2\beta_2} \cdots \left(\frac{e^{n_n}}{n_n}\right)^{q_n\beta_n} = C' \ .$$

If q_1, q_2, ... q_n are positive numbers, n_1, n_2, ... n_n will also be positive, and, therefore,

$$\frac{e^{n_r}}{n_r} \geq e$$

from which

$$\left(\frac{e^{n_r}}{n_r}\right)^{q_r\beta_r} \leq \frac{C' \, e^{q_r \beta_r}}{e^{\Sigma_h q_h \beta_h}} = K \, e^{q_r \beta_r} \ .$$

Setting

$$K = \frac{C'}{e^{\Sigma_h q_h \beta_h}}$$

we then have

(36)
$$\frac{e^{n_r}}{n_r} \leq e \sqrt[q_r\beta_r]{K}$$

This shows that n_r must always be bound between two positive numbers, one larger, and the other smaller than one. This can also be seen through very simple geometric considerations. Consider, in fact, the curve which has the equation (see Figure 20)

$$y = \frac{e^x}{x} \qquad \begin{array}{l} 0 < x < \infty \\[4pt] e < y < \infty \end{array} \ .$$

The ordinate y will be at a minimum, e , when $x = 1$, and it will tend toward infinity for $x = 0$ and $x = \infty$. Let us draw a straight line parallel to the x axis at a distance $y_0 > e$ from it. The line will intersect the curve at two points, A and B , which correspond to the points x^0 and x' on the abscissa.

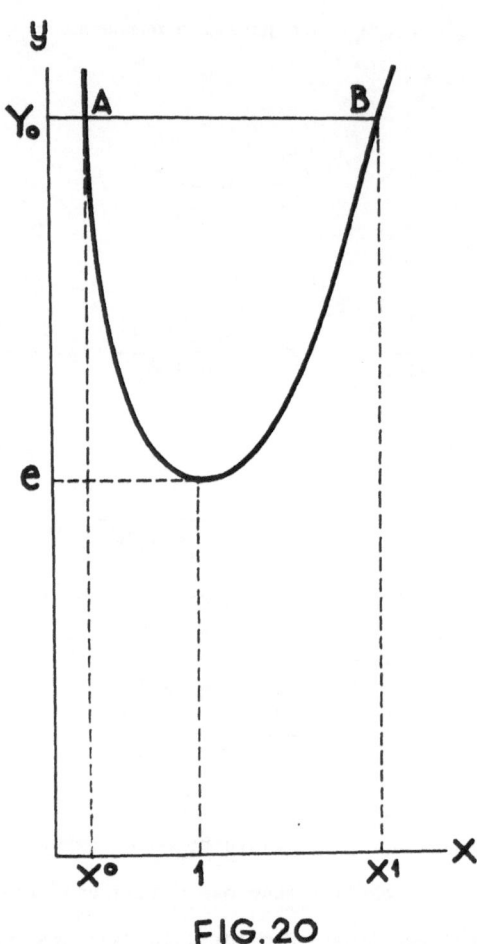

FIG. 20

If $\dfrac{e^x}{x} < y_0$, we have

$$0 < x^0 < 1 < x' < \infty \; ; \quad x^0 < x < x' \quad .$$

It then follows from (36) that $n_r^0 < n_r < n_r'$, in which both n_r^0 and n_r' are positive, the former being smaller and the latter larger than one. Setting

$$n_r^0 q_r \;=\; N_r^0 \, , \; n_r' q_r \;=\; N_r' \, ,$$

we have $N_r^0 < N_r < N_r'$, in which N_r^0 and N_r' are positive numbers, **respectively smaller than and larger than** q_r.

The fact that the q_r's are positive can be interpreted as follows. From Equations (B) we see that the necessary and sufficient conditions for the N_r's to be constant are that (see §2, Number 4)

(B') $$\varepsilon_r \beta_r + \sum_s^n {}_1 a_{sr} N_s \;=\; 0 \, , \quad (r = 1, 2, \ldots n) \, ,$$

namely, that the N_s's , which are the solutions of this system, are positive. These solutions are (see Equation 34)

(34') $$N_s \;=\; -\sum_r^n {}_1 A_{sr} \varepsilon_r \beta_r = \sum_r^n {}_1 A_{rs} \varepsilon_r \beta_r \;=\; q_s \, .$$

It follows that an equilibrium will exist if and only if the q_s's are positive. It is possible, however, that *the equilibrium may be reached through the extinction of some of the species.*

We can then state the following theorem: *If an equilibrium exists for all the species, and if the initial conditions are different from this state, the numbers of individuals in each species will remain bounded, each between two positive numbers.*

2. With this premise, it is convenient to introduce a number of definitions through which a number of properties can be stated without ambiguities. If

N(t) denotes the number of individuals of a species and it is always bound between two positive numbers, we will say that the species has a *limited variation between positive numbers*. If N(t) tends toward zero we will say that the species goes *extinct*, or that its *variation* results in *extinction*. If N(t) is bound between two positive numbers, and if it goes through maxima and minima during any amount of time $t > t_0$, we say that there are *fluctuations* in N .

We will say that there are *damped fluctuations* if the amplitude of the oscillations in N(t) at $t > t_0$ can be made as small as we please by choosing a sufficiently large t_0 . In this, and only in this case N tends toward a precise and finite limit as t tends toward infinity. We will say that N(t) varies *asymptotically* and that it *tends asymptotically* to the limit q if N(t) does not fluctuate, and if it tends toward the specific and finite limit q as t tends toward infinity.

3. We are now in a position to state:

If there exists an equilibrium, and if the initial state is different from it, there will be fluctuations in the numbers of at least some of the species, and these fluctuations will not dampen.

To prove this theorem notice that two cases are possible: all the N_1, N_2, ... N_n tend toward precise and finite limits as time increases indefinitely,[9] or, alternatively, some of the numbers oscillate, but the amplitudes of their oscillations remain larger than a positive number. There are, then, fluctuations, and some of them cannot dampen.

Now, it is not possible for all the N_1 , N_2 , ... N_n to tend toward q_1 , q_2 , ... q_n , namely, that the n_1 , n_2 , ... n_n all tend toward 1. This is so because the smallest value which the constant C' can assume is

$$m = e^{q_1\beta_1 + q_2\beta_2 + \dots + q_n\beta_n}$$

---

[9] The case in which some of the N_r's are constant is included in this case

and this happens only when all the n_r's are equal to 1. It is enough that at least one of these quantities is positive and different from 1 for C' to be larger than m. Therefore, if the initial state is not the same as the one in which n_1, n_2, ... n_n are all equal to 1 (in which case they would always maintain this value), we must have $C' > m$. If n_1, n_2, ... n_n were all to tend toward 1, the left-hand side of (35) should tend toward m. It always maintains, instead, the same value, $C' > m$.

Also, it is not possible for the N_1, N_2, ... N_n all to tend toward other limits q_1', q_2', ... q_n' which are completely or partially different from q_1, q_2, ... q_n. In fact, in this case the $\beta_r \frac{dN_r}{dt}$ would tend toward the following finite limits

$$(\varepsilon_r \beta_n + \sum_1^n {}_s a_{sr} q_s') q_r' \quad .$$

But, if N_r and $\frac{dN_r}{dt}$ tend toward precise and finite limits as t tends toward ∞, all the $\frac{dN_r}{dt}$ would tend toward zero, and therefore we should have

$$\varepsilon_r \beta_r + \sum_1^n {}_s a_{sr} q_s' = 0 \quad .$$

Solving these equations with respect to the q_s's, and remembering that the determinant (C) is non-zero, we find (see Equation 34)

$$q_s' = q_s , \quad (s = 1, 2, \ldots n)$$

which contradicts the premise. It is therefore necessary that some of the N_1, N_2, ... N_n continue to oscillate without dampening as time increases indefinitely, which proves the theorem.

Let us consider again Equations (B), and integrate them between two times, t_0 and t. We have

$$\frac{\beta_r}{T} \ln \frac{N_r}{N_r^0} = \varepsilon_r \beta_r + \sum_1^n {}_s a_{sr} \frac{1}{T} \int_{t_0}^t N_s dt$$

in which N_r^0 is the value of N_r at $t = t_0$, and $T = t - t_0$. Setting

$$\frac{1}{T} \int_{t_0}^{t} N_r dt = N_s \text{ ,}$$

N_s represents the average of the values assumed by N_s in the time interval (t_0, t). In terms of these quantities, the equations above become

$$\frac{\beta_r}{T} \ell n \frac{N_r}{N_r^0} = \varepsilon_r \beta_r + \sum_1^n a_{sr} N_s \text{ .}$$

By choosing a sufficiently large T, one can make

$$\frac{\beta_r}{T} \ell n \frac{N_r}{N_r^0} = \sigma_r$$

as small as one pleases since N_r and N_r^0 are bound between two determinate, positive numbers. The values of the N_s's which satisfy the preceding equations can then be made as close to the values of the q_s's as one wishes. That is to say

$$\lim_{T=\infty} N_r = q_r \text{ .}$$

We then have the following theorem:

Under the same conditions as in the preceding theorems, the limits of the average values of the N_r during a time interval (t_0, t) tend toward the quantities q_r as t tends toward ∞. Such limits will be called *asymptotic averages.*

As a corollary to the above theorem:

The asymptotic averages of the N_r do not depend on their initial values (Cf. the theorem on the *conservation of the averages* in Part I, §2, Number 9).

All these results can be summarized in the following general proposition:

I) *If there exists an equilibrium for a biological association as in (B),
the numbers of individuals in each species will be bound between positive numbers,
there will always be undamped oscillations, and the asymptotic averages of these
numbers will be the same as their equilibrium values.*[10]

As we have seen above, the fact that n_1, n_2, ... n_n all take the value 1
must necessarily imply that $N_1 = q_1$, $N_2 = q_2$, ... $N_n = q_n$, and vice versa. If
the initial values in (35) are chosen to be sufficiently close to 1, we can then
make C' as close as we please to $e^{\Sigma_h q_h \beta_h}$, namely, K as close as we please to
1. But at any instant, because of (36), we have

$$e \leq \frac{e^{n_r}}{n_r} \leq e^{q_r \beta_r} \sqrt{K} \quad .$$

It is then possible to **make** n_r remain as close to 1 as we please.

We thus conclude:

II) *The deviations from equilibrium can be made to remain arbitrarily
small by chosing an initial state which is sufficiently close to the equilibrium
point.*

In other words:

The equilibrium is always stable.[11]

4. It is easy to study small oscillations in the neighborhood of the
equilibrium point. From (34) one has, in fact,

$$(34'') \qquad\qquad \epsilon_r \beta_r = - \sum_1^n{}_s a_{sr} q_s$$

---

[10] The study of small oscillations in Number 4 will give a precise idea of the
behavior of these fluctuations.

[11] Here, as before (Cf. Part I, §2), stability is used in much the same way as
in mechanics. (It is also called neutrality to differentiate it from asymptotic
stability. Editors' note)

from which (B) becomes

$$\beta_r \frac{dN_r}{dt} = \sum_1^n {}_s a_{sr} (N_s - q_s)N_r \quad ,$$

that is,

$$\beta_r \frac{dn_r}{dt} = \sum_1^n {}_s a_{sr} q_s (n_s - 1)n_r \ .$$

Setting $n_r = 1 + \nu_r$ the preceding equations take the form

$$\frac{\beta_r d\nu_r}{dt} = \sum_1^n {}_s a_{sr} q_s \nu_s (1 + \nu_r) \ .$$

If the initial values of the ν_r's are sufficiently small, the theorem on the stability of the equilibrium guarantees that they will remain as small as one pleases.

Neglecting the 2nd order terms in ν_r , the preceding equations reduce to

$$(37) \qquad\qquad \beta_r \frac{d\nu_r}{dt} = \sum_1^n {}_s a_{sr} q_s \nu_s \quad .$$

Setting $\nu_r = A_r e^{-xt}$, (37) is written as

$$(38) \qquad\qquad \beta_r A_r x + \sum_1^n {}_s a_{sr} q_s A_s = 0 \ , \ (r = 1, 2 \ldots n) \ .$$

Eliminating the A_r's leads to the equation

$$(39) \qquad
\begin{vmatrix}
\frac{\beta_1}{q_1} x, & a_{21} , & a_{31} , & \cdots\cdots & a_{n1} \\[2mm]
a_{12} , & \frac{\beta_2}{q_2} x, & a_{32} , & \cdots\cdots & a_{n2} \\[2mm]
a_{13} , & a_{23} , & \frac{\beta_3}{q_3} x, & \cdots\cdots & a_{n3} \\[2mm]
\multicolumn{5}{c}{\cdots\cdots\cdots\cdots\cdots\cdots\cdots} \\[2mm]
a_{1n} , & a_{2n} , & a_{3n} , & \cdots\cdots & \frac{\beta_n}{q_n} x
\end{vmatrix} = 0 \ ,$$

which has only imaginary roots.

If there were, in fact, a real root x , we could assume that the A_r's were real, and from (38) we would obtain

$$x \sum_{1}^{n}{}_r \beta_r q_r A_r^2 + \sum_{1}^{n}{}_r \sum_{1}^{n}{}_s a_{sr} q_s q_r A_s A_r = 0 \ .$$

But the double summation is zero, so that

$$x \sum_{1}^{n} \beta_r q_r A_r^2 = 0 \ .$$

Since β_r and q_r are positive, we would have $x = 0$. But Equation (39) cannot have a zero root since, by assumption, the determinant (C) differs from zero.

If (39) had a complex root $a + ib$ it would also have its conjugate $a - ib$, and the A_r's corresponding to the first root would generally be complex numbers. Denote by A_r' their conjugates. They will correspond to the second root so that

$$(a + ib)\beta_r A_r + \sum_{1}^{n}{}_s a_{sr} q_s A_s = 0$$

$$(a - ib)\beta_r A_r' + \sum_{1}^{n}{}_s a_{sr} q_s A_s' = 0 \ ,$$

from which

$$(a + ib) \sum_{1}^{n}{}_r \beta_r q_r A_r A_r' + \sum_{1}^{n}{}_r \sum_{1}^{n}{}_s a_{sr} q_r q_s A_s A_r' = 0$$

$$(a - ib) \sum_{1}^{n}{}_r \beta_r q_r A_r' A_r + \sum_{1}^{n}{}_r \sum_{1}^{n}{}_s a_{sr} q_r q_s A_s' A_r = 0 \ .$$

Summing term by term, one has

$$2 a \sum_{1}^{n}{}_r \beta_r q_r A_r A_r' + \sum_{1}^{n}{}_r \sum_{1}^{n}{}_s (a_{sr} + a_{rs}) q_r q_s A_s A_r' = 0 \ ,$$

namely

$$a \sum_{1}^{n}{}_r \beta_r q_r A_r A_r' = 0 \ ,$$

and, therefore, $a = 0$, showing that the roots are pure imaginary.

To exchange x with $-x$ in (39) is the same as to exchange the rows with the columns. Therefore, if x is a root of (39), $-x$ will also be a root. Suppose that all such roots are different. We can denote them by

$$ib', ib'', \ldots ib^{\left(\frac{n}{2}\right)}, - ib', - ib'', \ldots - ib^{\left(\frac{n}{2}\right)}$$

and, clearly, we can write

$$ib^{(h)} = \frac{2\pi i}{T^{(h)}} .$$

Let us also write the coefficients A_r which correspond to this root in the form

$$M_r^{(h)} e^{\frac{2\pi i}{T^{(h)}} a_r^{(h)}}$$

in which the $M_r^{(h)}$ are the moduli, and the $\dfrac{2\pi a_r^{(h)}}{T^{(h)}}$ are the arguments. The coefficients A_r corresponding to the root $-ib^{(h)} = - \dfrac{2\pi i}{T^{(h)}}$ will be

$$M_r^{(h)} e^{- \frac{2\pi i}{T^{(h)}} a_r^{(h)}} .$$

From this we have the roots of Equations (37)

$$\nu_r'^{(h)} = M_r^{(h)} e^{- \frac{2\pi i}{T^{(h)}} (t-a_r^{(h)})} , \quad (r = 1,2\ldots n)$$

and their conjugates

$$\nu_r'^{(h)} = M_r^{(h)} e^{\frac{2\pi i}{T^{(h)}} (t-a_r^{(h)})} , \quad (r = 1,2\ldots n) .$$

Combining them through the conjugate constants

$$\frac{1}{2} c^{(h)} e^{\frac{2\pi i}{T^{(h)}} \alpha^{(h)}} \quad , \quad \frac{1}{2} c^{(h)} e^{-\frac{2\pi i}{T^{(h)}} \alpha^{(h)}}$$

we obtain a first real solution of Equations (37) which depends on the two arbitrary constants $c^{(h)}$ and $\alpha^{(h)}$,

$$v_r^{\prime\prime(h)} = c^{(h)} M_r^{(h)} \cos \frac{2\pi}{T^{(h)}} \left(t - a_r^{(h)} - \alpha^{(h)} \right).$$

This fluctuation is periodic, and of period $T^{(h)}$. Its amplitude and phase depend on the two arbitrary constants.

The general integral of (37) is then

$$(\text{III}) \qquad v_r = \sum_{1}^{\frac{n}{2}} c^{(h)} M_r^{(h)} \cos \frac{2\pi}{T^{(h)}} \left(t - a_r^{(h)} - \alpha^{(h)} \right), \quad (r = 1, 2 \dots n)$$

having n arbitrary constants

$$c', c'', \dots c^{\left(\frac{n}{2}\right)}; \quad \alpha', \alpha', \dots \alpha^{\left(\frac{n}{2}\right)}.$$

We have, therefore, the following general theorem:

III) *The small fluctuations of n coexisting species can be obtained by the superposition of $\frac{n}{2}$ undamped fluctuations, each with its own period.*

Since the periods $T^{(h)}$ will not generally be commensurable, the resulting fluctuations will not generally be periodic. Notice that we have an even number of species, and that the number of periods, $T^{(h)}$, is half that of the number of species. In conclusion, theorems I, II, and III give the *three general properties of the variations in an even number of coexisting species.*

5. From the above treatment we can derive a number of properties of the fixed points, and of the fluctuations associated with them. Thus, from (34") we have

$$(34''') \qquad \epsilon_r = \frac{1}{\beta_r} \sum_1^n s \; a_{rs} q_s \; .$$

The $\frac{a_{rs}}{\beta_r}$ (see §2, Number 1) represent interactions during the encounters between individuals of the various species. Choosing arbitrary positive numbers for the q_s's , *the equations* (34"') *determine all the possible growth rates of the species which are compatible with the existence of a stable equilibrium, and of the associated fluctuations.*

From (32) we have

$$\sum_1^n r \; \beta_r N_r - \sum_1^n r \; \beta_r N_r^0 = \sum_1^n r \; \epsilon_r \beta_r \int_0^t N_r dt \; ,$$

in which N_r^0 denotes the initial value of N_r . If there exists an equilibrium in which all species are present, all the N_r's must remain bounded between two positive numbers. Let g be a positive number smaller than the minimum of these numbers. If all the ϵ's were positive, we would then have

$$\sum_1^n r \; \beta_r N_r > \sum_1^n r \; \beta_r N_r^0 + (\sum_1^n r \; \epsilon_r \beta_r) g t$$

and if all the ϵ's were negative

$$\sum_1^n r \; \beta_r N_r < \sum_1^n r \; \beta_r N_r^0 + (\sum_1^n r \; \epsilon_r \beta_r) g t \; .$$

These two inequalities cannot hold simultaneously since their left-hand sides must always be limited while their right-hand sides must, in time, diverge to $+\infty$ and to $-\infty$ respectively. We have then the following proposition: *For the stable coexistence of all the species to be possible, it is necessary that some of their growth rates, ε_r, be positive, and that some be negative.* That is: *If all the species tend indefinitely to increase when alone, or if they all tend toward extinction, a steady state is not possible.*

In fact, the Equations (B') cannot have all positive roots if the ε_r's have the same sign. This can also be recognized directly, since from (B') we have

$$\sum_1^n {}_r \; \varepsilon_r \beta_r q_r \; = \; 0$$

and this equation cannot hold if all the q_r's are positive, and all the ε_r's have the same sign.

§4. <u>An odd number of coexisting species.</u>

1. We now consider the case in which an odd number of species coexist. The fundamental determinant

(C)
$$\begin{vmatrix} 0 & , & a_{21}, & a_{31}, & \ldots\ldots & a_{n1} \\ a_{12}, & 0 & , & a_{32}, & \ldots\ldots & a_{n2} \\ a_{13}, & a_{23}, & 0 & , & \ldots\ldots & a_{n3} \\ \cdot & \cdot & \cdot & \cdot & \cdots\cdots & \cdot \\ a_{1n}, & a_{2n}, & a_{3n}, & \ldots\ldots & 0 \end{vmatrix}$$

is antisymmetric (Cf. §2, Number 4) and of odd order, so that it is zero. The minors of order $n-1$ corresponding to any of the columns are proportional to the square roots of the minors corresponding to the diagonal elements. Such minors are antisymmetric determinants of even order, and, therefore, they are squares. We suppose that they are all non-zero, that is, that they are positive.

Denote by R_1, R_2, ... R_n the square roots proportional to the minors of

order n − 1 . By arbitrarily choosing the sign of one of them, the signs of all
the others will also be determined. We have

$$\sum_{1}^{n}{}_{h}\ a_{rh}R_{h} \ = \ 0 \ , \quad (r = 1,2\ldots n)$$

and, therefore, from (B) we get

$$\sum_{1}^{n}{}_{r}\ \frac{\beta_r}{N_r}\,R_r\,\frac{dN_r}{dt} \ = \ \sum_{1}^{n}{}_{r}\ \varepsilon_r\beta_r R_r \ .$$

Integrating, and passing from logarithms to numbers,

(40)
$$N_1^{\beta_1 R_1}\ N_2^{\beta_2 R_2}\ \ldots\ N_n^{\beta_n R_n} \ = \ C\ e^{Lt}$$

in which C is a positive constant, and $L = \sum_{1}^{n}{}_{r}\ \varepsilon_r\beta_r R_r$.

The probability that L will ever equal zero is vanishingly small, and we
may then assume that it is either a positive or a negative constant. Therefore,
as time grows indefinitely, the right-hand side of the previous equation will tend
either to zero or to ∞. Since we do not know, *a priori*, the sign of the R_h's we
can only conclude that at least one of the N_h's must assume either arbitrarily
small or arbitrarily large values. This does not exclude the possibility that
some of the numbers could become indefinitely small **and others** become indefinitely
large. Excluding the case L = 0 , we then have the proposition: *If the number*
of species is odd it is not possible for all the numbers of their individuals
to remain bounded between two positive numbers.

The meaning of this theorem is properly understood only by considering it
as a purely hypothetical result. Notice, to begin with, that if one of the species
tends toward extinction we will go back to the case of an even number of species
which we treated in the previous Section.

If the number of individuals in a species increases without bounds, it is easy to see that Equations (B) no longer hold. In fact, we have assumed that the ε_r's are constant, and independent of the numbers of individuals in the association. This is not unreasonable as long as the numbers in a species do not exceed a certain limit. Beyond that limit this can no longer hold, and some of the rates of growth must be modified to prevent an indefinite increase (Cf. §6).

2. Let us consider again the general Equations (B)

(B)
$$\beta_r \frac{dN_r}{dt} = (\varepsilon_r \beta_r + \sum_{1}^{n} s \; a_{sr} N_s) N_r$$

in the case of an odd nubmer of species. Suppose that the system of $n - 1$ equations

$$\varepsilon_r \beta_r + \sum_{2}^{n} s \; a_{sr} N_s = 0 , \quad (r = 2,3.....n)$$

has the positive roots

$$Q_2, \; Q_2, \; ..... \; Q_n \; .$$

We can then write (B) in the form

$$\beta_1 \frac{dN_1}{dt} = (\varepsilon_1 \beta_1 + \sum_{2}^{n} s \; a_{s1} N_s) N_1$$

$$\beta_r \frac{dN_r}{dt} = \left(\sum_{2}^{n} s \; a_{sr} (N_s - Q_s) + a_{1r} N_1 \right) N_r, \quad (r=2,3...n) \; .$$

Setting

$$N_1 = Q_1 v_1 , \quad N_r = Q_r(1+v_r) , \quad (r = 2, 3 ... n) \; ,$$

in which Q_1 is a positive constant, and neglecting the second order terms in

ν_1, ν_2, ... ν_n , we have

(41)
$$\beta_1 \frac{d\nu_1}{dt} = (\varepsilon_1\beta_1 + \sum_2^n{}_s a_{s1}Q_s)\nu_1$$

(42)
$$\beta_r \frac{d\nu_r}{dt} = \sum_2^n{}_s a_{sr}Q_s\nu_s + a_{1r}Q_1\nu_1 , \quad (r = 2,3 \dots n) .$$

3. Consider the antisymmetric determinant of even order

$$D' = \begin{vmatrix} 0 & , a_{23}, & a_{24}, & \dots\dots & a_{2n} \\ a_{32}, & 0 & , a_{34}, & \dots\dots & a_{3n} \\ a_{42}, & a_{43}, & 0 & , \dots\dots & a_{4n} \\ \cdot & \cdot & \cdot & \dots\dots & \cdot \\ a_{n2}, & a_{n3}, & a_{n4}, & \dots\dots & 0 \end{vmatrix} .$$

We will have

$$Q_s = - \frac{\sum_2^n{}_r \varepsilon_r\beta_r D'_{rs}}{D'}$$

in which D'_{rs} is the reciprocal of a_{rs} in the preceding determinant, so that

$$\sum_2^n{}_s a_{s1}Q_s = - \frac{\sum_2^n{}_r \varepsilon_r\beta_r \sum_2^n{}_s a_{s1}D'_{rs}}{D'} .$$

Consider now the fundamental antisymmetric determinant of odd order

$$D = \begin{vmatrix} 0 & , a_{12}, & a_{13}, & \cdots\cdots & a_{1n} \\ a_{21}, & 0 & , a_{23}, & \cdots\cdots & a_{2n} \\ a_{31}, & a_{32}, & 0 & , \cdots\cdots & a_{3n} \\ \cdot\cdot\cdot & \cdot\cdot\cdot & \cdot\cdot\cdot & \cdot\cdot\cdot & \cdot \\ a_{n1}, & a_{n2}, & a_{n3}, & \cdots\cdots & 0 \end{vmatrix}$$

and let D_{rs} be the conjugate element of a_{rs} . We will have

$$D' = D_{11}$$

$$\sum_{2}^{n} {}_s \ a_{s1} D'_{rs} = - \sum_{2}^{n} {}_s \ a_{1s} D'_{rs} = - D_{r1}$$

and, therefore,

$$\sum_{2}^{n} {}_s \ a_{s1} Q_s = \frac{\sum_{2}^{n} {}_r \ \epsilon_r \beta_r D_{r1}}{D_{11}}$$

$$\epsilon_1 \beta_1 + \sum_{2}^{n} {}_s \ a_{s1} Q_s = \frac{\sum_{1}^{n} {}_r \ \epsilon_r \beta_r D_{r1}}{D_{11}} \ .$$

One has, however,

$$D_{r1} = \sqrt{D_{rr}} \ \sqrt{D_{11}} \ ,$$

from which

$$\epsilon_1 \beta_1 + \sum_{2}^{n} {}_s \ a_{s1} Q_s = \frac{1}{R_1} \sum_{1}^{n} {}_r \ \epsilon_r \beta_r R_r = \frac{L}{R_1}$$

setting, as in Number 1, $\sum_{1}^{n} {}_r \ \epsilon_r \beta_r R_r = L$, and denoting by R_r the $\sqrt{D_{rr}}$. As was stated there, the signs of the R_r's are determined by choosing the sign of only one of them. We therefore choose $R_1 > 0$.

4. Suppose that L is negative, and write $\frac{L}{\beta_1 R_1} = -m$. In this case
Equation (41) is integrated by

$$\nu_1 = A_1 e^{-mt} \; .$$

This justifies having previously neglected the second order terms. It also shows,
a posteriori, that these terms cannot be neglected when L is positive.

Let us integrate the equations

$$\beta_r \frac{d\nu_r}{dt} = \sum_2^n s \; a_{sr} Q_s \nu_s \; , \quad (r = 2, 3, \ldots n)$$

which have the same form as (37). Their solutions, u_r , have the same form as
the integrals (III) derived in §3, Number 4. The integrals of (42) are then given
by

$$\nu_r = u_r + A_r e^{-mt}$$

in which the coefficients A_r are easily calculated by the standard theory of
linear differential equations.

As in Number 1 we see that when $R_1 > 0$ and L < 0 : *The variation is
obtained by superimposing on the fluctuations of species 2, 3, ... n a decrease
in all the species. In other words the first species tends asymptotically to
extinction while all the others tend to fluctuate around their fixed point, as
in the conservative associations with an even number of species.* (Cf. §7, Number 6
for a formal definition of a conservative association.)

§5. <u>Extensions of the three fundamental properties of the fluctuations.</u>

1. In §2 of Part I three fundamental properties of fluctuations were
derived for the case of two coexisting species. We now attempt to extend them
to the general case of n species.

In §3 it was shown that: *If the fundamental determinant corresponding to*

*an association of an even number of species is different from zero, and if there
exists an equilibrium in which all are present, the numbers of all the species will
always fluctuate without dampening between two positive bounds.* This extends the
first property to the general case in which the fluctuation persists, but the
periodicity has been lost.

2. Assume that the above conditions are satisfied. Concerning the
extension of the second property, on the constancy of the average numbers of
individuals, notice that the time span on which the averages must be computed is
undetermined because there is no periodicity. We know, however, that, as the time
span on which the averages are computed tends toward infinity, these averages
tend toward the roots of the equilibrium conditions. These roots do not depend
on the initial conditions, so that: *The second property is not changed if we
substitute for the averages their limits as the time interval over which they are
computed tends toward infinity (asymptotic averages).*

3. Assuming again that the conditions above are satisfied, we want to
extend the third property on the perturbation of the averages to the general case,
as applied to the asymptotic averages. For the third property (Part I, §2, Number 9)
we distinguish between predator and prey. With more than two species it might
as well be that the individuals of one species are preyed upon by another species,
and that at the same time they feed upon a third species. To preserve an absolute
distinction between predator and prey we have to assume that if a species A is
a predator, it is not preyed upon by any other species. Similarly, if a species
B is a prey, it does not in turn feed on any other species which is a member of
the association.

4. We can now show that, under the assumptions given above: *The number
of predator species must be the same as the number of prey species.* Suppose, in
fact, that the number of prey species is different from that of the predators.
Let 1, 2, ... p be the predator species, and p + 1, p + 2, ... p + q the

prey species with $q \gtrless p$. The equilibrium conditions are (Cf. §3, Number 4):

$$\begin{cases}
\beta_1 \varepsilon_1 = a_{1,p+1} N_{p+1} + a_{1,p+2} N_{p+2} + \cdots + a_{1,p+q} N_{p+q} \\
\beta_2 \varepsilon_2 = a_{2,p+1} N_{p+1} + a_{2,p+2} N_{p+2} + \cdots + a_{2,p+q} N_{p+q} \\
\cdots \cdots \cdots \cdots \cdots \cdots \cdots \cdots \cdots \cdots \cdots \\
\beta_p \varepsilon_p = a_{p,p+1} N_{p+1} + a_{p,p+2} N_{p+2} + \cdots + a_{p,p+q} N_{p+q}
\end{cases}$$

$$\begin{cases}
\beta_{p+1} \varepsilon_{p+1} = a_{p+1,1} N_1 + a_{p+1,2} N_2 + \cdots + a_{p+1,p} N_p \\
\beta_{p+2} \varepsilon_{p+2} = a_{p+2,1} N_1 + a_{p+2,2} N_2 + \cdots + a_{p+2,p} N_p \\
\cdots \cdots \cdots \cdots \cdots \cdots \cdots \cdots \cdots \cdots \cdots \\
\beta_{p+q} \varepsilon_{p+q} = a_{p+q,1} N_1 + a_{p+q,2} N_2 + \cdots + a_{p+q,p} N_p
\end{cases}\ .$$

In these equations the $a_{1,p+1}, \cdots a_{p,p+q}$ are negative while the $a_{p+1,1}$, $\cdots a_{p+q,p}$ are positive. Correspondingly, the $\varepsilon_1, \varepsilon_2, \cdots \varepsilon_p$ are negative, and the $\varepsilon_{p+1}, \varepsilon_{p+2}, \cdots \varepsilon_{p+q}$ are positive since we assume that none of them can be zero.

If $q \gtrless p$, the antisymmetric determinant of the a_{rs}'s can easily be seen to be zero, which contradicts the premise. We must then have $q = p$, that is, the number of prey species must equal the number of predator species.

5. Let us now increase the

$$|\varepsilon_1|, \ |\varepsilon_2| \cdots |\varepsilon_p|$$

and at the same time decrease the

$$|\varepsilon_{p+1}|, \ |\varepsilon_{p+2}| \cdots |\varepsilon_{p+q}| \ .$$

In this way we destroy both predator and prey species in proportion to the numbers of their individuals. Some of the

$$N_{p+1}, \; N_{p+2}, \; \dots \; N_{p+q}$$

which satisfy the equilibrium conditions must then increase, and some of the

$$N_1, \; N_2, \; \dots \; N_p$$

must decrease. Conversely, if none of the $N_{p+1}, \; \dots \; N_{p+q}$ decreases, but all or some of them increase, none of the $N_1, \; \dots \; N_p$ will increase, but all or some will decrease, and the $|\varepsilon_{p+1}|, \; |\varepsilon_{p+2}|, \; \dots \; |\varepsilon_{p+q}|$ must decrease, and the $|\varepsilon_1|, \; |\varepsilon_2|, \; \dots \; |\varepsilon_p|$ must increase. This means that individuals in both the predator and prey species are destroyed in proportion to their numbers.

We have thus extended to the general case the third property, which may be formulated in the following way:

Consider a conservative association of even order, with non-zero determinant, for which an equilibrium exists, and in which it is possible to distinguish between predator and prey species. By uniformly destroying individuals in all the species in proportion to their numbers, the asymptotic averages of some of the prey species, possibly all of them, will increase, and the asymptotic averages of some of the predator species, possibly all of them, will decrease.

Obviously this property holds only up to some limit for the intensity of destruction (Cf. Part I, §4 and §5). Recall also that the roots of the equilibrium conditions must be positive.

§6. The case in which the growth rate of each given species depends on the number of its own individuals

1. We have several times had the occasion to consider cases leading to an indefinite increase in the numbers of one or more species. We warned, as in §4, Number 1, that such results should be considered as purely abstract ones. As the number of individuals increases the fundamental equations loose their validity. In particular, the growth rates must be modified as the number of individuals grows without bounds.

For these reasons it is necessary to take into account the effect of the number of individuals in a species on the species' own rate of growth. Such an effect can be neglected as long as the number of individuals remains within certain bounds, but it becomes essential when growth proceeds without bounds. Let us see how this **can be done**.

Consider first the case of **a** single species having a constant rate of increase, ε , so that

$$\frac{dN}{dt} = \varepsilon N \quad .$$

The number of individuals, N , will be

$$N = N_0 e^{\varepsilon t}$$

in which N_0 is the initial number. If ε is positive, N will increase without bounds.

Suppose now that the growth rate is given by $\varepsilon - \lambda N$, in which ε and λ are positive constants. In this case

$$\frac{dN}{dt} = (\varepsilon - \lambda N)N \ ,$$

from which

$$\varepsilon dt = \frac{\varepsilon dN}{N(\varepsilon - \lambda N)} = \frac{dN}{N} + \frac{\lambda dN}{\varepsilon - \lambda N} \quad .$$

Integrating, and passing from logarithms to numbers, we get

$$C \ e^{\varepsilon t} = \frac{N}{\varepsilon - \lambda N}$$

in which C is a constant. From this

$$N = \frac{C\varepsilon e^{\varepsilon t}}{1 + C\lambda e^{\varepsilon t}}$$

and, therefore,

$$\lim_{t=\infty} N = \frac{\varepsilon}{\lambda} \quad .$$

Denoting by N_0 the initial number of individuals, one has

$$C = \frac{N_0}{\varepsilon - N_0 \lambda}$$

and, therefore,

$$N = \frac{\varepsilon N_0 e^{\varepsilon t}}{\varepsilon + N_0 \lambda (e^{\varepsilon t} - 1)} \quad .$$

The number of individuals cannot increase without bounds and remains bounded between N_0 and $\frac{\varepsilon}{\lambda}$.

Similarly, in the case of n coexisting species, suppose that the coefficients of increase are given by

$$\varepsilon_r - \lambda_r N_r$$

in which the λ_r's are either positive constants or zero. The Equations (B) then become

(D) $$\beta_r \frac{dN_r}{dt} = (\varepsilon_r \beta_r - \lambda_r \beta_r N_r + \sum_1^n{}_s a_{sr} N_s) N_r \quad .$$

If the equations

$$\varepsilon_r \beta_r - \lambda_r \beta_r N_r + \sum_1^n{}_s a_{sr} N_s = 0$$

have all positive roots, $N_r = q_r$, we can write Equations (D) as

$$\beta_r \frac{dN_r}{dt} = (-\lambda_r \beta_r (N_r - q_r) + \sum_1^n{}_s a_{sr}(N_s - q_s)) N_r \quad .$$

Setting $N_r = q_r n_r$ we derive

$$\beta_r \frac{dn_r}{dt} = (-\lambda_r \beta_r q_r (n_r - 1) + \sum_1^n{}_s a_{sr} q_s (n_s - 1)) n_r$$

and

$$\beta_r \frac{d \ln n_r}{dt} = - \lambda_r \beta_r q_r n_r + \lambda_r \beta_r q_r + \sum_1^n {}_s a_{sr} q_s n_s - \sum_1^n {}_s a_{sr} q_s \ .$$

From these it follows that

$$\sum_1^n {}_r \beta_r q_r \frac{dn_r}{dt} = - \sum_1^n {}_r \lambda_r \beta_r q_r^2 n_r^2 + \sum_1^n {}_r \lambda_r \beta_r q_r^2 n_r - \sum_1^n {}_r \sum_1^n {}_s a_{sr} q_r q_s n_r$$

$$\sum_1^n {}_r \beta_r q_r \frac{d \ln n_r}{dt} = - \sum_1^n {}_r \lambda_r \beta_r q_r^2 n_r + \sum_1^n {}_r \lambda_r \beta_r q_r^2 + \sum_1^n {}_r \sum_1^n {}_s a_{sr} q_s q_r n_s$$

and, subtracting side by side, we get

$$\frac{d}{dt} \Sigma_r \beta_r q_r (n_r - \ln n_r) = - \Sigma_r \lambda_r \beta_r q_r^2 (n_r^2 - 2n_r + 1)$$

$$= - \Sigma_r \lambda_r \beta_r q_r^2 (n_r - 1)^2 \ .$$

Finally, integrating, and passing from logarithms to numbers, we get

$$\left(\frac{e^{n_1}}{n_1} \right)^{\beta_1 q_1} \left(\frac{e^{n_2}}{n_2} \right)^{\beta_2 q_2} \cdots \cdots \left(\frac{e^{n_n}}{n_n} \right)^{\beta_n q_n} = Ce^{-\int_0^1 \Sigma_r \lambda_r \beta_r q_r^2 (n_r - 1)^2 dt}$$

in which C is a positive constant.

From this one then has (Cf. Part II, §3, Number 1)

$$\frac{e^{n_r}}{n_\smile} \leq e^{\sqrt[q_r \beta_r]{G}}$$

in which

$$G = \frac{C}{e^{\Sigma_h \beta_h q_h}} \; e^{-\int_0^t \Sigma_r \lambda_r \beta_r q_r^2 (n_r - 1)^2 dt} \; .$$

2. This treatment has a number of consequences:

1) Each of the n_1, n_2, ... n_n, and, therefore, each of the N_1, N_2, ... N_n must remain bounded between two positive numbers.

The proof is similar to that in Part II, §3, Number 1 where G is substituted for K. G is obtained when a factor analogous to K is multiplied by a power of e which, because it has a negative exponent, is smaller than 1. Notice also that, since

$$e \le \frac{e^{n_r}}{n_r} \le e \, G^{\frac{1}{q_r \beta_r}} \; ,$$

G cannot tend toward zero as time increases indefinitely.

2) If λ_r differs from zero, N_r must tend toward q_r unless they already coincide. Furthermore, *if it does not tend monotonically toward this limit, the fluctuation of species r must dampen.*

In other words, by choosing an arbitrarily small and positive σ, there must be a time t after which one has $|n_r - 1| < \sigma$. Suppose, in fact, that there exists an arbitrarily large value of t such that $|n_r - 1| \ge \sigma$. Since, from Equations (D'), $\frac{dn_r}{dt}$ is a limited quantity, and we can find a number ϑ such that $\vartheta > \left| \frac{dn_r}{dt} \right|$, then, in some time interval corresponding to an arbitrarily large time, greater than $\frac{\sigma}{2\vartheta}$, we should be able to find $|n_r - 1| \ge \frac{\sigma}{2}$. This would imply that

$$\int_0^t \sum_1^n \lambda_r \beta_r q_r^2 (n_r - 1)^2 \, dt$$

could be made arbitrarily large by choosing a sufficiently large time, which is absurd since G cannot decrease indefinitely toward zero. Notice that the values q_r of the N_r's are equilibrium values which excludes, as in §3, Number 1, that some of them might be zero. The following theorem, then, holds:

Suppose that the growth rates of one or more species decreases linearly with the number of their respective individuals, while the growth rates of other species remain constant. If an equilibrium state exists, and the initial state is different from that state, the species with linearly decreasing growth rates will always tend to equilibrium, either monotonically, or through dampened fluctuations. If all the growth rates decrease in this way the entire system will tend toward its equilibrium state.[12]

A decrease in the growth rate of a species with the increase of its own numbers has much the same effect as internal friction in mechanical systems, that is, it dampens fluctuations.

3. Consider again the Equations (D), and set $\lambda_r \beta_r = a_{rr}$ instead of $a_{rr} = 0$, while preserving the conditions $a_{rs} = -a_{sr}$ whenever $s \gtreqless r$. These equations then become

$$\beta_r \frac{dN_r}{dt} = \varepsilon_r \beta_r N_r - \sum_1^n s \ a_{rs} N_s N_r \ .$$

Considering the quadratic form

(43)
$$F(N_1, N_2, \ldots N_n) = \sum_1^n r \sum_1^n s \ a_{sr} N_s N_r = \sum_1^n r \ a_{rr} N_r^2$$

_____

[12]The system can tend towards its equilibrium state even if some of the λ_r's are zero. By studying special cases it is easy to see that this can also happen either monotonically, or through dampened fluctuations (Cf. §5, Number 4).

we have

$$\frac{d}{dt} \sum_{1}^{n} r \ \beta_r N_r = - F(N_1, N_2 \ldots N_n) + \sum_{1}^{n} r \epsilon_r \beta_r N_r \ .$$

Suppose that from a certain time on the various causes of increase and decrease of the species begin to operate in such a way that all the ϵ_r become zero. Then,

$$\sum_{1}^{n} r \ \epsilon_r \beta_r N_r$$

can be neglected. F will then measure the decrease in time of $\sum_{1}^{n} r \ \beta_r N_r$. As in the first rough approximation in §2, Number 1 consider the β_r in terms of weight. F then measures the decrease, through time, in the total weight of the individuals of all the species, which would take place if the constant factors of growth and decline within each species were removed.

§7. Conservative and dissipative biological associations

1. We can extend the considerations in the previous section and thereby develop a basic distinction between classes of systems. To this end suppose that the growth rates of each species depend linearly on the respective numbers of individuals within each species, and, through the effects of encounters, on the numbers of individuals of other species as well. We assume only that the effect of encounters is proportional, according to constants, to the number of encounters, whether or not the hypothesis in §2, Number 1 is satisfied.

The Equations (B) and (D) then assume the general form

(E)
$$\frac{dN_r}{dt} = \left(\epsilon_r - \sum_{1}^{n} s \ p_{rs} N_s \right) N_r$$

in which the ϵ_r's and the p_{rs}'s are general, constant coefficients. We can consider the ϵ_r's as depending on constant effects of increase or decrease of the species, and the other terms as depending on the reciprocal actions of individuals. This greatly extends the notion of interactions among individuals as compared with the one considered previously.

The ε_r's are the growth rates which each species would have it it were in isolation, whereas the

$$\varepsilon_r - \sum_{1\ s}^{n} p_{rs} N_s$$

are the growth rates which result from their coexistence. The latter will be called the *net growth rates;* the ε_r's will be called the *gross growth rates* or just the *growth rates* when no confusion can arise.

2. First of all the proof in §2, Number 3, also applies, namely: *It is impossible for all of the species to go extinct if at least one species maintains a positive rate of growth.*

Let, then, α_1, α_2, ... α_n be positive quantities, and set

$$F(N_1, N_2, \ldots N_n) = \sum_{1\ r}^{n} \sum_{1\ s}^{n} \alpha_r p_{rs} N_s N_r \ .$$

The following theorem holds:

If the quadratic form F is positive definite, there will exist a number N such that none of the numbers N_1, N_2, ... N_n can remain larger than N from a certain time on.

It follows from Equations (E), in fact, that

$$\sum_{1\ r}^{n} \alpha_r \frac{dN_r}{dt} = \sum_{1\ r}^{n} \alpha_r \varepsilon_r N_r - F(N_1, N_2, \ldots, N_n) \ .$$

Set $N_r = 1$ and denote by m_r the lower limit of the values of F for all the possible values of N_1, N_2, ..., N_{r-1}, N_{r+1}, ..., N_n . Clearly, $m_r > 0$. Let m be the smallest of the numbers m_1, m_2, ..., m_r , and also let

$$\sum_{1\ r}^{n} |\alpha_r \varepsilon_r| < E \ .$$

Suppose that, from a certain time t_1, N_r remains larger than

$$\frac{E + 1}{m} = N .$$

Denote by $M(t_2)$ the largest of the numbers $N_1(t_2)$, $N_2(t_2)$, ... $N_n(t_2)$, where $t_2 > t_1$. We have

$$F(N_1, N_2 ... N_n)_{t=t_2} > m M^2(t_2)$$

$$\sum_1^n r \ \alpha_r \epsilon_r N_r(t_2) < EM(t_2) ,$$

from which

$$\left(\sum_1^n r \ \alpha_r \frac{dN_r}{dt} \right)_{t=t_2} > (E - m M(t_2))M(t_2) .$$

Now

$$M(t_2) > \frac{E + 1}{m}$$

and, therefore,

$$\left(\sum_1^n r \ \alpha_r \frac{dN_r}{dt} \right)_{t=t_2} < - \frac{E + 1}{m} , \quad (t_2 > t_1) .$$

It then follows that, from a given time on, some of the N_1, N_2, ... N_n would have to become negative. But this is clearly absurd since the N_r's can only be positive. This also follows from Equations (E), which give

$$N_r = N_r^0 e^{\int_0^t (\epsilon_r - \Sigma_s p_{rs} N_s) dt}$$

in which N_r^0 is the value of N_r at $t = 0$. Then, since N_r^0 must be positive, N_r will also remain positive. Therefore, there exists a number $N = \frac{E+1}{m}$ such that none of the N_r can remain larger after a given time t_1.

Combining the two theorems proved above leads to the following proposition:

If at least one of the growth rates is positive, and the quadratic form F is positive definite, the biological association will be stable.

By *stability* we mean that it is impossible for the entire association to become extinct, and that none of the species can increase without bounds (Cf. Part I, §4, Numbers 8, 9 and Part II, §2, Number 3). Since the N_i's are always positive, the theorem above may be extended to the case in which the form F is positive for all positive values of the N_i's , and zero only when all the $N_i = 0$.

3. It is easy to recognize that the determinant formed with the p_{rs}'s cannot be zero when the form F is positive definite.[13] If this determinant were zero, there would be a set of numbers $N_1, N_2, \ldots N_n$ (positive, negative, or zero, but not all zero) such that

$$\sum_1^n s \; p_{rs} N_s \; = \; 0 \, .$$

We would then have

$$0 \; = \; \sum_1^n r \sum_1^n s \, \alpha_r p_{rs} N_s N_r \; = \; F(N_1, N_2, \ldots N_n)$$

which contradicts the hypothesis that the form F is positive definite.

With this premise, suppose that the Equations

$$(E') \qquad\qquad \varepsilon_r - \sum_1^n s \; p_{rs} N_s \; = \; 0$$

when solved with respect to N_s have non-zero solutions, q_s . Then, from the identity

---

[13] In Part III, §1, Number 6 we will demonstrate that the above mentioned determinant is positive when the form F is positive.

$$\varepsilon_r = \sum_1^n {}_s \, p_{rs} q_s \ ,$$

equations (E) become

$$\frac{dN_r}{dt} = - \sum_1^n {}_s \, p_{rs}(N_s - q_s)N_r \ .$$

Setting

$$\frac{N_r}{q_r} = n_r$$

we have

(44)
$$\frac{dn_r}{dt} = - \sum_1^n {}_s \, p_{rs} q_s (n_s - 1)n_r \ .$$

From these equations, it follows that

$$\frac{1}{n_r} \frac{dn_r}{dt} = - \sum_1^n {}_s \, p_{rs} q_s (n_s - 1) \ .$$

Denoting by α_r some positive constants, it also follows that

$$\Sigma_r \alpha_r q_r \ \frac{n_r - 1}{n_r} \frac{dn_r}{dt} = - \sum_1^n {}_r \sum_1^n {}_s \, p_{rs} \alpha_r q_r q_s (n_s - 1)(n_r - 1) \ .$$

Setting

(45)
$$\frac{1}{2} \ (p_{rs}\alpha_r + p_{sr}\alpha_s) = m_{rs} = m_{sr} \ ,$$

(46)
$$F(x_1, x_2 \ldots x_n) = - \Sigma_r \Sigma_s m_{rs} x_r x_s \ ,$$

the preceding equation becomes

$$\frac{d}{dt} \ \Sigma_r \alpha_r q_r (n_r - \ln n_r) = - F(x_1, x_2, \ldots x_n)$$

in which

$$x_r = (n_r-1)q_r \ .$$

Integrating, and passing from logarithms to numbers, we have

$$\left(\frac{e^{n_1}}{n_1}\right)^{\alpha_1 q_1} \left(\frac{e^{n_2}}{n_2}\right)^{\alpha_2 q_2} \cdots \cdots \left(\frac{e^{n_n}}{n_n}\right)^{\alpha_n q_n} = Ce^{-\int_0^t Fdt}$$

in which C is a positive constant.

Suppose that the q_1, q_2, ... q_n are all positive, that is, *there exists an equilibrium state*, and suppose that we can choose positive constants α_1, α_2, ... α_n such that the quadratic form (46) becomes *identically zero*[14] (see also Part II, §3, Number 1). Then, *the numbers of individuals of the various species will all remain bounded between positive numbers, and there will be undamped fluctuations*. If there exists an equilibrium, and we can choose positive constants α_1, α_2, ... α_n such that form (46) is *positive*, then *the variations of the numbers of individuals of each species will be bounded between positive numbers.*[15] If this form is *positive definite,* in addition to the above property, we can state that: *the variations in the single species will be either monotonic, or they will exhibit damped fluctuations, so that the biological association will tend toward its equilibrium state.*

The proof of this last statement is analogous to that in §6, Number 2. Furthermore, *whenever form* (46) *is positive, the limits of the averages of* N_1, N_2, ... N_n *as the time intervals increase without bounds will be* q_1, q_2, ... q_n .

The expression F may be written as

$$(47) \qquad F = \sum_1^n {}_r \sum_1^n {}_s \alpha_r p_{rs}(N_r-q_r)(N_s-q_s) \ .$$

---

[14] In this case, for the determinant of the p_{rs} to be non-zero, n must be even.

[15] From this property it is easy to recognize that in this case, as in §3, Number 3, the averages of the N_1, N_2, ... N_n tend to q_1, q_2, ... q_n as the time interval tends to infinity.

In §6, Number 2 the damping of the fluctuations was compared to internal friction in mechanical systems. F, written as above, can be taken as a measure of this damping action, that is, *of the tendency of all species to proceed toward the equilibrium state.*

We shall call the from $F(x_1, x_2, \ldots x_n)$ the *fundamental form,* and we shall call the Equations (E') as well as (B') the *equilibrium equations.* As in §3, Number 2 of this Part, we ignore the possibility of zero roots, **namely** that the equilibrium might be reached through the extinction of some of the species.

4. It is easy to work out the necessary and sufficient conditions which the p_{rs}'s must satisfy in order for form (46) to be identically zero. In fact, from (45) it follows that

$$\frac{p_{rs}}{p_{sr}} = - \frac{\alpha_s}{\alpha_r} \, ,$$

from which

(48) $$p_{rs}p_{sg}p_{gr} + p_{rg}p_{gs}p_{sr} = 0 \, ,$$

for all three-by-three combinations of the indices 1, 2, ... n .

Since the α_r's are all positive, the above conditions, as well as the condition that p_{rs} and p_{sr} be of oppositve sign, and therefore that all $p_{rr} = 0$, are *necessary conditions.* It is easy to verify that they also are *sufficient conditions.*

In fact, from (48) we have

$$\left(- \frac{p_{rs}}{p_{sr}} \right) = \left(- \frac{p_{rg}}{p_{gr}} \right) : \left(- \frac{p_{sg}}{p_{gs}} \right)$$

from which, setting

$$- \frac{p_{rs}}{p_{sr}} = w_{rs} \, ,$$

we have $w_{rs} = w_{rg} : w_{sg}$. Since the first member does not depend on g we have

$$\frac{w_{rg}}{w_{sg}} = \frac{w_r}{w_s}$$

in which w_r and w_s depend respectively only on r and s . Therefore,

$$w_{rg} = \frac{w_r}{\left(\dfrac{w_s}{w_{sg}}\right)}$$

from which it follows that $\dfrac{w_s}{w_{sg}}$ must be independent of s .

We may then set

$$w_{rg} = \frac{\alpha_g}{\alpha_r} \ .$$

If for each p_{rg} the corresponding p_{gr} is of opposite sign, the w_{rg}'s will all be positive, and we may therefore choose positive α_r's. We then have

$$-\frac{p_{rg}}{p_{gr}} = \frac{\alpha_g}{\alpha_r} \ ,$$

that is, $\alpha_r p_{rg} + \alpha_g p_{gr} = 0$. These conditions, therefore, are also sufficient.[16]

5. The samll monotonic variations and the small damped fluctuations can be studied by setting $n_s = 1 + \nu_s$ in Equations (44), and by neglecting second order terms. We assume, of course, that an equilibrium exists, and that F is positive definite. We have then

---

[16] Miss Elena Freda has remarked, and it also follows from the way the proof was carried out, that such conditions hold only if all the p_{rs} $(r \gtrless s)$ are non-zero. When some of them are zero, the analogous conditions no longer involve three terms, as Miss Freda has pointed out through a number of examples.

(49)
$$\frac{d\nu_r}{dt} = - \sum_1^n s \; p_{rs} q_s \nu_s \quad .$$

Setting $\nu_r = \gamma_r e^{xt}$ we find the equations

(49')
$$\gamma_r x + \sum_1^n s \; p_{rs} q_s \gamma_s = 0$$

and the equation in x

(50)
$$\begin{vmatrix} p_{11} + \dfrac{x}{q_1} , & p_{12} & p_{13} & , & p_{1n} \\[2mm] p_{21} & , p_{22} + \dfrac{x}{q_2} , & p_{23} & , & p_{2n} \\[2mm] p_{31} & , p_{32} & , p_{33} + \dfrac{x}{q_3} , & \ldots & p_{3n} \\[1mm] \cdots\cdots\cdots\cdots\cdots\cdots\cdots\cdots\cdots\cdots\cdots \\[1mm] p_{n1} & , p_{n2} & , p_{n3} & , & p_{nn} + \dfrac{x}{q_n} \end{vmatrix} = 0$$

The roots of this equation have negative real parts. In fact, if

$$x = x' + ix''$$

is a root and $x' - ix''$ its conjugate, and if

$$(\gamma_r' + i\gamma_r'')e^{(x'+ix'')t}, \; (\gamma_r' - i\gamma_r'')e^{(x'-ix'')t}, \; (r=1,2,\ldots n)$$

satisfy Equations (49'), we have

$$\Sigma_r \alpha_r q_r (\gamma_r'^2 + \gamma_r''^2) x' + F(q_1\gamma_1',\ldots q_n\gamma_n') + F(q_1\gamma_1'',\ldots q_n\gamma_n'') = 0 \quad ,$$

showing that x' is negative.

The general·case we have just considered comprises all of the special cases which we have already examined, and it incorporates all the methods we have already employed.

6. We are now in the position to reexamine the hypothesis in §2, Number 1, and to better understand its meaning. Suppose we assign to each individual of species r a positive value α_r . The value of the whole association will then be $V = \sum_1^n r \; \alpha_r N_r$. From the equalities (E) it follows that

$$dV = \sum_1^n r \; \alpha_r \varepsilon_r N_r dt - \sum_1^n r \sum_1^n s \; p_{rs} \alpha_r N_r N_s dt \; .$$

In any time interval dt the change in the value of the association consists then of two parts

$$dV_1 = \sum_1^n \alpha_r \varepsilon_r N_r dt$$

$$dV_2 = \sum_1^n r \sum_1^n s \; p_{rs} \alpha_r N_r N_s dt \; .$$

The first part takes into account the constant properties of increase and decrease within each species, as determined by the ε_r's . The second part arises from the interactions among individuals within the general context defined above. If we can choose the α_r's so that dV_2 is zero for any $N_1, N_2, \ldots N_n$, the individual interactions will not affect the value of the association.

A biological association in which it is possible to assign to the individuals values such that the total value of their association is not affected by their interactions is called a conservative association. The hypothesis in Part II, §2, Number 1 is clearly satisfied by a conservative system. Conversely, if this hypothesis is verified, and the interactions within each species are neglegible, the association will be conservative. Fully conservative biological associations are probably ideal entities, which can only approximate natural situations. Examples are: the case of two species treated in Part I, §3, and those cases considered in Part II, §2, Number 2.

Suppose that we can assign, to single individuals, values which are the same for all individuals within a given species and such that the *fundamental form* F *is positive definite*. In this case the interactions among individuals tend to decrease the total value of the association. Such associations can then be called *dissipative*. It would seem that dissipative associations more closely approximate many natural situations.

As we saw in Number 4, the necessary and sufficient conditions for a system to be conservative are

$$p_{rs}p_{sg}p_{gr} \; + \; p_{rg}p_{gs}p_{sr} \; = \; 0$$

for all three by three combinations of the indices, with p_{rs} and p_{sr} having opposite sign, and $p_{rr} = 0$.[17]

Suppose that in the preceding equations the relationship (48) holds, but only for three by three combinations without repetition of the indices. Then, if all the p_{rr} are positive, the association will still be a dissipative one. This case is of special interest since it corresponds to the hypothesis in §2, Number 1 on the interactions of individuals of the different species where the growth rate of each species decreases with its own numbers (Cf. §6, Numbers 2,3).

---

[17] A dissipative association is characterized by positive p_{rr} , and it is therefore clearly distinct from a conservative association. Recall that the preceding conditions hold when the p_{rs} $(r \gtrless s)$ are non-zero.

PART THREE

FURTHER DEVELOPMENTS AND APPLICATIONS OF THE GENERAL THEORY

§1. General theorems on conservative and dissipative associations

1. We may restate one of the propositions in Part II, §2, Number 3: *The value of a conservative biological association will 1) tend toward zero if and only if all the growth rates are negative, and 2) tend toward infinity if they are all positive.*[18]

The *first part* of this proposition *also applies to dissipative systems.* From Equations (E), and taking into account (45), it follows that

$$\sum_{1}^{n} {}_{r} \; \alpha_{r} \frac{dN_{r}}{dt} = \sum_{1}^{n} {}_{r} \; \alpha_{r} \epsilon_{r} N_{r} - \sum_{1}^{n} {}_{r} \sum_{1}^{n} {}_{s} \; m_{rs} N_{r} N_{s} \; .$$

If the form

$$\sum_{1}^{n} {}_{r} \sum_{1}^{n} {}_{s} \; m_{rs} N_{r} N_{s}$$

is positive, whether or not it is positive definite, it will be

$$\sum_{1}^{n} {}_{r} \; \alpha_{r} \frac{dN_{r}}{dt} \le \sum_{1}^{n} {}_{r} \; \alpha_{r} \epsilon_{r} N_{r} \; .$$

If, furthermore

$$\epsilon_{r} < - \epsilon \; , \quad (r=1,2....n)$$

in which ϵ is a positive quantity, it then follows from Part II, §2, Number 3 that

$$\sum_{1}^{n} {}_{r} \; \alpha_{r} N_{r} < \sum_{1}^{n} {}_{r} \; \alpha_{r} N_{r}^{0} \; e^{-\epsilon t}$$

---

[18] In this proposition we have excluded the case of zero growth rates, and we shall also exclude it in the extention of the proposition. Editor's note: another peculiar property of conservative systems is discussed in Appendix, §3.

This shows that the value of an association will tend to zero if all the growth rates are negative. As mentioned in Part II, §7, Number 2, it is enough that one of these growth rates remain positive for the association not to become extinct.

Regarding the *second part* of the proposition above, we have from the proof in Part II, §7, Number 2 that: *The value of a dissipative biological association remains bounded.*

2. Let us again consider the Equations (B) (see Part II, §2, Number 2) for conservative systems

$$\beta_r \frac{dN_r}{dt} = (\epsilon_r \beta_r + \sum_1^n {}_s a_{sr} N_s) N_r \ , \ (r=1,2,\dots n)$$

$$a_{rs} = -a_{sr} \ , \qquad a_{rr} = 0 \ .$$

From these equations it follows that

$$\sum_1^n {}_r \beta_r \frac{dN_r}{dt} = \sum_1^n {}_r \epsilon_r \beta_r N_r \ .$$

Let ρ be larger than the largest $|\epsilon_r|$; we shall have

$$- \rho \sum_1^n {}_r \beta_r N_r < \sum_1^n {}_r \beta_r \frac{dN_r}{dt} < \rho \sum_1^n \beta_r N_r \ ,$$

and, therefore,

$$- \rho < \frac{\frac{d}{dt} \left(\sum_1^n {}_r \beta_r N_r \right)}{\sum_1^n {}_r \beta_r N_r} < \rho \ .$$

From this

$$\sum_1^n {}_r \beta_r N_r^0 e^{-\rho t} < \sum_1^n {}_r \beta_r N_r < \sum_1^n {}_r \beta_r N_r^0 e^{\rho t}$$

in which N_r^0 denotes the value of N_r at $t = 0$, and we have: *A conservative association can neither go to extinction nor increase without bounds in a finite time.*

3. It is also possible to prove something more, that is: *No single species can go to ∞ or to extinction in a finite time.*

The first part of this proposition is a direct consequence of the theorem proved above. Regarding the second part, let us first remark that from the general theorems of differential equations, starting from finite initial values of the N_1, N_2, ... N_n, these values will be *analytical function of time.* Therefore, in any neighborthood of t_0 where N_1, N_2, ... N_n assume finite values they can be developed in a power series of $t - t_0$.

We can now prove the theorem:

If at a certain instant t_0 in which N_1, N_2, ... N_n are finite, one has $N_h = 0$, then N_h will be zero at all later times, and it must also have been zero at all previous times.

If $(N_h)_{t=t_0} = 0$ one has, in fact, from Equations (B) $\left(\dfrac{dN_h}{dt}\right)_{t=t_0} = 0$, and the higher derivatives also give

$$\left(\frac{d^2 N_h}{dt^2}\right)_{t=t_0} = 0 \ , \ \left(\frac{d^3 N_h}{dt^3}\right)_{t=t_0} = 0,\ldots$$

That is to say, all the derivatives of N_h with respect to t at time t_0 are zero, and, therefore, the analytic function N_h is a zero constant.

Suppose that a species, the h-th, becomes extinct in a finite time. That is, starting from a positive initial number of individuals it reaches zero at a time t_0. We would then have $(N_h)_{t=t_0} = 0$ while the other species would have, as they had at the beginning, a finite number of individuals. Therefore, N_h should always have been zero, which is contrary to the premise. This proves the second part of the proposition.

4. Let us now consider the case of dissipative associations. We know that the number of individuals in each species is limited and we can also prove, much in the same way as above, that their numbers cannot become zero in a finite time. It then follows that:

In a dissipative association no species can go extinct in a finite time, and the number of individuals in each species is limited.

5. We can then prove the following theorem:

If the number of individuals in each species of an association (dissipative, or conservative of even order with non-zero determinant) remains bounded between two positive numbers, all the roots of the equilibrium equations are strictly positive.

From the general equations,

$$(E) \qquad \frac{dN_r}{dt} = (\varepsilon_r - \sum_1^n {}_s \ p_{rs} N_s) N_r \ ,$$

it follows, in fact, that

$$\frac{d \ \ln N_r}{dt} = \varepsilon_r - \sum_1^n {}_s \ p_{rs} N_s \ .$$

Integrating between 0 and T, with $N_r(0) = N_r^0$, we have

$$\ln \frac{N_r}{N_r^0} = \varepsilon_r T - \sum_1^n {}_s \ p_{rs} \int_0^T N_s dt \ ,$$

that is,

$$\frac{1}{T} \ \ln \frac{N_r}{N_r^0} = \varepsilon_r - \sum_1^n {}_s \ p_{rs} \frac{1}{T} \int_0^T N_s dt \ .$$

If the N_r's always remain bounded between two positive limits, as T increases without bounds, the left-hand sides will tend to zero. Therefore, there exists a

$$\lim_{T=\infty} \frac{1}{T} \int_0^T N_s dt = N_s \; ,$$

and we have

$$0 = \epsilon_r - \sum_1^n s \; p_{rs} \; N_s \; .$$

The N_r's are, therefore, the roots of the equilibrium condition. But the N_r's are positive numbers since they are averages of the N_s's , and, therefore, the roots of the equilibrium condition can be neither zero nor negative. From these results, and from the proof in Part II, §3, and §7, we may conclude that:

A necessary and sufficient condition for the numbers of each species to remain bounded between positive numbers is that the roots of the equilibrium equation are positive. This applies to dissipative associations, and to conservative associations of even order and with non-zero determinant.

6. We had to explicitly state above for a conservative system that its determinant is non-zero. This is not necessary in the case of dissipative systems. For such a system the form

$$F = \sum_1^n r \; \sum_1^n s \; \alpha_r p_{rs} N_r N_s \; ,$$

in which the α_1, α_2, ... α_n are positive quantities, is positive definite and, therefore, the determinant of the p_{rs}'s is always positive.

We have shown in Part II, §7, Number 2 that this determinant is non-zero. We will now prove that it cannot be negative. Set, in fact,

$$\frac{1}{2} \left(\alpha_r p_{rs} + \alpha_s p_{sr} \right) = m_{rs} = m_{sr} \; ,$$

$$\omega_{rs} = \frac{m_{rs} + h_{rs}}{\alpha_r}$$

in which

$$h_{rs} = - h_{sr} .$$

We will have

$$F = \sum_{1}^{n} {}_r \sum_{1}^{n} {}_s \; \alpha_r p_{rs} N_r N_s = \sum_{1}^{n} {}_r \sum_{1}^{n} {}_s \; \alpha_r \frac{m_{rs}}{\alpha_r} N_r N_s =$$

$$\sum_{1}^{n} {}_r \sum_{1}^{n} {}_s \; \alpha_r \omega_{rs} N_r N_s \; ,$$

and, therefore, the constants ω_{rs}'s and the constants $\dfrac{m_{rs}}{\alpha_r}$'s will correspond, as well as the p_{rs}'s , to dissipative systems.

To change the ω_{rs}'s from the values $\dfrac{m_{rs}}{\alpha_r}$ to the values p_{rs} it will be enough to increase the h_{rs}'s from zero to some values $H_{rs} = -H_{sr}$. If the determinant of the p_{rs}'s were negative, the h_{rs}'s would have to go through values such that the determinant of the ω_{rs}'s would be zero. This holds since the determinant of the m_{rs}'s , and, therefore, that of the $\dfrac{m_{rs}}{\alpha_r}$'s is positive. But from what was previously shown, the fact that the determinant of the ω_{rs}'s should be zero is inconsistent with dissipative systems.

7. In Part I, §2, Number 1 we remarked that the integration procedure employed there would also apply if the growth rates of the two species, instead of being linear and of the form

$$\epsilon_1 - \gamma_1 N_2 \; , \qquad -\epsilon_2 + \gamma_2 N_1$$

were arbitrary functions, $F_1(N_2)$ and $F_2(N_1)$ of N_2 and N_1 respectively. We will now attempt to extend the equations (E) in Part II, §7, Number 1 with the condition that an integral be conserved, in analogy with what was found in Part II, §3, Number 1 for conservative associations with even numbers of species.

8. To this end we exchange the polynomials

$$\varepsilon_r - \sum_1^n s \; p_{rs} N_s$$

in the equations (E) with the functions

$$f_r(N_1, N_2, \ldots N_n) .$$

That is to say, the growth rate of each species is affected in a general, arbitrary way by the interactions of individuals within the same species.

Equations (E) then become

(F) $$\frac{dN_r}{dt} = f_r(N_1, N_2, \ldots N_n)N_r .$$

Suppose we can find some $\phi_1(N_1), \phi_2(N_2), \ldots \phi_n(N_n)$ which satisfy the condition

$$\sum_1^n r \; \phi_r(N_r)f_r(N_1, N_2, \ldots N_n) = 0 .$$

We then have

$$\sum_1^n r \; \frac{\phi_r(N_r)}{N_r} dN_r = 0 ,$$

and integrating,

$$\sum_1^n r \int \frac{\phi_r(N_r)}{N_r} dN_r = C ,$$

in which C is a constant. This integral can also be written as

$$e^{\psi_1(N_1)} e^{\psi_2(N_2)} \ldots e^{\psi_n(N_n)} = \text{Constant}$$

having set

$$\psi_r(N_r) = \int \frac{\phi_r(N_r)}{N_r} dN_r .$$

The condition above is satisfied if and only if

$$f_r(N_1, N_2, \ldots N_n) = \sum_1^n {}_s F_{rs}(N_1, N_2, \ldots N_n)\phi_s(N_s)$$

in which the functions $F_{rs}(N_1, N_2, \ldots N_n)$ are such that

$$F_{rs} = -F_{sr} \, , \quad F_{rr} = 0 \, .$$

The Equations (F) then become

$$\frac{dN_r}{dt} = N_r \sum_1^n {}_s F_{rs}(N_1, N_2, \ldots N_n)\phi_s(N_s) \, .$$

Suppose that the lower limit of ψ_r is finite, and

$$\lim_{N_r=0} \psi_r(N_r) = \infty \, , \quad \lim_{N_r=\infty} \psi_r(N_r) = \infty \, , \quad (r = 1,2,\ldots n) \, .$$

Then, by the same reasoning as in Part II, 3, Number 1 we see that the variation of each N_r is bounded between positive numbers.

A property of conservative systems can then be extended in the following way:

Suppose that the interactions among species result in growth rates of the form

$$\sum_1^n {}_s F_{rs}(N_1, N_2, \ldots N_n) \, \phi_s(N_s)$$

in which

$$F_{rs} = -F_{sr} \, , \quad F_{rr} = 0 \, .$$

The differential equations describing the changes in numbers of each species,

$$\frac{dN_r}{dt} = N_r \sum_1^n {}_s F_{rs}(N_1, N_2, \ldots N_n)\phi_s(N_s) \, ,$$

will have the integral

$$\sum_{1}^{n} r \ \psi_r(N_r) \ = \ constant$$

in which $\psi_r = \int \dfrac{\phi_r(N_r)}{N_r} \ dN_r$. *Furthermore, if the lower bound of* ψ_r *is finite,*

and $\lim\limits_{N_r=0} \phi_r(N_r) = \infty$, $\lim\limits_{N_r=\infty} \phi_r(N_r) = \infty$, *the variation of each* N_r *will be*

bounded between positive numbers. (Editor's note: cf. also Appendix, § 4.)

§2. The principle of superposition of proper and forced fluctuations

1. In Equations (E) of Part II, §7, which may be considered as the most general equations we have treated so far, and which summarize all those previously studied, we had assumed that the growth rates were constant. In reality, however, the growth rates are not constant, rather they change according to one or more periodic variations. In all practical cases, in fact, one has to consider an annual periodicity due to the change of the seasons and to meteorological conditions. Of course, other periodical changes may also be present.

To take into account the periodic perturbations of the growth rates, we replace ε_r by

$$\varepsilon_r + g_r' \ \cos kt + g_r'' \ \sin kt$$

in which g_r' , g_r'' and k are constant quantities. The Equations (E) then become

$$(G) \qquad \frac{dN_r}{dt} \ = \ (\varepsilon_r + g_r' \ \cos kt + g_r'' \sin kt - \sum_{1}^{n} s \ p_{rs} N_s) N_r$$

and the Equations (44) become

$$\frac{dn_r}{dt} \ = \ (g_r' \ \cos kt + g_r'' \ \sin kt - \sum_{1}^{n} s \ p_{rs} q_s (n_s - 1)) n_r \ ,$$

in which the q_s's are assumed to be positive.

The small fluctuations are studied setting

$$n_s = 1 + v_s$$

and considering g'_r, g''_r, and v_r as infinitesimal quantities of the 1st order. By neglecting higher order terms, (49) takes the form

(51)
$$\frac{dv_r}{dt} = g'_r \cos kt + g''_r \sin kt - \sum_1^n {}_s \, p_{rs} q_s v_s \; .$$

Setting

$$g'_r - i g''_r = G_r$$

we have

$$G_r \, e^{ikt} = g'_r \cos kt + g''_r \sin kt + i(g'_r \sin kt - g''_r \cos kt) \; ,$$

Writing the differential equations

(52)
$$\frac{dv_r}{dt} = G_r \, e^{ikt} - \sum_1^n {}_s \, p_{rs} q_s v_s$$

it is enough to integrate the real parts of (52) in order to integrate (51). Let us write

(53)
$$v_r = A_r \, e^{ikt} + \gamma_r \, e^{xt}$$

in which the terms $\gamma_r \, e^{xt}$ denote the integrals of (49). We then have

(54)
$$ik \, A_r + \sum_1^n \, p_{rs} q_s A_s = G_r \; .$$

The determinant of the coefficients of the A_r's in the preceding equations is

$$q_1 \; q_2 \cdots\cdots q_n \quad \begin{vmatrix} \dfrac{ik}{q_1} + p_{11} \, , \; p_{12} & , \; p_{13} \cdots p_{1n} \\[2ex] p_{21} & , \; \dfrac{ik}{q_2} + p_{22} \, , \; p_{23} \cdots p_{2n} \\[2ex] \cdots\cdots\cdots\cdots\cdots\cdots\cdots\cdots \\[2ex] p_{n1} & , \; p_{n2} & , \; p_{n3} \cdots \dfrac{ik}{q_n} + p_{nn} \end{vmatrix}$$

Therefore, if ik does not coincide with any of the roots of Equation (50) we can calculate the A_r's from the Equations (54). From them we can then derive the integrals (53), in which the real parts may be separated from the imaginary parts. The terms $\gamma_r \, e^{xt}$ correspond to the *proper variations* of the association, and the terms $A_r \, e^{ikt}$ to the *forced fluctuations*.

Noting that the ε_r's are the mean values of the growth rates during the period $\dfrac{2\pi}{k}$, the following theorem holds:

Suppose that the growth rates are periodic, and their values do not differ much from their average values. Suppose also that by taking the average values of the growth rates we get monotonic variations, damped fluctuations, or undamped fluctuations in the neighborhood of the equilibrium state (proper variations). Then, for small fluctuations corresponding to periodic growth rates, the super-position principle holds. That is, whenever the period of the forced fluctuations does not coincide with the periods of possible proper variations, the total variation is obtained by superimposing on the proper variations forced variations whose period is the same as that of the growth rates.

§3. Variations between positive limits superimposed on extinction.

1. Consider again the Equations (E) in which we have distinguished between the terms ε_r and $\varepsilon_r - \sum\limits_1^n p_{rs} N_s$. The former are the *gross rates of growth* which do not take into account any of the interactions among individuals. By taking these interactions into account we obtain the latter, which we have

called the *net rates of growth* (Part II, §7, Number 1).

Suppose that in an association of n species the first m species are missing, that is $N_1 = N_2 = \ldots N_m = 0$. The net rates of growth of the remaining n - m species are then

$$\epsilon_r - \sum_{m+1}^{n} s \; p_{rs} N_s \; , \; (r = m+1 \ldots\ldots n) \; .$$

It is meaningless to talk about increase or decrease in the first m species. Nevertheless, the expressions

$$\epsilon_r - \sum_{m+1}^{n} p_{rs} N_s \; , \; (r = 1, 2, \ldots m)$$

assume specific values, and are therefore called the *virtual growth rates* of the first m species (cf. Part II, §7, number 1).

2. Let us now prove the following theorem:

Suppose that in a conservative or dissipative association of n species an equilibrium exists for a set of n - m of them. Suppose also that the virtual growth rates of the remaining m species assume negative values at this equilibrium. Then the small variations of the entire association result from superimposing the variation between positive limits of the n - m species on a decline in all the species.

Let $q_{m+1} , \ldots q_n$ be the positive values of the $N_{m+1} , \ldots N_n$ for which the n - m species are at equilibrium. Setting

$$N_r = q_r + v_r \; , \; (r = m+1, \ldots n)$$

the Equations (E) can be written as

$$(55) \qquad \frac{dN_i}{dt} = \left(\epsilon_i - \sum_{1}^{m} s \; p_{is} N_s - \sum_{m+1}^{n} \ell \; p_{i\ell} (q_\ell + v_\ell) \right) N_i,$$

$$(i = 1, 2 \ldots m) ,$$

$$(56) \quad \frac{dv_r}{dt} = \left(- \sum_{m+1}^{n} \ell \; P_{r\ell} v_\ell - \sum_{1}^{m} s \; P_{rs} N_s \right) (q_r + v_r) \; , \; (r = m+1, \ldots n) \; .$$

By retaining only the 1st order terms they simplify to

$$(57) \qquad \frac{dN_i}{dt} = \left(\varepsilon_i - \sum_{m+i}^{n} \ell \; P_{i\ell} q_\ell \right) N_i \; (i=1,2 \ldots .m) \; ,$$

$$(58) \quad \frac{dv_r}{dt} = - \sum_{m+1}^{n} \ell \; P_{r\ell} q_r v_\ell - \sum_{i}^{m} s \; P_{rs} q_r N_s \; , \; (r=m+1, \ldots .n) \; .$$

The virtual growth rates,

$$\varepsilon_i - \sum_{m+1}^{n} P_{i\ell} q_\ell = - n_i \; , \quad (i = 1,2, \ldots .m) \; ,$$

are negative, which justifies having neglected 2nd order terms. We have then

$$N_i = N_i^0 \; e^{-n_i t} \; , \; (i = 1,2 \ldots m)$$

in which N_i^0 denote the initial values of the N_i. Equations (58) then become

$$(59) \qquad \frac{dv_r}{dt} = - \sum_{m+1}^{n} P_{r\ell} q_r v_\ell - \sum_{i}^{m} P_{rs} q_r N_s^0 \; e^{-n_s t} \; , \; (r=m+1, \ldots n) \; .$$

Since the entire association is either conservative or dissipative, this is also the case for the association of the $n - m$ species alone. Therefore, if $v_r = f_r(t)$ are the solutions of the equations

$$\frac{dv_r}{dt} = - \sum_{m+1}^{n} P_{r\ell} q_r v_\ell \; , \; (r=m+1, \ldots n) \; ,$$

$q_r + f_r$ will give the damped or undamped fluctuations, or the monotonic variation of these $n - m$ species in the neighborhood of their equilibrium. But the solutions of the Equations (59) have the form

$$(60) \qquad v_r = f_r(t) + \sum_1^m i \, A_{ri} \, e^{-n_i^t} \,, \quad (r=m+1,\ldots n)$$

and, since $N_r = q_r + v_r$, the theorem is proved.

Therefore, the variation we have considered leads to *a monotonic decrease to extinction of the 1st m species, and to variations of the remaining ones in a neighborhood of their equilibrium state.*

3.   If an equilibrium state exists for a whole association, whether conservative or dissipative, all the variations will be bounded between positive limits.   This excludes the possibility that one or more species might decrease to extinction.   We will try to show directly that, *if there exists an equilibrium for the whole association, and also one for part of it, the virtual growth rates of the remaining species cannot all be negative at the partial equilibrium.*

Let $q_1, q_2, \ldots q_n$ be the positive values of $N_1, N_2, \ldots N_n$ at the equilibrium for the entire association, and $q'_{m+1}, \ldots q'_n$ of $N_{m+1}, \ldots N_n$ at the steady state of species $m+1, \ldots n$.   We have

$$\varepsilon_i - \sum_1^n s \, p_{is} N_s = \sum_1^n s \, p_{is}(q_s - N_s) \,, \quad (i=1,2,\ldots n)$$

$$\varepsilon_r - \sum_1^n s \, p_{rs} N_s = \sum_1^n s \, p_{rs}(q_s - N_s) = \sum_{m+1}^n \ell \, p_{r\ell}(q'_\ell - N_\ell) - \sum_1^m s \, p_{rs} N_s \,,$$

$$(r = m + 1, \ldots n) \ .$$

Setting

$$F(x_1, x_2, \ldots x_n) = \sum_1^n i \sum_1^n s \, p_{is} \alpha_i x_i x_s \,,$$

in which the α_i's are positive, we have

$$F(q_1 - N_1, \ldots, q_n - N_n) = \sum_1^m {}_i \; \alpha_i (\varepsilon_i - \sum_1^n {}_s \; p_{is} N_s)(q_i - N_i)$$

$$+ \sum_{m+1}^n \; \alpha_r \left[\sum_{m+1}^n {}_\ell \, p_{r\ell}(q_\ell' - N_\ell) - \sum_1^m {}_s \; p_{rs} N_s \right] (q_r - N_r) \; .$$

By choosing $N_1 = N_2, \ldots N_m = 0$, $N_{m+1} = q_{m+1}'$, $\ldots N_n = q_n'$ we obtain

$$F = \sum_1^m \; \alpha_i \left(\varepsilon_i - \sum_{m+1}^n {}_s \; p_{is} q_s' \right) q_i \; .$$

Since the q_i's are positive, if the form F is zero or positive it follows that the virtual growth rates

$$\varepsilon_i - \sum_{m+1}^n {}_s \; p_{is} q_s' \; , \quad (i = 1, 2, \ldots m)$$

cannot all be negative. This proves the theorem.

§4. <u>The perturbation produced in a stable biological association</u>
<u>by the addition of a new species.</u>

    1. Consider the equilibrium conditions

$$\varepsilon_r - \sum_1^n {}_s \; p_{rs} N_s = 0 \; , \quad (r = 1, 2, \ldots n)$$

for the unknowns N_1, N_2, $\ldots N_n$. Let

$$q_1, q_2, \ldots \ldots q_n$$

be the solutions, and suppose that $q_1 < 0$.

    Now consider the equations

$$\varepsilon_r - \sum_{2}^{n} s \ p_{rs} N_s \ = \ 0 \quad (r = 2,3 \ldots n)$$

for the unknowns N_2, N_3, ... N_n , and let

$$q_2', \ q_3', \ldots \ldots q_n'$$

be their positive roots. Suppose then that the form

$$F(x_1, x_2, \ldots x_n) = \sum_{1}^{n} r \ \sum_{1}^{n} s \ m_{rs} x_r x_s \ ,$$

in which

$$m_{rs} = \frac{1}{2} (p_{rs} \alpha_r + p_{sr} \alpha_s)$$

and the α_1, α_2, ... α_n are positive quantities, is positive definite. We will have

$$\varepsilon_r - \sum_{1}^{n} s \ p_{rs} N_s = \sum_{1}^{n} s \ p_{rs} (q_s - N_s), \quad (r = 1, 2 \ldots n)$$

$$\varepsilon_r - \sum_{2}^{n} s \ p_{rs} N_s = \sum_{2}^{n} s \ p_{rs} (q_s' - N_s), \quad (r = 2, 3 \ldots n)$$

from which

$$\varepsilon_r - \sum_{1}^{n} s \ p_{rs} N_s = \sum_{2}^{n} s \ p_{rs} (q_s' - N_s) - p_{r1} N_1 =$$

$$= \sum_{1}^{n} s \ p_{rs} (q_s - N_s), \quad (r = 2, 3 \ldots n)$$

$$\varepsilon_1 - \sum_{1}^{n} s \ p_{1s} N_s = \sum_{1}^{n} s \ p_{1s} (q_s - N_s) \ .$$

It then follows that

$$\sum_{1}^{n} r \sum_{1}^{n} s \, p_{rs}(q_s - N_s)(q_r - N_r)\alpha_r = F(q_1 - N_1, \ldots q_n - N_n)$$

$$= \sum_{2}^{n} r \left(\sum_{2}^{n} s \, p_{rs}(q_s' - N_s) - p_{r1}N_1 \right)(q_r - N_r)\alpha_r$$

$$+ \left(\varepsilon_1 - \sum_{1}^{n} s \, p_{1s}N_s \right)(q_1 - N_1)\alpha_1 \, .$$

2. If we now take

$$N_1 = 0, \, N_2 = q_2', \ldots N_n = q_n'$$

we will have

$$F(q_1, q_2 - q_2' \ldots q_n - q_n') = (\varepsilon_1 - \sum_{2}^{n} s \, p_{1s}q_s')\alpha_1 q_1 > 0 \, .$$

But $q_1 < 0$ whereas the left hand side is positive. It follows, therefore, that

$$\varepsilon_1 - \sum_{2}^{n} s \, p_{1s}q_s' < 0 \, .$$

Taking into account the results of the preceding section we can then state the following theorem:

If there exists an equilibrium state for a given biological association, the possibility for an equilibrium will be lost by the addition of a new species if there results a negative root, corresponding to the number of individuals of the new species, in the equilibrium equations of the whole association. The small variations of the whole association (assumed to be dissipative) will then consist of the decline toward extinction of the new species, superimposed on a variation of the original association in the neighborhood of its equilibrium state.[19]

---

[19]It is useful to compare this theorem with the one in §1, Number 5.

The new species will tend to extinction while the others tend to a variation in a neighborhood of their equilibrium state. Namely, the addition of such a new species produces a perturbation which tends to die out.

§5. Appendix to Part III; The special case of a biological association of three species.

1. As an example of the treatment carried out so far, we will examine a special case which may be worked out fully and explicitly. Let us suppose that three species coexist in a limited habitat, such as an island. Suppose also that the first species feeds upon the second, and the second upon the third, but not vice versa. We can take, for example, the closed association of a carnivore, an herbivore, and a plant. The same would also apply to an herbivorous host-specific insect and its parasite. In such cases we may assume that the same treatment we have applied to animals, also applies to the plant species we are considering.

2. Assume, at first, that the association is *conservative* (Cf., Part II, §7). Denoting by N_1, N_2, N_3 the numbers of individuals in the three species, we obtain the equations (See (B), Part II, §2, Number 2)

$$\beta_1 \frac{dN_1}{dt} = (\beta_1 \epsilon_1 + a_{21} N_2 + a_{31} N_3) N_1$$

$$\beta_2 \frac{dN_2}{dt} = (\beta_2 \epsilon_2 + a_{12} N_1 + a_{32} N_3) N_2$$

$$\beta_3 \frac{dN_3}{dt} = (\beta_3 \epsilon_3 + a_{13} N_1 + a_{23} N_2) N_3 \quad .$$

In the present case we must choose

$$\varepsilon_1 = -\ell < 0 \ , \quad a_{21} = a > 0 \ , \quad a_{31} = 0$$

$$\varepsilon_2 = -m < 0 \ , \quad a_{12} = -a < 0 \ , \quad a_{32} = b > 0$$

$$\varepsilon_3 = k > 0 \ , \quad a_{13} = 0 \quad , \quad a_{23} = -b < 0$$

in which β_1, β_2, β_3 , a, ℓ, m, k are constant quantities. We have, therefore,

$$(\mathrm{H}_1) \qquad\qquad \beta_1 \frac{dN_1}{dt} = (-\beta_1 \ell + aN_2)N_1$$

$$(\mathrm{H}_2) \qquad\qquad \beta_2 \frac{dN_2}{dt} = (-\beta_2 m - a_1 N_1 + bN_3)N_2$$

$$(\mathrm{H}_3) \qquad\qquad \beta_3 \frac{dN_3}{dt} = (\beta_3 k - bN_2)N_3 \ .$$

Since we are dealing with an odd number of species, we can use the integral (40) in Part II, §4, Number 1, and obtain

$$(61) \qquad\qquad N_1^{\beta_1 b} \, N_3^{\beta_3 a} = C \, e^{(\beta_3 ka - \beta_1 \ell b)t}$$

We will distinguish between two cases, according to whether the binomial

$$(62) \qquad\qquad \beta_3 ka - \beta_1 \ell b$$

is positive or negative.

3. Suppose, at first, that this binomial is negative. We must exclude the possibility that N_3 might tend toward zero since, from (H_2), N_2 would then progressively decrease. But, if N_2 were to become smaller than $\dfrac{\beta_3 k}{b}$, N_3 would increase indefinitely which is contrary to the initial assumption. Therefore, N_1 will have to become smaller than an arbitrary positive quantity as time grows indefinitely.

Let us set

$$\frac{\beta_1 \ell}{a} = Q_1 \; ; \quad \frac{\beta_3 k}{b} = Q_2 \; ; \quad \frac{\beta_2 m}{b} = Q_3 \; .$$

From the assumption that the binomial (52) should be negative we have $Q_1 > Q_2$, and the Equations (H_1), (H_2), and (H_3) become

$$\beta_I \frac{dN_1}{dt} = - a(Q_1 - N_2)N_1$$

$$\beta_2 \frac{dN_2}{dt} = (-aN_1 + b(N_3 - Q_3))N_2$$

$$\beta_3 \frac{dN_3}{dt} = - b(N_2 - Q_2)N_3 \; .$$

Setting

$$N_1 = Q_1 \nu_1 ; \quad N_2 = Q_2(1 + \nu_2); \quad N_3 = Q_3(1 + \nu_3)$$

we obtain the equations

$$\beta_1 \frac{d\nu_1}{dt} = - a(Q_1 - Q_2)\nu_1 + aQ_2\nu_2\nu_1$$

$$\beta_2 \frac{d\nu_2}{dt} = - aQ_1\nu_1 + bQ_3\nu_3 - aQ_1\nu_1\nu_2 + bQ_3\nu_2\nu_3$$

$$\beta_3 \frac{d\nu_3}{dt} = - bQ_2\nu_2 - bQ_2\nu_2\nu_3 \; .$$

Neglecting second order terms, these equations become

$$\beta_1 \frac{d\nu_1}{dt} = - a(Q_1-Q_2)\nu_1$$

$$\beta_2 \frac{d\nu_2}{dt} = - aQ_1\nu_1 + bQ_3\nu_3$$

$$\beta_3 \frac{d\nu_3}{dt} = - bQ_2\nu_2 \ .$$

The general integral of this system is

$$\nu_1 = \frac{\beta_2(\rho^2 + km)}{\beta_1 \ell} \ C_1 e^{-\rho t}$$

$$\nu_2 = \rho C_1 e^{-\rho t} + C_2\sqrt{m} \ \sin (\sqrt{km} \ t + C_3)$$

$$\nu_3 = kC_1 e^{-\rho t} + C_2\sqrt{k} \ \cos (\sqrt{km} \ t + C_3)$$

in which $\rho = \dfrac{a(Q_1-Q_2)}{\beta_1}$, and C_1, C_2, C_3 are three constants. The negative

exponent $-\rho t$ justifies our having neglected the second order terms.

Setting

$$\nu_1' = \frac{\beta_2(\rho^2+km)}{\beta_1\ell} \; C_1 e^{-\rho t} \; , \; \nu_2' = \rho C_1 e^{-\rho t}, \; \nu_3' = k C_1 e^{-\rho t}$$

$$\nu_1'' = 0, \; \nu_2'' = C_2\sqrt{m} \, \sin{(\sqrt{km} \, t + C_3)}, \; \nu_3'' = C_2\sqrt{k} \, \cos{(\sqrt{km} \, t + C_3)}$$

$$N_1' = Q_1\nu_1', \quad N_2' = Q_2\nu_2' \quad , \quad N_3' = Q_3\nu_3'$$

$$N_1'' = 0 \quad , \quad N_2'' = Q_2(1+\nu_2''), \quad N_3'' = Q_3(1+\nu_3'')$$

we have

$$N_1 = N_1' + N_1'' \, , \qquad N_2 = N_2' + N_2'' \, , \qquad N_3 = N_3' + N_3'' \quad .$$

The variation in the association will then consist of the superposition of the variation N_1', N_2', N_3' on the variation N_1'', N_2'', N_3'' . The former variation consists of the tendency toward extinction of the three species, the latter of undamped fluctuations of the second and third species. In other words: *The first species declines monotonically to extinction while the other two species tend to the state of undamped fluctuation studied in §2 of Part I.*

The fact that the binomial (62) is negative means that the amount of food provided by the plant, through the herbivores, is not enough to offset the death rate of the carnivores. The assumption that the association is conservative holds since the plant has a constant growth rate, k , and its numbers remain bounded between two positive constants.

4. The case in which the binomial (62) is positive is very different since, from (61), N_1 would grow without bounds, and, from (H_2) and (H_3) the same would also occur for the other two species. It is clear, however, that the number of plants, N_3 , cannot grow beyond a limit imposed by the area of the island. In this case, then, it would be absurd to assume that the association is conservative, that is that k is a constant. We must then assume that the growth rate of the plant decreases with its numbers. We will then have a *dissipative*

system, to which the calculations in §6 of Part II apply.

By substituting $\beta_3 k - \lambda N_3$ for $\beta_3 k$ in Equation (H_3), in which λ is positive, we obtain

(63')
$$\beta_1 \frac{dN_1}{dt} = (-\beta_1 \ell + aN_2)N_1$$

(63")
$$\beta_2 \frac{dN_2}{dt} = (-\beta_2 m - aN_1 + bN_3)N_2$$

(63"')
$$\beta_3 \frac{dN_3}{dt} = (\beta_3 k - \lambda N_3 - bN_2)N_3 \ .$$

Solving the equilibrium conditions,

$$- \beta_1 \ell + aN_2 = 0$$

$$- \beta_2 m - aN_1 + bN_3 = 0$$

$$\beta_3 k - \lambda N_3 - bN_2 = 0 \ ,$$

the values

$$q_1 = \frac{ab\beta_3 k - b^2\beta_1 \ell - a\lambda\beta_2 m}{a^2\lambda} = \frac{b(a\beta_3 k - b\beta_1 \ell) - a\lambda\beta_2 m}{a^2\lambda}$$

$$q_2 = \frac{\beta_1 \ell}{a}$$

$$q_3 = \frac{a\beta_3 k - b\beta_1 \ell}{a\lambda}$$

are obtained for N_1, N_2, and N_3 respectively.

If we assume that

$$ab\beta_3 k - b^2\beta_1 \ell - a\lambda\beta_2 m > 0$$

the three roots q_1, q_2, q_3 will be positive so that there is an equilibrium state. Let us set

$$N_1 = q_1(1 + \nu_1)$$

$$N_2 = q_2(1 + \nu_2)$$

$$N_3 = q_3(1 + \nu_3) \ ,$$

and neglect the second order terms in the Equations (63'), (63"), and (63"'). These assume the form

(63$'_1$)
$$\beta_1 \frac{d\nu_1}{dt} = aq_2\nu_2$$

(63$''_1$)
$$\beta_2 \frac{d\nu_2}{dt} = - aq_1\nu_1 + bq_3\nu_3$$

(63$'''_1$)
$$\beta_3 \frac{d\nu_3}{dt} = - bq_2\nu_2 - \lambda q_3\nu_3 \ .$$

Setting, in these equations,

$$\nu_1 = A_1 e^{xt}$$

$$\nu_2 = A_2 e^{xt}$$

$$\nu_3 = A_3 e^{xt}$$

they become

$$A_1\beta_1 x = aq_2 A_2$$

$$A_2\beta_2 x = -aq_1 A_1 + bq_3 A_3$$

$$A_3\beta_3 x = -bq_2 A_2 - \lambda q_3 A_3 \ .$$

Eliminating A_1, A_2, A_3 we obtain the equation

$$\begin{vmatrix} -\beta_1 x \, , & aq_2 & 0 \\ -aq_1 \, , & -\beta_2 x \, , & bq_3 \\ 0 \, , & -bq_2 \, , & -\lambda q_3 - \beta_3 x \end{vmatrix} = 0 \ .$$

This equation can only have negative real roots, or complex roots with negative real parts. Let x' be the conjugate of x, and A_1', A_2', A_3' be the conjugates of A_1, A_2, A_3. We then have

$$A_1'\beta_1 x' = aq_2 A_2'$$

$$A_2'\beta_2 x' = -aq_1 A_1' + bq_3 A_3'$$

$$A_3'\beta_3 x' = -bq_2 A_2' - \lambda q_3 A_3'$$

from which follows

$$(A_1 A_1' \beta_1 q_1 + A_2 A_2' \beta_2 q_2 + A_3 A_3' \beta_3 q_3)(x+x') = -2\lambda q_3^2 A_3 A_3' \quad '$$

Unless $A_3 = 0$, which would entail $A_1 = A_2 = 0$, $x + x'$ is, therefore, negative. The equation in x above can then have only negative roots, or complex roots with negative real parts. This proves that *the variation will either be monotonic or exhibit damped oscillations, and, therefore, the system will tend to its equilibrium state.*[20]

5. We are now left with the case in which the binomial (62) is positive, and

(64) $$ab\beta_3 k - b^2\beta_1 \ell - a\lambda\beta_2 m < 0 \ .$$

Rewrite this expression in the form

$$a(b\beta_3 k - \lambda\beta_2 m) - b^2\beta_1 \ell \quad ,$$

---

[20]Cf., the note in Part II, §6, Number 3.

and distinguish, together with inequality (64), the two subcases:

I.   (65) $b\beta_3 k - \lambda\beta_2 m < 0$,

which directly entails inequality (64), and

II.   (66) $b\beta_3 k - \lambda\beta_2 m > 0$.[21]

I. Set

$$\frac{\beta_3 k}{\lambda} = g$$

$$N_3 = g(1 + \nu_3) \quad .$$

Equations (63'), (63"), (63"') then become

$$\beta_1 \frac{dN_1}{dt} = - \beta_1 \ell N_1 + aN_1 N_2$$

$$\beta_2 \frac{dN_2}{dt} = (- \beta_2 m + bg)N_2 - aN_1 N_2 + bgN_2 \nu_3$$

$$\beta_3 \frac{d\nu_3}{dt} = (- \lambda g \nu_3 - bN_2)(1 + \nu_3) \quad .$$

Neglecting second order terms, one has

$$\frac{dN_1}{dt} = - \ell N_1$$

$$\frac{dN_2}{dt} = - \omega N_2$$

$$\frac{d\nu_3}{dt} = \frac{-b}{\beta_3} N_2 - \frac{\lambda g}{\beta_3} \nu_3$$

---

[21] We have not considered the cases

$$b\beta_3 k - \lambda\beta_2 m = 0$$

$$ab\beta_3 k - b^2 \beta_1 \ell - a\lambda\beta_2 m = 0 \quad .$$

in which, from (65),

$$\omega = \frac{\beta_2 m - bg}{\beta_2} > 0 .$$

The general integral of this sytem is

$$N_1 = C_1 e^{-\ell t}$$

$$N_2 = C_2 e^{-\omega t}$$

$$\nu_3 = C_2 \frac{b}{\beta_3(\omega-k)} e^{-\omega t} + C_3 e^{\frac{-\lambda g}{\beta_3} t}$$

The negative exponents justify our having neglected second order terms. In this case, then, both the first and the second species go to extinction while the numbers of the plant species tend to $g = \frac{\beta_3 k}{\lambda}$.

II. Set $\frac{\beta_2 m}{b} = f_3$ so that, from (66)

(67)
$$\frac{\beta_2 k - \frac{\lambda \beta_2 m}{b}}{b} = f_2 > 0 .$$

Set also $N_2 = f_2(1 + \nu_2)$, $N_3 = f_3(1 + \nu_3)$. Equations (63'), (63''), (63''') become

$$\beta_1 \frac{dN_1}{dt} = (-\beta_1 \ell + af_2)N_1 + af_2 \nu_2 N_1$$

$$\beta_2 \frac{d\nu_2}{dt} = (bf_3 \nu_3 - aN_1)(1 + \nu_2)$$

$$\beta_3 \frac{d\nu_3}{dt} = (-\lambda f_3 \nu_3 - bf_2 \nu_2)(1 + \nu_3) .$$

Neglecting second order terms we have

$$\frac{dN_1}{dt} = - \frac{\rho_1}{\beta_1} N_1$$

$$\frac{d\nu_2}{dt} = - \frac{a}{\beta_2} N_1 + \frac{bf_3}{\beta_2} \nu_3$$

$$\frac{d\nu_3}{dt} = - \frac{bf_2}{\beta_3} \nu_2 - \frac{\lambda f_3}{\beta_3} \nu_3$$

in which, from (64)

$$- \rho_1 = - \beta_1 \ell + af_2 = \frac{ab\beta_3 k - a\lambda\beta_2 m - b^2\beta_1\ell}{b^2} < 0 .$$

The signs of the coefficients again justify our approximation. Without going through the same reasoning as above it is easy to recognize that N_1 declines monotonically to zero while ν_2 and ν_3 tend to an equilibrium state through damped oscillitations.

6. The results of all the possible cases can be summarized as follows:

1. $\qquad\qquad \beta_3 ka - \beta_1 \ell b < 0$

Even if the plant species should increase without bounds, the amount of food which reaches the carnivores through the herbivores would not be enough to support them. The carnivores tend to extinction while the herbivores and plants tend to exhibit undamped, periodic fluctuations.

2. $\qquad\qquad \beta_3 ka - \beta_1 \ell b > 0 .$

If the plant species were to have a constant growth rate, its numbers would increase without bounds. It becomes convenient then to assume that its growth rate should decrease linearly with the number of individuals.

$2_a \qquad\qquad b\beta_3 k - \lambda\beta_2 m < 0 .$

The quantity of food provided by the plant is not enough to support the herbivore. Both the herbivore and carnivore species tend to extinction, whereas the number of individuals in the plant species tends to a constant value.

$$b\beta_3 k - \lambda\beta_2 m > 0$$

$$2_b$$

$$ab\beta_3 k - b^2\beta_1 m - a\lambda\beta_2 m < 0 \ .$$

The plant species provides enough food for the herbivore, but the latter does not provide enough for the carnivore. The carnivore species then tends to extinction while the hebivore and plant species tend to a stable equilibrium through damped fluctuations.

$$2_c \qquad ab\beta_3 k - b^2\beta_1 \ell - a\lambda\beta_2 m > 0 \ .$$

There is enough nourishment for the three species, which tend to a stable equilibrium, either monotonically or through damped fluctuations.

We have now identified and discussed all the meaningful situations, while disregarding those which merely correspond to equalities, rather than inequalities among the constants, and have a vanishingly small probability.

PART FOUR

THE STUDY OF DELAYED EFFECTS

§1. Extension of the first fundamental property of the fluctuations to the case

of historical actions.

1. In most biological phenomena the past influences contemporary as well

as future events. I have several times had the opportunity to deal with historical

actions through integro-differential equations, and equations with functional

derivatives.[22] In the preceding parts of this essay I have dealt with the

problems of biological fluctuations only through simple differential equations.

Delayed effects had been neglected in this first rudimentary approach. Wishing

to move one step closer to reality, it becomes necessary to make use of integro-

differential equations, although they will still be in their simplest possible

form.[23]

2. Consider the case of two coexisting species whose rates of growth, if

alone, would be $\varepsilon_1 > 0, - \varepsilon_2 < 0$. In this case, corresponding to the second

species feeding on the first species, we have derived the differential equations

(Cf., Part I, §2 and §4)

$$(A_1) \quad \frac{dN_1}{dt} = N_1(\varepsilon_1 - \gamma_1 N_2) , \qquad (A_2) \quad \frac{dN_2}{dt} = N_2(-\varepsilon_2 + \gamma_2 N_1)$$

in which N_1 and N_2 denote the numbers in the two species, and γ_1 , γ_2 are

two positive constants.

---

[22] Cf., Volterra, Lecons sur les fonctions de lignes, Paris, Gauthier-Villars 1913,
and Saggi scientifici, Bologna, Zanichelli 1920.

[23] In his previously cited work Dr. Lotka mentioned the possibility of lags and
leads in epidemics, without dealing with them analytically. However, in his paper
with Sharpe, Contribution to the analysis of Malaria epidemiology: IV Incubation
lag, he explicitly considered lags, but in a quite different way from my own
analysis of historical actions and integro-differential equations.

Let us go back to the reasoning in Part I, §4, Number 1 by which these equations were derived. If the first species were alone, its numbers would grow by the amount $\varepsilon_1 N_1 dt$ in a time interval dt . In the same time interval, however, there would be a number of encounters between the two species proportional to $N_1 N_2 dt$. The number of individuals of the first species eaten by the second in the time interval dt would also be proportional to this quantity, according to the constant γ_1 . We may then write

$$dN_1 = \varepsilon_1 N_1 dt - \gamma_1 N_1 N_2 dt$$

from which Equation (A_1) follows.

Considering now the second species, we notice that if it were alone the number of its individuals would decrease by the amount $-\varepsilon_2 N_2 dt$ in a time interval dt . The individuals of the second species feed only on the individuals of the first species which they encounter. As a first level of approximation, we had assumed that the number of individuals of the second species increased in a time interval dt in proportion to the number of encounters in this time interval according to the constant coefficient γ_2 . We then have

$$dN_2 = - \varepsilon_2 N_2 dt + \gamma_2 N_1 N_2 dt ,$$

from which Equation (A_2) follows.

The reasoning leading to Equation (A_1), although schematic, does not pose serious difficulties. One must, however, take into account a very serious objection to the reasoning by which (A_2) was derived. The two cases, in fact, lack the symmetry which they might at first sight appear to have. In fact, the destruction of individuals of the first species during an interval dt can be reasonably considered as due to the individuals of the second species which are present at that very time. However, the nourishment which contributes to the growth rate of the predator in the time interval dt is not the nourishment obtained during this time interval, but rather the nourishment obtained at previous intervals.

Although this will not often be strictly true, suppose for simplicity that the age structure of the predator does not change with time. Denote by $\phi(\xi)d\xi$ the proportion of individuals in the age interval $(\xi, \xi+d\xi)$. Then the fraction of individuals of age larger than $t-\tau$ will be

$$\int_{t-\tau}^{\infty} \phi(\xi)d\xi = f(t-\tau) \ .$$

It follows that the proportion of the $N_2(t)$ individuals present at time t which were already alive at an earlier time τ will be $f(t-\tau)N_2(t)$. The quantity of food they ingested in the time interval $(\tau, \tau+ d\tau)$ can then be expressed by

$$\gamma f(t-\tau)N_2(t)N_1(\tau)d\tau$$

in which γ is a positive constant.

This nourishment will affect the growth rate of the predator at time t in different measures, depending on the length of elapsed time, τ . To take into account this effect we can multiply the preceding expression by a positive function, $\phi(t-\tau)$ obtaining

$$\gamma\phi(t-\tau)f(t-\tau)N_2(t)N_1(\tau)d\tau = F(t-\tau)N_2(t)N_1(\tau)d\tau \ .$$

These quantities must be summed up, so to speak, over all the time intervals prior to the instant t , giving us

$$\int_{-\infty}^{t} F(t-\tau)N_2(t)N_1(\tau)d\tau \ .$$

Therefore, Equation (A_2) must be substituted by

$$\frac{dN_2}{dt} = N_2(t)\left(- \epsilon_2 + \int_{-\infty}^{t} F(t-\tau)N_1(\tau)d\tau\right).$$

Then, instead of (A_1) and (A_2) we will have the simultaneous equations

$$(IV) \qquad \frac{dN_1}{dt} = N_1(t)(\varepsilon_1 - \gamma_1 \mathring{N}_2(t)) \ ,$$

$$(V) \qquad \frac{dN_2}{dt} = N_2(t)\left(-\varepsilon_2 + \int_{-\infty}^{t} F(t-\tau)N_1(\tau)d\tau\right),$$

the former being an ordinary differential equation, and the latter an integro-differential equation.

Regarding the function $F(t-\tau)$ we must assume that as its argument increases without bound it becomes infinitesimal of such an order that its integral converges. It is also quite reasonable to suppose that this function will remain zero for all values of its argument larger than a certain value T_0 .

3. To make the treatment more symmetrical, we shall consider, in place of Equations (IV) and (V)

$$(K) \qquad \frac{dN_1}{dt} = N_1(t)\left(\varepsilon_1 - \gamma_1 N_2(t) - \int_{-\infty}^{t} F_1(t-\tau)N_2(\tau)d\tau\right),$$

$$(L) \qquad \frac{dN_2}{dt} = N_2(t)\left(-\varepsilon_2 + \gamma_2 N_1(t) + \int_{-\infty}^{t} F_2(t-\tau)N_1(\tau)d\tau\right),$$

in which γ_1 and γ_2 are two positive constants (γ_2 can also be zero), and in which F_1 and F_2 are finite, continuous, positive functions which are zero for values of the argument equal to or greater than a $T_0 > 0$ (F_1 can also be zero). The integral terms in the equation for the prey could be interpreted in much the same way as for the predator.

The functions $N_1(t)$ and $N_2(t)$ can be arbitrarily chosen in the time interval $(t_0 - T_0, t_0)$, except for being finite, continuous, and positive. We can then extend them in a $t_0 \leq t < t_1$, employing for example the method of successive approximations, so that they are finite, and they satisfy Equations (K) and (L). The values of $N_1(t)$ and $N_2(t)$ in the time interval $(t_0 - T_0, t_0)$ will **join** in a continuous manner with those in the interval (t_0, t_1) for $t = t_0$. This, however, will not generally hold for their derivatives.

4. THEOREM I. *The integrals of the Equations (K) and (L) are positive for any* $t_0 \leq t < t_1$.

Setting $N_1(t_0) = N_1^0$, $N_2(t_0) = N_2^0$ we have, in fact,

$$N_1(t) = N_1^0\, e^{P_1(t)}\ , \qquad N_2(t) = N_1^0\, e^{P_2(t)}\ , \qquad t_0 \leq t < t_1\ ,$$

in which

$$P_1(t) = \int_{t_0}^{t} \left(\varepsilon_1 - \gamma_1 N_2(\theta) - \int_{-\infty}^{\theta} F_1(\theta-\tau) N_2(\tau) d\tau \right) d\theta\ ,$$

$$P_2(t) = \int_{t_0}^{t} \left(-\varepsilon_2 + \gamma_2 N_1(\theta) + \int_{-\infty}^{\theta} F_2(\theta-\tau) N_1(\tau) d\tau \right) d\theta\ ,$$

so that N_1 and N_2, being exponentials, will always remain positive.

THEOREM II. *For any* $t_0 < t < t_1$ *we will have*

(68)
$$N_1(t) < N_1^0\, e^{\varepsilon_1(t-t_0)} < N_1^0\, e^{\varepsilon_1(t_1-t_0)} = N_1(t_1-t_0)$$

(68')
$$N_2(t) < N_2^0\, e^{\frac{\gamma_2+\Gamma_2}{\varepsilon_1} N_1^0 e^{\varepsilon_1(t-t_0)}} < N_2^0\, e^{\frac{\gamma_2+\Gamma_2}{\varepsilon_1} N_1^0 e^{\varepsilon_1(t_1-t_0)}} = N_2(t_1-t_2)$$

$$(69) \qquad \left| \frac{dN_1(t)}{dt} \right| \; < \; N_1(t_1-t_0)(\varepsilon_1 + (\gamma_1+\Gamma_1) \; N_2(t_1-t_0))$$

$$(69') \qquad \left| \frac{dN_2(t)}{dt} \right| \; < \; N_2(t_1-t_0)(\varepsilon_2 + (\gamma_2+\Gamma_2) \; N_1(t_1-t_0))$$

in which

$$(70) \qquad \begin{cases} \Gamma_1 \; = \; \int_0^{T_0} F_1(\xi)d\xi \; , \qquad \Gamma_2 \; = \; \int_0^{T_0} F_2(\xi)d\xi \; . \\[4mm] N_1(t) \; = \; N_1^0 e^{\varepsilon_1 t} \; , \qquad N_2(t) \; = \; N_2^0 e^{\frac{\gamma_2+\Gamma_2}{\varepsilon_1} N_1^0 e^{\varepsilon_1 t}} \; . \end{cases}$$

In fact, from the previous theorem, $P_1(t) < \varepsilon_1$ from which (68) follows, and, therefore,

$$P_2(t) \; < \; \int_{t_0}^t \left(\gamma_2 N_1^0 \; e^{\varepsilon_1(\theta-t_0)} + e^{\varepsilon_1(\theta-t_0)} \int_{-\infty}^\theta F_1(\theta-\tau)d\tau \right) d\theta \; .$$

But

$$\int_{-\infty}^\theta F_1(\theta-\tau)d\tau \; = \; \int_0^\infty F_1(\xi)d\xi \; = \; \int_0^{T_0} F_1(\xi)d\xi \; = \; \Gamma_1 \; ,$$

from which the relation (68') easily follows. Equations (K) and (L) then lead immediately from (68) and (68') to (69) and (69').

THEOREM III. $N_1(t)$ and $N_2(t)$, $\dfrac{dN_1}{dt}$ and $\dfrac{dN_2}{dt}$ tend toward determinate and finite limits as t tends toward t_1.

From (68), (68'), and Theorem I the upper bounds of the absolute values of N_1 and N_2 are finite in the interval (t_0, t_1). Therefore, N_1 and N_2 cannot tend to ∞ as t tends to $t = t_1$. If there were no limits as $t = t_1$ the upper limits of the magnitude of $\dfrac{dN_1}{dt}$ and $\dfrac{dN_2}{dt}$ would be ∞. In fact, the oscillations of N_1 and N_2 in time intervals $(t_1 - \alpha, t_1)$ would remain larger than a $\sigma > 0$, no matter how small α. But this would contradict inequalities (69) and (69'), thus proving the theorem.

THEOREM IV. *If the integrals of Equations (K) and (L) exist for any* $t_0 < t < t_1$, *one can find a* $t_2 > t_1$ *such that the integrals also exist for any* $t_0 < t < t_2$.

This follows from the fact that the integrals and their derivatives will tend toward precise and finite limits as $t = t_1$. Then, proceeding from t_1 as we did starting from t_0, we can extend the integrals to an interval (t_1, t_2) in which $t_2 > t_1$. At time t_1 the same integrals and their derivatives that were previously determined in the intervals (t_0, t_1) and (t_1, t_2) will join continuously, thus proving the theorem.

By appropriately specifying the process of successive approximations we can then deduce that *the integrals of the Equations (K) and (L) can be extended to any* $t_0 < t < \infty$, *and they will always remain positive.*

5. THEOREM V. *There exist constant values of* N_1 *and* N_2 *which satisfy the Equations (K) and (L).*

It is, in fact, enough to take

$$N_1 = K_1 \quad , \qquad N_2 = K_2$$

such that

$$\epsilon_1 - K_2 \left(\gamma_1 + \int_0^{T_0} F_1(\xi)d\xi \right) = 0$$

$$\epsilon_2 - K_1 \left(\gamma_2 + \int_0^{T_0} F_2(\xi)d\xi \right) = 0 \quad ,$$

and, therefore

$$K_1 = \frac{\epsilon_2}{\gamma_2 + \Gamma_2} \quad , \qquad K_2 = \frac{\epsilon_1}{\gamma_1 + \Gamma_1} \quad .$$

THEOREM VI. *If* α *is any positive quantity one cannot then have, at all values of time over a given limit,*

(71) $$N_1 > K_1 + \alpha$$

or

(71') $$N_1 < K_1 - \alpha \ .$$

Analogously, one cannot have, at all values of time over a given limit,

(72) $$N_2 > K_2 + \alpha$$

or

(72') $$N_2 < K_2 - \alpha \ .$$

Suppose, in fact, that, from a certain value of time t_1 onward, (71) is always satisfied. It would follow from Equation (K) that, if $t > t_1 + T_0$,

$$\frac{dN_2}{dt} > N_2(t)(\gamma_2 + \Gamma_2)\alpha \ .$$

Denoting by N_2' the value of N_2 at time t_1 , we will then have

$$N_2(t) > N_2' \; e^{\alpha(\gamma_2 + \Gamma_2)(t - t_1)} \quad ,$$

namely, $N_2(t)$ increases indefinitely with time.

There will then be a value of time, t_2 , such that for any $t > t_2$

$$N_2(t) > K_2 + \alpha \quad ,$$

and then, from (K), if $t > t_2 + T_0$,

$$\frac{dN_1}{dt} < - N_1(t)\alpha(\gamma_1 + \Gamma_1) \quad .$$

Setting $N_1(t_2) = N_1''$ it follows that

$$N_1(t) < N_1'' \; e^{-\alpha(\gamma_1 + \Gamma_1)(t - t_2)} \quad ,$$

namely, N_1 will tend to zero as time increases without bounds.

There will then be a value of time, $t > t_1$, for which

$$N_1(t) < K_1 + \alpha \quad ,$$

which contradicts the initial assumption. The other parts of the theorem may be proved in analogous fashion.

COROLLARY: N_1 *cannot tend toward any value different from* K_1 , *nor can* N_2 *tend toward any value different from* K_2 . *In particular the numbers of individuals of the two species cannot tend to zero or to* ∞ .

6. THEOREM VII. N_1 *cannot tend monotonically towards* K_1 *nor can* N_2 *tend monotonically towards* K_2 .

Suppose, for instance, that, from a certain time onward, N_1 tends monotonically toward K_1 , passing through values larger than K_1 . Since (L) can be written as

$$\frac{dN_2}{dt} = N_2(t) \left(\gamma_2(N_1(t)-K_1) + \int_{-\infty}^{t} F_2(t-\tau)(N_1(\tau)-K_1)d\tau \right) \quad ,$$

at least for a sufficiently large time we will have

$$\frac{dN_2}{dt} > 0 \quad .$$

Therefore, from a certain time on, N_2 will continue to increase, and as a consequence of the theorem above it must tend to K_2 through smaller values. But Equation (K) can be written as

$$\frac{dN_1}{dt} = N_1(t) \left(\gamma_1(K_2-N_2(t)) + \int_{-\infty}^{t} F_1(t-\tau)(K_2-N_2(\tau))d\tau \right) \quad ,$$

from which

$$\frac{dN_1}{dt} > 0 \quad ,$$

so that N_1 will increase from a certain time on, which contradicts the initial assumption. Similarly, one can prove that N_1 cannot increase toward K_1 . An analogous statement can be proved for N_2 .

COROLLARY: N_1 and N_2 *must oscillate as time increases indefinitely, passing through an infinite number of maxima and minima.*

THEOREM VIII. N_1 and N_2 *must cross infinitely often the values* K_1 K_2 , *respectively, as time increases without bounds.*

If N_1 were to remain, from a certain time on, always larger than K_1 ,

$$- \varepsilon_2 + \gamma_2 N_1 + \int_{-\infty}^{t} F_2(t-\tau) N_1(\tau) d\tau$$

would remain positive. That is to say it could never become zero, and N_2 could not have maxima or minima, which contradicts the corollary above. One can prove in the same manner that N_1 cannot remain smaller than K_1 from a certain time on. Therefore, both N_1 and N_2 must cross the values K_1 and K_2 respectively infinitely often as time increases without bounds.

7. The preceding theorem extends to the case of historical actions the property of *the persistent fluctuations of two species about their equilibrium values* (Cf., Part I, §2, Number 9). In this case, however, the fluctuations are not periodic.

§2. <u>Extension of the second and third fundamental properties of the fluctuations to the case of historical actions.</u>

1. In the case of historical actions the properties of the conservation of the averages and of the perturbation of the averages remain, essentially, unchanged. As we shall see in this section, however, it is far more difficult to prove that these properties hold.

2. Let us consider again the Equations (K) and (L) of the preceding section. Taking into account that

$$F_1(\tau) = F_2(\tau) = 0$$

for $\tau > T_0$, the equations can be written as

(K')
$$\frac{dN_1}{dt} = N_1(t) \left(\varepsilon_1 - \gamma_1 N_2(t) - \int_0^{T_0} F_1(\tau) N_2(t-\tau) d\tau \right)$$

$$(L') \qquad \frac{dN_2}{dt} = N_2(t) \left(-\varepsilon_2 + \gamma_2 N_1(t) + \int_0^{T_0} F_2(\tau) N_1(t-\tau) d\tau \right) \ ,$$

or also as

$$(K'') \qquad \frac{dN_1}{dt} = N_1(t) \left(\varepsilon_1 - \gamma_1 N_2(t) - \int_{t-T_0}^t F_1(t-\tau) N_2(\tau) d\tau \right)$$

$$(L'') \qquad \frac{dN_2}{dt} = N_2(t) \left(-\varepsilon_2 + \gamma_2 N_1(t) + \int_{t-T_0}^t F_2(t-\tau) N_1(\tau) d\tau \right) \ ,$$

in which ε_1 and ε_2 are positive, and γ_1, γ_2, F_1, F_2 are never negative.

From these considerations it follows that if at least one of the quantities, γ_1 and Γ_1 , is non-zero, and at least one of the quantities, γ_2 and Γ_2, is also non-zero, then, since N_1 and N_2 are always positive:

$$\varepsilon_1 > \frac{1}{N_1} \frac{dN_1}{dt} > -\gamma_1 N_2(t) - \int_0^{T_0} F_1(\xi) N_2(t-\xi) d\xi$$

$$-\varepsilon_2 < \frac{1}{N_2} \frac{dN_2}{dt} < \gamma_2 N_1(t) + \int_0^{T_0} F_2(\xi) N_1(t-\xi) d\xi \qquad .$$

Integrating between τ and t , with $\tau < t$,

$$(73) \qquad e^{-\varepsilon_1(t-\tau)} < \frac{N_1(\tau)}{N_1(t)} < e^{\gamma_1 \int_\tau^t N_2(\xi) d\xi + \int_\tau^t d\xi \int_0^{T_0} F_1(\eta) N_2(\xi-\eta) d\eta}$$

$$(73') \qquad e^{-\varepsilon_2(t-\tau)} > e^{-\gamma_2 \int_\tau^t N_1(\xi) d\xi - \int_\tau^t d\xi \int_0^{T_0} F_2(\eta) N_1(\xi-\eta) d\eta} \qquad .$$

From (73) we have

$$\gamma_1 \int_\tau^t N_2(\xi)d\xi + \int_\tau^t d\xi \int_0^{T_0} F_1(\eta)N_2(\xi-\eta)d\eta \quad <$$

$$< N_2(t) \int_\tau^t e^{\varepsilon_2(t-\xi)}d\xi \left(\gamma_1 + \int_0^{T_0} F_1(\eta)e^{\varepsilon_2\eta}d\eta \right) < N_2(t) \frac{e^{\varepsilon_2(T_0+t-\tau)}}{\varepsilon_2} (\gamma_1+\Gamma_1) .$$

Equations (73) and (73') can then be written as

(74) $\qquad N_1(t)e^{-\varepsilon_1(t-\tau)} < N_1(\tau) < N_1(t)e^{N_2(t) \frac{e^{\varepsilon_2(T_0+t-\tau)}}{\varepsilon_2}(\gamma_1+\Gamma_1)}$

(74') $\quad N_2(t)e^{\varepsilon_2(t-\tau)} > N_2(\tau) > N_2(t)e^{-\gamma_2 \int_\tau^t N_1(\xi)d\xi - \int_\tau^t d\xi \int_0^{T_0} F_2(\eta)N_1(\xi-\eta)d\eta}$.

If $t-\tau \leq T_0$, it follows from the preceding equations that

(75) $\qquad N_1(t)e^{-\varepsilon_1 T_0} < N_1(\tau) < N_1(t)e^{N_2(t) \frac{e^{2\varepsilon_2 T_0}}{\varepsilon_2}(\gamma_1+\Gamma_1)}$

(75') $\qquad N_2(t)e^{\varepsilon_2 T_0} > N_2(\tau) > 0$.

Then, if $t - \tau \leq T_0$, and at the same time $N_2 < K_2 = \dfrac{\varepsilon_1}{\gamma_1+\Gamma_1}$, one has as a consequence of (75)

(76) $\qquad N_1(t)e^{-\varepsilon_1 T_0} < N_1(\tau) < N_1(t)e^{\varepsilon_1 m}$

in which

$$m = \frac{1}{\epsilon_2} e^{2\epsilon_2 T_0} \quad .$$

3. Suppose that N_1 attains a maximum or a minimum at time t . From (K') we will have

$$\epsilon_1 - \gamma_1 N_2(t) - \int_0^{T_0} F_1(\tau)N_2(t-\tau)d\tau = 0 \quad .$$

If $F_1 > 0$, it follows from (75') that

$$\epsilon_1 - \gamma_1 N_2(t) > 0$$

$$\epsilon_1 - \gamma_1 N_2(t) - N_2(t)e^{\epsilon_2 T_0} \int_0^{T_0} F_1(\tau)d\tau < 0 \quad .$$

Therefore, if $\gamma_1 > 0$, $\Gamma_1 > 0$ we will have

(77)
$$\frac{\epsilon_1}{\gamma_1 + \Gamma_1 e^{\epsilon_2 T_0}} < N_2(t) < \frac{\epsilon_1}{\gamma_1}$$

while, if $\gamma_1 > 0$, $\Gamma_1 = 0$ we will have $N_2(t) = \frac{\epsilon_1}{\gamma_1}$.

Now Theorem VIII of the preceding section guarantees that, as time increases without bounds, N_1 must pass through an infinite number of minima lower than K_1 , and N_2 through an infinite number of minima lower than K_2 . Let t be the time at which N_2 assumes a minimum value. From Equation (L') we have

$$\epsilon_2 - \gamma_2 N_1(t) - \int_0^{T_0} F_2(\tau)N_1(t-\tau)d\tau = 0 \quad .$$

If $F_2 > 0$ it follows from (76) that

$$\varepsilon_2 - \gamma_2 N_1(t) - N_1(t) e^{-\varepsilon_1 T_0} \Gamma_2 > 0$$

$$\varepsilon_2 - \gamma_2 N_1(t) - N_1(t) e^{\varepsilon_1 m} \Gamma_2 < 0$$

from which

(77')
$$\frac{\varepsilon_2}{\gamma_2 + e^{-\varepsilon_1 T_0} \Gamma_2} > N_1(t) > \frac{\varepsilon_2}{\gamma_2 + e^{\varepsilon_1 m} \Gamma_2} .$$

We are then in a position to state the following properties:

LEMMA I. *If* $\gamma_1 > 0$, *the maxima and minima of* N_1 *correspond to values of* N_2 *within the following limits*

$$\frac{\varepsilon_1}{\gamma_1 + \Gamma_1 e^{\varepsilon_1 T_0}} \leq N_2 \leq \frac{\varepsilon_1}{\gamma_1} .$$

The stronger inequalities correspond to $\Gamma_1 > 0$, *the weaker ones to* $\Gamma_1 = 0$.

LEMMA II. *If at least one of the numbers* γ_2 *and* Γ_2 *is greater than zero, the minimum values of* N_2 *smaller than* K_2 *correspond to values of* N_1 *within the following limits,*

$$\frac{\varepsilon_2}{\gamma_2 + e^{-\varepsilon_1 T_0} \Gamma_2} \geq N_1 \geq \frac{\varepsilon_2}{\gamma_2 + e^{\varepsilon_1 m} \Gamma_2} .$$

The stronger inequalities correspond to $\Gamma_2 > 0$, *the weaker ones to* $\Gamma_2 = 0$.

LEMMA III. *If* $N_1(t_1)$ *and* $N_2(t_2)$ *are minima of* $N_1(t)$ *and* $N_2(t)$, *and are smaller than* K_1 *and* K_2 *respectively, we will have*

$$(t_1 - T_0 \leq \tau \leq t_1) , \quad N_1(\tau) < p_1 K_1 , \quad p_1 = e^{\left(\frac{\varepsilon_1}{\varepsilon_2} \frac{\gamma_1 + \Gamma_1}{\gamma_1} e^{2\varepsilon_2 T_0} \right)}$$

$$(t_2 - T_0 \leq \tau \leq t_2) , \quad N_2(\tau) < p_2 K_2 , \quad p_2 = e^{\varepsilon_2 T_0} ,$$

assuming that $\gamma_1 > 0$, *and that at least one of the numbers* γ_2 *and* Γ_2 *is greater than zero.*

4. We will now prove the following THEOREM: *Assume that* $\gamma_1 > 0$, *and that at least one of the numbers* γ_2 *and* Γ_2 *is larger than zero. Consider a time interval from any initial time to the time at which* N_1 *reaches a minimum lower than* K_1 . *The average value of* N_1 *will then tend towards* K_1 *as this time interval increases without bounds. Similarly, as the time interval in which* N_2 *reaches a minimum smaller than* K_2 *increases without bounds, the average value of* N_2 *will tend toward* K_2 .

Let us remark first of all that the Equations (K") and (L") can both be written in the form

$$(M) \qquad (-1)^{i+1} \frac{dN_1}{dt} = N_i(t) \left(\varepsilon_i - \gamma_i N_{i+1}(t) - \int_{t-T_0}^{t} F_i(t-\tau) N_{i+1}(\tau) d\tau \right)$$

with the convention that we substitute for i and $i+1$ the numbers 1 or 2 according to whether the subscripts are odd or even.

With this premise, let us divide both sides of (M) by $N_i(t)$, and integrate between the initial time t_0 and the time $t_{i+1} > t_0 + T_0$. We will have

$$(78) \qquad (-1)^{i+1} \ell n \; \frac{N_i(t_{i+1})}{N_i(t_0)} \; = \; \varepsilon_i(t_{i+1}-t_0) - \gamma_i \int_{t_0}^{t_{i+1}} N_{i+1}(t)dt \; -$$

$$- \int_{t_0}^{t_{i+1}} dt \int_{t-T_0}^{t} F_i(t-\tau)N_{i+1}(\tau)d\tau \; .$$

The last integral can be written as

$$I_i = \int_{t_0}^{t_{i+1}} dt \int_{t_0-T_0}^{t_0} F_i(t-\tau)N_{i+1}(\tau)d\tau + \int_{t_0}^{t_{i+1}} dt \int_{t_0}^{t} F_i(t-\tau)N_{i+1}(\tau)d\tau \; .$$

By then transforming the second of these integrals, one has[24]

$$I_i = \int_{t_0}^{t_{i+1}} dt \int_{t_0-T_0}^{t_0} F_i(t-\tau)N_{i+1}(\tau)d\tau + \int_{t_0}^{t_{i+1}} N_{i+1}(\tau)d\tau \int_{\tau}^{t_{i+1}} F_i(t-\tau)dt$$

$$= \int_{t_0-T_0}^{t_0} N_{i+1}(\tau)d\tau \int_{t_0}^{t_{i+1}} F_i(t-\tau)dt + \int_{t_0}^{t_{i+1}} N_{i+1}(\tau)d\tau \int_{0}^{t_{i+1}-\tau} F_i(\xi)d\xi \; =$$

---

[24]The transformation performed here is also known as "Dirichlet's principle", that is,

$$\int_{a}^{b} dx \int_{a}^{x} F(x,y)dy \; = \; \int_{a}^{b} dy \int_{y}^{b} F(x,y)dx.$$

$$= \int_{t_0-T_0}^{t_0} N_{i+1}(\tau)d\tau \int_{t_0}^{t_{i+1}} F_i(t-\tau)dt \; + \; \int_{t_0}^{t_{i+1}-T_0} N_{i+1}(\tau)d\tau \int_0^{t_{i+1}-\tau} F_i(\xi)d\xi \; +$$

$$+ \; \int_{t_{i+1}-T_0}^{t_{i+1}} N_{i+1}(\tau)d\tau \int_0^{t_{i+1}-\tau} F_i(\xi)d\xi \; .$$

Notice now that in the first and last terms of the preceding sum we have

$$\int_{t_0}^{t_{i+1}} F_i(t-\tau)dt \le \Gamma_i \; , \qquad \int_0^{t_{i+1}-\tau} F_i(\xi)d\xi \le \Gamma_i$$

while in the second term

$$\int_0^{t_{i+1}-\tau} F_i(\xi)d\xi \; = \; \int_0^{T_0} F_i(\xi)d\xi \; = \; \Gamma_i$$

since in this term the smallest value of $t_{i+1}-\tau$ is T_0 , and $F_i(\xi)$ is zero when $\xi > T_0$.

We then have

$$I_i = \Gamma_i \left\{ \theta' \int_{t_0-T_0}^{t_0} N_{i+1}(\tau)d\tau \; + \theta'' \int_{t_{i+1}-T_0}^{t_{i+1}} N_{i+1}(\tau)d\tau \; + \int_{t_0}^{t_{i+1}-T_0} N_{i+1}(\tau)d\tau \right\},$$

in which θ' and θ'' denote numbers between 0 and 1 . We can also write

$$(79) \quad I_i = \Gamma_i \left\{ \theta' \int_{t_0-T_0}^{t_0} N_{i+1}(\tau)d\tau \; - (1-\theta'') \int_{t_{i+1}-T_0}^{t_{i+1}} N_{i+1}(\tau)d\tau \; + \int_{t_0}^{t_{i+1}} N_{i+1}(\tau)d\tau \right\} .$$

Suppose that N_{i+1} reaches a minimum lower than K_{i+1} at the time t_{i+1}. From Lemma III we have

$$\int_{t_{i+1}-T_0}^{t_{i+1}} N_{i+1}(\tau)d\tau < p_{i+1}K_{i+1}T_0 \ .$$

Denote by M_{i+1} the maximum reached by $N_{i+1}(\tau)$ in the time interval (t_0-T_0,t_0). This will take a finite value independent of t_{i+1}, and we will have

$$\int_{t_0-T_0}^{t_0} N_{i+1}(\tau)d\tau < M_{i+1}T_0 \ .$$

Let θ''' and θ'''' be two numbers respectively between 0 and θ', and 0 and $1 - \theta''$. We can replace Equations (79) by

$$I_i = \Gamma_i \left\{ \theta''' \, M_{i+1}T_0 - \theta^{iv}p_{i+1}K_{i+1}T_0 + \int_{t_0}^{t_{i+1}} N_{i+1}(\tau)d\tau \right\} \ .$$

Substituting this value of the integral in (78), we obtain

$$(-1)^{i+1} \ln \frac{N_i(t_{i+1})}{N_i(t_0)} = \varepsilon_1(t_{i+1}-t_0) - (\gamma_i+\Gamma_i) \int_{t_0}^{t_{i+1}} N_{i+1}(\tau)d\tau -$$

$$- T_0\Gamma_i \, \theta''' \, M_{i+1} - \theta^{iv} p_{i+1}K_{i+1} \} \ ,$$

and, therefore,

$$\frac{1}{t_{i+1}-t_0} \int_{t_0}^{t_{i+1}} N_{i+1}(\tau)d\tau = \frac{\varepsilon_1}{\gamma_i+\Gamma_i} \quad +$$

$$+ \frac{1}{(\gamma_i+\Gamma_i)(t_{i+1}-t_0)} \left\{ (-1)^i \, \ell n \, \frac{N_i(t_{i+1})}{N_i(t_0)} \quad - \right.$$

$$\left. - \, T_0\Gamma_i(\theta''' \, M_{i+1}-\theta^{iv} P_{i+1}K_{i+1}) \right\} \, .$$

Let us consider now the minima smaller than K_{i+1} at ever increasing times t_{i+1}. From the Lemmas I and II, $N_i(t_{i+1})$ will remain bounded between two positive limits which are independent of t_{i+1}. From this $\ell n \, \dfrac{N_i(t_{i+1})}{N_i(t_0)}$ will always remain smaller than a fintie limit; the previous equation will then give

$$\lim_{t_{i+1}=\infty} \frac{1}{t_{i+1}-t_0} \int_{t_0}^{t_{i+1}} N_{i+1}(\tau)d\tau = \frac{\varepsilon_i}{\gamma_i+\Gamma_i} = K_{i+1}$$

which proves the theorem.

5. The fluctuations in the case of historical actions are described by Equations (IV) and (V) of §1, Number 2. They correspond to the Equations (K) and (L) in which $F_1 = 0$ and $\gamma_2 = 0$, while γ_1 and F_2, and therefore Γ_2, remain positive. The conditions under which the previous theorem holds are also satisfied in this special case.

The number of individuals, N, of a species will oscillate around a constant value K, which is an equilibrium value. We can define the *asymptotic average* of this number as the limit of the average in a time interval between any arbitrary time and a time corresponding to a minimum of N smaller than K , when the length of this interval increases without bounds (Cf., Part II, §3). The asymptotic averages of the numbers of individuals, N_1 and N_2 , of the two species would be, from this theorem, K_1 and K_2 .

All the results obtained so far for the case of historical actions can be summarized in the following three properties:

1) PROPERTY OF THE FLUCTUATIONS. *The numbers of the individuals of the two species forever oscillate around values corresponding to the equilibrium values. As time increases indefinitely they pass through an infinite number of maxima and minima.*

2) CONSERVATION OF THE AVERAGES. *The asymptotic averages of the numbers of the two species do not depend on the initial state, and they coincide with the equilibrium values.*

3) PERTURBATION OF THE AVERAGES. *If one removes individuals of the two species uniformly, and in proportion to their numbers, the asymptotic average of the prey increases, and that of the predator decreases.*

Since $\gamma_2 = \Gamma_1 = 0$, the asymptotic averages of the two species are ε_2/Γ_2 and ε_1/γ_1 respectively, the same as the equilibrium values. Also a uniform removal of individuals in proportion to their numbers is the same as increasing ε_2 and decreasing ε_1 , while Γ_2 and γ_1 remain unchanged. There-fore the former average value increases, and the latter decreases.

§3. <u>Non-periodicity of the small fluctuations in the case of historical actions.</u>

1. In the present section I would like to show that it is not possible to have small periodic fluctuations around an equilibrium in the case of delayed

effects. To this end, consider again the equations in the form (K'), (L') (Cf., §2, Number 2), that is,

$$(K') \qquad \frac{dN_1}{dt} = N_1(t)\left(\varepsilon_1 - \gamma_1 N_2(t) - \int_0^{T_0} F_1(\tau)N_2(t-\tau)d\tau\right)$$

$$(L') \qquad \frac{dN_2}{dt} = N_2(t)\left(-\varepsilon_2 + \gamma_2 N_1(t) + \int_0^{T_0} F_2(\tau)N_1(t-\tau)d\tau\right).$$

Their equilibrium values are given by (Cf., formulas (70) and Theorem V in §1)

$$N_1 = K_1 = \frac{\varepsilon_2}{\gamma_2 + \Gamma_2} \quad , \qquad N_2 = K_2 = \frac{\varepsilon_1}{\gamma_1 + \Gamma_1}$$

$$\Gamma_1 = \int_0^{T_0} F_1(\tau)d\tau \quad , \qquad \Gamma_2 = \int_0^{T_0} F_2(\tau)d\tau \quad .$$

Set

$$N_1 = K_1 + n_1 \; , \qquad N_2 = K_2 + n_2 \; .$$

Equations (K') and (L') take the form

$$\frac{dn_1}{dt} = -(K_1 + n_1)\left(\gamma_1 n_2(t) + \int_0^{T_0} F_1(\tau)n_2(t-\tau)d\tau\right)$$

$$\frac{dn_2}{dt} = (K_2 + n_2)\left(\gamma_2 n_1(t) + \int_0^{T_0} F_2(\tau)n_1(t-\tau)d\tau\right).$$

Let us also write

$$\frac{n_1}{K_1} = \nu_1 \ , \quad \frac{n_2}{K_2} = \nu_2 \ , \quad \gamma_1 K_1 = \alpha_1 \ , \quad \gamma_2 K_2 = \alpha_2$$

$$K_1 F_1(t) = \Phi_1(t) \ , \quad K_2 F_2(t) = \Phi_2(t) \ .$$

If the second degree terms in ν_1 and ν_2 can be neglected, the previous equations become

$$(80) \qquad \frac{d\nu_1}{dt} + \alpha_1 \nu_2(t) + \int_0^{T_0} \Phi_1(\tau)\nu_2(t-\tau)d\tau = 0$$

$$(80') \qquad \frac{d\nu_2}{dt} - \alpha_2 \nu_1(t) - \int_0^{T_0} \Phi_2(\tau)\nu_1(t-\tau)d\tau = 0 \ .$$

2. Suppose now that the preceding equations had periodic solutions of equal periodicity. Developing these solutions in a Fourier series, in which terms corresponding to $m = 0$ must not occur, we would get

$$\nu_1 = \sum_{1}^{\infty} m \ (a_m' \sin m\lambda t + b_m' \cos m\lambda t)$$

$$\nu_2 = \sum_{1}^{\infty} m \ (a_m'' \sin m\lambda t + b_m'' \cos m\lambda t) \ .$$

Assume also that this series can be derived term by term. Substituting this derivation in the preceding equation we have

$$\sum_{1}^{\infty} m \ \left\{ \left[a_m' m\lambda + \alpha_1 b_m'' - a_m'' \int_0^{T_0} \Phi_1(\tau) \sin m\lambda\tau d\tau + b_m'' \int_0^{T_0} \Phi_1(\tau) \cos m\lambda\tau d\tau \right] \cos m\lambda t \ + \right.$$

$$+ \left[-b_m'm\lambda + \alpha_1 a_m'' + a_m'' \int_0^{T_0} \Phi_1(\tau)\cos m\lambda\tau d\tau + b_m'' \int_0^{T_0} \Phi_1(\tau)\sin m\lambda\tau d\tau\right] \sin m\lambda t \Bigg\} = 0$$

$$\sum_1^\infty m \left\{ \left[a_m''m\lambda - \alpha_2 b_m' + a_m' \int_0^{T_0} \Phi_2(\tau)\sin m\lambda\tau d\tau - b_m' \int_0^{T_0} \Phi_2(\tau)\cos m\lambda\tau d\tau\right] \cos m\lambda t \right. +$$

$$+ \left. \left[-b_m''m\lambda - \alpha_2 a_m' - a_m' \int_0^{T_0} \Phi_2(\tau)\cos m\lambda\tau d\tau - b_m' \int_0^{T_0} \Phi_2(\tau)\sin m\lambda\tau d\tau\right] \sin m\lambda t \right\} = 0 \; .$$

From these equations we derive

$$a_m'm\lambda \qquad\qquad - a_m''M_{11}^{(m)} \qquad + b_m''(\alpha_1 + M_{12}^{(m)}) = 0 \; ,$$

$$\qquad - b_m'm\lambda \qquad + a_m''(\alpha_1 + M_{12}^{(m)}) + b_m'' M_{11}^{(m)} \qquad = 0 \; ,$$

$$a_m'M_{21}^{(m)} - b_m'(\alpha_2 + M_{22}^{(m)}) \qquad + a_m'' m\lambda \qquad\qquad = 0 \; ,$$

$$- a_m'(\alpha_2 + M_{22}^{(m)}) - b_m'M_{21}^{(m)} \qquad\qquad - b_m'' m\lambda \qquad = 0 \; ,$$

in which

$$M_{11}^{(m)} = \int_0^{T_0} \Phi_1(\tau)\sin m\lambda\tau d\tau \quad , \qquad M_{12}^{(m)} = \int_0^{T_0} \Phi_1(\tau)\cos m\lambda\tau d\tau$$

$$M_{21}^{(m)} = \int_0^{T_0} \Phi_2(\tau)\sin m\lambda\tau d\tau \quad , \qquad M_{22}^{(m)} = \int_0^{T_0} \Phi_2(\tau)\cos m\lambda\tau d\tau \quad .$$

Eliminating a_m' , b_m' , a_m'' , b_m'' from the four linear equations above, we have, if these are not all zero,

$$
(N) \qquad
\begin{vmatrix}
m\lambda & , & 0 & , & -M_{11}^{(m)} & , & \alpha_1 + M_{12}^{(m)} \\[2ex]
0 & , & -m\lambda & , & \alpha_1 + M_{12}^{(m)} & , & M_{11}^{(m)} \\[2ex]
M_{21}^{(m)} & , & -(\alpha_2 + M_{22}^{(m)}) & , & m\lambda & , & 0 \\[2ex]
-(\alpha_2 + M_{22}^{(m)}) & , & -M_{21}^{(m)} & , & 0 & , & -m\lambda
\end{vmatrix} = 0 ,
$$

and the development of this determinant gives

$$
(N') \qquad
\begin{aligned}
&\left[m^2\lambda^2 - (\alpha_1 + M_{12}^{(m)})(\alpha_2 + M_{22}^{(m)}) \right]^2 + M_{11}^2(\alpha_2 + M_{22}^{(m)} + \\
&+ M_{21}^2 (\alpha_1 + M_{12}^{(m)})^2 + M_{11}^{(m)^2} M_{21}^{(m)^2} + 2\, m^2\lambda^2\, M_{11}^{(m)}\, M_{21}^{(m)} = 0 .
\end{aligned}
$$

3. Through the same analysis previously employed in another problem concerning delayed effects,[25] one can recognize that $M_{11}^{(m)}$ and $M_{21}^{(m)}$ will be positive quantities if $\Phi_1(\tau)$ and $\Phi_2(\tau)$ are positive functions, decreasing with $0 \leq \tau < T_0$. Suppose, in fact, that

$$
\frac{2(h-1)\pi}{m\lambda} < T_0 \leq \frac{2h\,\pi}{m\lambda}
$$

in which h is a positive integer, and set $\frac{2\pi}{m\lambda} = \omega$. We will have

$$
\int_0^\omega \Phi_1(\tau)\, \sin m\lambda\tau d\tau > 0 , \quad \int_\omega^{2\omega} \Phi_1(\tau)\, \sin m\lambda\tau d\tau > 0 , \quad \ldots \quad \int_{(h-1)\omega}^{T_0} \Phi_1(\tau)\, \sin m\lambda\tau d\tau > 0 ,
$$

―――――――
[25] Vibrazioni elastiche nel caso delle eredità, "Rend. della R. Acc. dei Lincei" vol. XXI, serie 5ª, 2º sem. fasc. 1º, luglio 1912.

in which $i = 1, 2$, from which it follows that

$$M_{11}^{(m)} = \int_0^{T_0} \Phi_1(\tau) \sin m\lambda\tau d\tau > 0 \ , \qquad M_{21}^{(m)} = \int_0^{T_0} \Phi_2(\tau) \sin m\lambda\tau d\tau > 0 \ .$$

Then, in order for the determinant in (N) to be positive, it is enough that, for $0 \leq \tau < T_0$, any one of the following three conditions be satisfied:

(VI) $\Phi_1(\tau) > 0$ and decreasing , $\Phi_2(\tau) > 0$ and decreasing

(VII) $\Phi_1(\tau) = 0$, $\Phi_2(\tau) > 0$ and decreasing, $\alpha_1 \gtrless 0$

(VIII) $\Phi_2(\tau) = 0$, $\Phi_1(\tau) > 0$ and decreasing, $\alpha_2 \gtrless 0$.

In these three cases the Equations (N) and (N') cannot be satisfied and, therefore, the a_m' , b_m' , a_m'' , b_m'' must all be zero, which is to say that there cannot be periodic solutions.

The original equations for delayed effects, (IV) and (V) in §1, are derived from (K') and (L') by setting $F_1 = 0$, $\gamma_2 = 0$, while $\gamma_1 > 0$, $\varepsilon_1 > 0$. Also, if the delayed effect of the interaction decreases with time, $F_2 > 0$ is decreasing. This satisfies the condition (VII) above which *precludes the existence of small periodic fluctuations in the case of delayed effects.*

4. This theorem can be proved in a different way, without using a Fourier expansion. From Equations (80) and (80') it follows that

$$\left(\alpha_2 \nu_1(t) + \int_0^{T_0} \Phi_2(\tau)\nu_1(t-\tau)d\tau \right)\frac{d\nu_1}{dt} + \left(\alpha_1 \nu_2(t) + \int_0^{T_0} \Phi_1(\tau)\nu_2(t-\tau)d\tau \right)\frac{d\nu_2}{dt} = 0 \ ,$$

which can also be written as

$$\frac{d}{dt}\left(\frac{1}{2}\alpha_2 v_1^2(t) + v_1(t)\int_0^{T_0}\Phi_2(\tau)v_1(t-\tau)d\tau + \frac{1}{2}\alpha_1 v_2^2(t) + v_2(t)\int_0^{T_0}\Phi_1(\tau)v_2(t-\tau)d\tau\right) =$$

$$= v_1(t)\int_0^{T_0}\Phi_2(\tau)\frac{d}{dt}v_1(t-\tau)d\tau + v_2(t)\int_0^{T_0}\Phi_1(\tau)\frac{d}{dt}v_2(t-\tau)d\tau$$

$$= -v_1(t)\int_0^{T_0}\Phi_2(\tau)\frac{d}{d\tau}\left[v_1(t-\tau)-v_1(t)\right]d\tau - v_2(t)\int_0^{T_0}\Phi_1(\tau)\frac{d}{d\tau}\left[v_2(t-\tau)-v_2(t)\right]d\tau .$$

Integrating by parts, and taking into account that

$$\Phi_2(T_0) = \Phi_1(T_0) = 0 ,$$

(Cf., §1, Number 3), the last member becomes

$$v_1(t)\int_0^{T_0}\Phi_2'(\tau)\left[v_1(t-\tau)-v_1(t)\right]d\tau + v_2(t)\int_0^{T_0}\Phi_1'(\tau)\left[v_2(t-\tau)-v_2(t)\right]d\tau .$$

The preceding equation is then written as

$$(81)\begin{cases}\frac{d}{dt}\left(\frac{1}{2}\alpha_2 v_1^2(t) + v_1(t)\int_0^{T_0}\Phi_2(\tau)v_1(t-\tau)d\tau + \frac{1}{2}\alpha_1 v_2^2(t) + v_2(t)\int_0^{T_0}\Phi_1(\tau)v_2(t-\tau)d\tau\right) = \\ = \int_0^{T_0}\Phi_2'(\tau)\left[v_1(t)v_1(t-\tau)-v_1^2(t)\right]d\tau + \int_0^{T_0}\Phi_1'(\tau)\left[v_2(t)v_2(t-\tau)-v_2^2(t)\right]d\tau .\end{cases}$$

Notice now that

$$\frac{1}{2} \frac{d}{dt} \left\{ \int_0^{T_0} \Phi_2(\tau) v_1^2(t-\tau) d\tau + \int_0^{T_0} \Phi_1(\tau) v_2^2(t-\tau) d\tau \right\} =$$

$$= -\frac{1}{2} \int_0^{T_0} \Phi_2(\tau) \frac{d}{d\tau} \left[v_1^2(t-\tau) - v_2^2(t) \right] d\tau -$$

$$- \frac{1}{2} \int_0^{T_0} \Phi_1(\tau) \frac{d}{d\tau} \left[v_2^2(t-\tau) - v_2^2(t) \right] d\tau =$$

$$= \frac{1}{2} \int_0^{T_0} \Phi_2'(\tau) \left[v_1^2(t-\tau) - v_1^2(t) \right] d\tau +$$

$$+ \frac{1}{2} \int_0^{T_0} \Phi_1'(\tau) \left[v^2(t-\tau) - v_2^2(t) \right] d\tau .$$

Subtracting, side by side, the preceding equations from (81),

$$\frac{1}{2} \frac{d}{dt} \left\{ \alpha_2 v_1^2 + \alpha_1 v_2^2 + \int_0^{T_0} \left[\Phi_2(\tau) v_1(t-\tau)(2v_1(t) - v_1(t-\tau)) + \right. \right.$$

$$\left. \left. + \Phi_1(\tau) v_2(t-\tau)(2v_2(t) - v_2(t-\tau)) \right] d\tau \right\} =$$

$$= - \frac{1}{2} \int_0^{T_0} \left[\Phi_2'(\tau)(v_1(t-\tau) - v_1(t))^2 + \Phi_1'(\tau)(v_2(t-\tau) - v_2(t))^2 \right] d\tau .$$

Setting

$$A_1 = \int_0^{T_0} \Phi_1(\tau)d\tau \ , \qquad A_2 = \int_0^{T_0} \Phi_2(\tau)d\tau \ ,$$

$$H = (\alpha_2 + A_2)\nu_1^2(t) + (\alpha_1 + A_1)\nu_2^2(t) - \int_0^{T_0} \Big[\Phi_2(\tau)(\nu_1(t) - \nu_1(t-\tau))^2 +$$

$$+ \ \Phi_1(\tau)(\nu_2(t) - \nu_2(t-\tau))^2\Big]d\tau \ ,$$

the preceding equation can be written as

$$\frac{dH}{dt} = - \int_0^{T_0} \Big[\phi_2'(\tau)(\nu_1(t) - \nu_1(t-\tau))^2 + \phi_1'(\tau)(\nu_2(t) - \nu_2(t-\tau))^2\Big] \ d\tau \ .$$

If one of the functions Φ_1 , Φ_2 is decreasing, and the other is also decreasing or zero, the right-hand side of the last equation will be positive, and therefore, H is an ever increasing function. But if ν_1 and ν_2 were periodic, each with the same period, H would also be periodic and, therefore, it could not be always increasing. This constitutes the required proof which *precludes that* ν_1 *and* ν_2 *are periodic with the same period.*

5. From the results of Theorems VI, VII, and VIII of §1, or directly from Equations (80) and (80'), and by taking into account the last result above, we have the following theorems:

THEOREM I. *The integrals,* $\nu_1(t)$ *and* $\nu_2(t)$, *of Equations (80) and (80') oscillate around zero, and cross it infinitely often, passing through infinite numbers of maxima and minima, as time increases without bounds. Yet, if one of the two functions* Φ_1 *and* Φ_2 *is decreasing, and the other is also decreasing or zero, the expression* H *in terms of* ν_1 *and* ν_2 *will vary in the same direction, increasing all the time.*

THEOREM II. *If at some instant* H *assumes a positive value, the oscillations of* ν_1 *and* ν_2 *will never dampen.*

In fact, if $\nu_1(t)$ and $\nu_2(t)$ were to tend toward zero as time increased without bounds, so also would H . But H must, on the contrary, continue to increase from the positive value it had initially assumed.

(Editors' note). The appendix which follows reports a number of details from publications which appeared prior to the present monograph. It has been added to the re-edition in V. Volterra, "Opere matematiche", Accademia Nazionale dei Lincei, Rome, (1962), Vol. V.

APPENDIX

1. <u>Computation of the period of oscillation in a conservative predator-prey system.</u>

To compute the period, T, of the system in Part I, §2 (Number 6), set

$$n_1 = 1 + v_1 \ , \quad n_2 = 1 + v_2$$

Equations (18) then become

$$v_1^2 \left(\frac{1}{1 \cdot 2} - \frac{2v_1}{1 \cdot 2 \cdot 3} + \frac{3v_1^2}{1 \cdot 2 \cdot 3 \cdot 4} - \cdots \right) = 1 - ex^{1/\varepsilon_2}$$

$$v_2^2 \left(\frac{1}{1 \cdot 2} - \frac{2v_2}{1 \cdot 2 \cdot 3} + \frac{3v_2^2}{1 \cdot 2 \cdot 3 \cdot 4} - \cdots \right) = 1 - e\left(\frac{x}{c}\right)^{-1/\varepsilon_1}$$

in which, setting

$$S(v) = \frac{1}{1 \cdot 2} - \frac{2v}{1 \cdot 2 \cdot 3} + \frac{3v^2}{1 \cdot 2 \cdot 3 \cdot 4} - \cdots \ ,$$

and conveniently choosing the signs of the radicals,

$$n_1 - 1 = v_1 = \sqrt{1 - ex^{1/\varepsilon_2}} \ \frac{1}{\sqrt{S(v_1)}}$$

$$n_2 - 1 = v_2 = \sqrt{1 - e\left(\frac{x}{c}\right)^{-1/c_1}} \ \frac{1}{\sqrt{S(v_2)}} \ .$$

Then each of the four integrals is written as

$$\int_{x_1}^{x_2} \frac{dx}{\varepsilon_1 \varepsilon_2 x \sqrt{\left(1-ex^{1/\varepsilon_2}\right)\left(1-e\left(\frac{x}{c}\right)^{1/\varepsilon_1}\right)}} \ \sqrt{S(v_1)S(v_2)}$$

in which

$$x_1 = Ce^{\varepsilon_1} \quad , \qquad x_2 = e^{-\varepsilon_2} \quad ,$$

that is

$$\int_{x_1}^{x_2} \frac{dx}{\varepsilon_1 \varepsilon_2 x \sqrt{\left(1-\left(\frac{x}{x_2}\right)^{1/\varepsilon_2}\right)\left(1-\left(\frac{x}{x_1}\right)^{-1/\varepsilon_1}\right)}} \sqrt{S(v_1)S(v_2)} \quad .$$

But, as is well known,

$$1 - \left(\frac{x}{x_2}\right)^{1/\varepsilon_2} = \sum_1^\infty {}_m (-1)^{m-1} \left(\frac{1}{\varepsilon_2}\right)_m \left(1 - \frac{x}{x_2}\right)^m$$

$$1 - \left(\frac{x}{x_1}\right)^{-1/\varepsilon_1} = \sum_1^\infty {}_m (-1)^{m-1} \left(-\frac{1}{\varepsilon_1}\right)_m \left(1 - \frac{x}{x_1}\right)^m$$

in which

$$\left(\frac{1}{\varepsilon_2}\right)_m \qquad \text{and} \qquad \left(\frac{-1}{\varepsilon_1}\right)_m$$

denote the binomial coefficients.

The preceding integral can then be written as

$$\int_{x_1}^{x_2} \frac{\sqrt{S(v_1)S(v_2)}\; dx}{\varepsilon_1\varepsilon_2 x \sqrt{\left\{\sum_1^\infty {}_m (-1)^{m-1} \left(\frac{1}{\varepsilon_2}\right)_m \left(1 - \frac{x}{x_2}\right)^m\right\}\left\{\sum_1^\infty {}_m (-1)^{m-1} \left(-\frac{1}{\varepsilon_1}\right)_m \left(1 - \frac{x}{x_1}\right)^m\right\}}}$$

Suppose that the fluctuations are small so that $S(v_1)$, $S(v_2)$ can be neglected in all the terms except the first ones in the series at the bottom of the preceding formula. This then becomes

$$\int_{x_1}^{x_2} \frac{dx}{2\varepsilon_1\varepsilon_2 x \sqrt{\dfrac{1}{\varepsilon_1\varepsilon_2}\left(1 - \dfrac{x}{x_2}\right)\left(\dfrac{x}{x_1} - 1\right)}} = \frac{1}{2\sqrt{\varepsilon_1\varepsilon_2}} \int_{x_1}^{x_2} \frac{dx}{x\sqrt{\left(1 - \dfrac{x}{x_2}\right)\left(\dfrac{x}{x_1} - 1\right)}}$$

which can be easily transformed to

$$\frac{1}{\sqrt{\varepsilon_1\varepsilon_2}} \int_0^\infty \frac{dt}{1+t^2} = \frac{\pi}{2\sqrt{\varepsilon_1\varepsilon_2}} \quad .$$

Then the four integrals over the arcs R_2S_1, S_1S_2, S_2R_1, R_1R_2 are approximately equal, and, therefore,

$$T = 4 \frac{\pi}{2\sqrt{\varepsilon_1\varepsilon_2}} = \frac{2\pi}{\sqrt{\varepsilon_1\varepsilon_2}} \quad .$$

2. The perturbation of the averages in the case of small oscillations.

A change in the predator-prey cycle (Part I, §2,8) can be better understood by considering the case of small fluctuations, through the approximate solution (22), and supposing furthermore that the variations in ε_1, ε_2, γ_1, γ_2 are small.

Denoting by N_1', N_2' the changed values of N_1, N_2 when the ε_1, ε_2, γ_1, γ_2, E,α are changed by $\delta\varepsilon_1$, $\delta\varepsilon_2$, $\delta\gamma_1$, $\delta\gamma_2$, δE, $\delta\alpha$, and taking into account only the first order terms, the (22) become

$$(22_1) \qquad N_1' = \frac{\epsilon_2}{\gamma_2} + \frac{\gamma_1}{\sqrt{\epsilon_1}} \, E \cos \Theta' + \frac{1}{\gamma_2} \, \delta\epsilon_2 - \frac{\epsilon_2}{\gamma_2^2} \, \delta\gamma_2$$

$$+ \left(-\frac{1}{2} \frac{\gamma_1}{\epsilon_1^{3/2}} \, E \, \delta\epsilon_1 + \frac{1}{\sqrt{\epsilon_1}} \, E \, \delta\gamma_1 + \frac{\gamma_1}{\sqrt{\epsilon_1}} \, \delta E \right) \cos \Theta' - \frac{\gamma_1}{\sqrt{\epsilon_1}} \, E \, \delta\alpha \cdot \sin \Theta',$$

$$(22_2) \qquad N_2' = \frac{\epsilon_1}{\gamma_1} + \frac{\gamma_2}{\sqrt{\epsilon_2}} \, E \sin \Theta' + \frac{1}{\gamma_1} \, \delta\epsilon_1 - \frac{\epsilon_1}{\gamma_1^2} \, \delta\gamma_1$$

$$+ \left(-\frac{1}{2} \frac{\gamma_2}{\epsilon_2^{3/2}} \, E \, \delta\epsilon_2 + \frac{1}{\sqrt{\epsilon_2}} \, E\delta\gamma_2 + \frac{\gamma_2}{\sqrt{\epsilon_2}} \, \delta E \right) \sin \Theta' + \frac{\gamma_2}{\sqrt{\epsilon_2}} \, E \, \delta\alpha \cdot \cos \Theta'$$

in which

$$\Theta' = \sqrt{(\epsilon_1 + \delta\epsilon_1)(\epsilon_2 + \delta\epsilon_2)} \, t + \alpha \, .$$

Clearly, it was not possible to compute $\cos \Theta'$ and $\sin \Theta'$ through the first order terms in $\delta\epsilon_1$, $\delta\epsilon_2$ since t can diverge. Suppose now that the change in the ϵ_1, ϵ_2, ... takes place at time 0 , and that at this time the values of N_1' and N_2' are

$$N_1 = \frac{\epsilon_2}{\gamma_2} + \frac{\gamma_1}{\sqrt{\epsilon_1}} \, E \cos \Theta \, , \qquad N_2 = \frac{\epsilon_1}{\gamma_1} + \frac{\gamma_2}{\sqrt{\epsilon_2}} \, E \sin \Theta$$

in which

$$\Theta = \sqrt{\epsilon_1 \epsilon_2} \, t + \alpha \, .$$

Noticing that for $t = 0$ one has $\Theta = \Theta' = \alpha$, we have the equations

$$\frac{\gamma_2}{\sqrt{\epsilon_2}} (\sin\alpha \, \delta E + \cos\alpha \, E \, \delta\alpha) = \frac{1}{\gamma_1} \, \delta\epsilon_1 + \frac{\gamma_2}{2\epsilon_2^{3/2}} \, E \sin\alpha \, \delta\epsilon_2 + \frac{\epsilon_1}{\gamma_1^2} \, \delta\gamma_1 - \frac{E \sin\alpha}{\sqrt{\epsilon_2}} \, \delta\gamma_2 \, ,$$

$$\frac{\gamma_1}{\sqrt{\epsilon_1}} (\cos\alpha \, \delta E - \sin\alpha \, E \, \delta\alpha) = -\frac{1}{\gamma_2} \, \delta\epsilon_2 + \frac{\gamma_1}{2\epsilon_1^{3/2}} \, E \cos\alpha \, \delta\epsilon_1 + \frac{\epsilon_2}{\gamma_2^2} \, \delta\gamma_2 - \frac{E \cos\alpha}{\sqrt{\epsilon_1}} \, \delta\gamma_1 \, ,$$

from which one obtains

$$(22_3) \qquad \delta E = \left(\frac{1}{2} \frac{E}{\epsilon_1} - M_1 \sin\alpha \right)\delta\epsilon_1 + \left(\frac{1}{2} \frac{E}{\epsilon_2} - M_2 \cos\alpha \right)\delta\epsilon_2 +$$

$$+ \left(- \frac{E}{\gamma_1} + P_1 \sin\alpha \right)\delta\gamma_1 + \left(+ \frac{E}{\gamma_2} + P_2 \cos\alpha \right)\delta\gamma_2$$

$$(22_4) \qquad E\delta\alpha = - M_1 \cos\alpha \, \delta\epsilon_1 + M_2 \sin\alpha \, \delta\epsilon_2 + P_1 \cos\alpha \, \delta\gamma_1 - P_2 \sin\alpha \, \delta\gamma_2$$

in which

$$M_1 = \frac{\sqrt{\epsilon_2}}{\gamma_1\gamma_2} + \frac{1}{2} \frac{E}{\epsilon_1} \sin\alpha \qquad ; \qquad M_2 = \frac{\sqrt{\epsilon_1}}{\gamma_1\gamma_2} + \frac{1}{2} \frac{E}{\epsilon_2} \cos\alpha \qquad ;$$

$$P_1 = \left(\frac{\sqrt{\epsilon_2}}{\gamma_1\gamma_2} + \frac{E}{\epsilon_1} \sin\alpha \right) \frac{\epsilon_1}{\gamma_1} \qquad ; \qquad P_2 = \left(\frac{\sqrt{\epsilon_1}}{\gamma_1\gamma_2} + \frac{E}{\epsilon_2} \cos\alpha \right) \frac{\epsilon_2}{\gamma_2} \qquad .$$

Substituting the values (22_3), (22_4), in Equations (22_1), (22_2) we obtain the perturbed values of the number of individuals in the two species, as long as the perturbation introduced at time 0 lasts. If we take the averages of N_1', N_2' over a period, all the terms containing $\cos \Theta'$ and $\sin \Theta'$ clearly disappear, and we have as average values

$$\frac{\epsilon_2}{\gamma_2} + \frac{\delta\epsilon_2}{\gamma_2} - \frac{\epsilon_2\delta\gamma_2}{\gamma_2^2} \quad , \qquad \frac{\epsilon_1}{\gamma_1} + \frac{\delta\epsilon_1}{\gamma_1} - \frac{\epsilon_1\delta\gamma_1}{\gamma_1^2}$$

which lead to the very properties we had previously found.

3. Time reversal in conservative systems.

The process of variation in the numbers of individuals of the species enjoys of a special symmetry which is worth discussing. Let us change t to $2t_0 - t$ in equations (B) of Part II, §2, Number 2. Setting

$$(50^*) \qquad\qquad N_r(2t_0 - t) \;=\; N'_r(t)$$

we obtain the equations

$$\beta_r \; \frac{dN'_r}{dt} \;=\; \left(- \varepsilon_r \beta_r + \sum_{1}^{n} s \, a_{rs} N'_s \right) N'_r \;.$$

These equations correspond to the variations in an association in which the coefficients ε_r are changed to $-\varepsilon_r$, and a_{rs} to $-a_{rs}$ (Cf., Part I, §4, Number 7). This may be called the *conjugate* association.

The fundamental theorem of conjugate conservative associations can be stated as follows:

At time t_0 *the two conjugate systems coincide, that is,*

$$N_1(t_0) = N'_1(t_0) \;, \quad N_2(t_0) = N'_2(t_0) \;, \;\ldots\;, \quad N_n(t_0) = N'_n(t_0) \;;$$

and furthermore,

$$N_r(t_0-t) = N'_r(t_0+t) \quad, \quad N_r(t_0+t) = N'_r(t_0-t) \;,$$

namely one of the two systems assumes, with time, all the values assumed by the other system in previous times, and in the reverse order.

This is immediately proved by setting $t = t_0$ in (50), or by exchanging $t_0 + t$ for t , or $t_0 - t$ for t . We may also state that *the variations of the two conjugate systems are symmetric with respect to the time* t_0 , *or that one is the mirror image of the other.* It is also immediately proved that: *The conjugate system of a conservative system is also conservative.*

Notice that the conjugate system of a dissipative association is neither conservative nor dissipative since the corresponding fundamental form is negative.

The consideration of conjugate systems permits the proof, in an equivalent manner, of the second part of the property in Part III, §1, Number 3. Assuming, in fact, that starting from an initial state in which all the N_1, N_2, ... N_n are finite and non-zero, say for example C_1, C_2, ... C_n , a conservative association may reach in a finite time T the extinction of a certain species h , $N_h = 0$.

Consider the conjugate association which coincides at the initial time $t = 0$ with the original one. At the time $-T$ the N_1', N_2', ... N_n' should have taken the same values as the N_1, N_2, ... N_n at the time T . Therefore, from a previous theorem N_1', N_2', ... N_n' will be finite and, according to the hypothesis, N_h' should be zero. Translating the time origin $-T$, we would then have that the time origin N_1', N_2', ... N_n' is finite, and $N_h' = 0$. From a previous theorem N_h' must always remain zero, and so it will be after a time T . But at this time it would have to coincide with C_h which is positive. This proves the absurdity of the hypothesis that the species h could become extinct in a finite time. There-fore: *In a conservative biological association none of the species can become extinct in a finite time.*

4. <u>A generalization of conservative systems.</u>

With reference to Equations (F) in Part III, §1, Number 8, set

$$\phi_r(N_r) = q_r - N_r \psi_r(N_r)$$

in which the q_r's are constant. We will have

$$\int \frac{\phi_r dN_r}{N_r} = q_r \ln N_r - \int \psi_r(N_r) dN_r = q_r \ln N_r - \theta_r(N_r) \ .$$

Therefore, taking the antilogarithm, the integral we had previously obtained becomes:

$$\left(\frac{e^{\theta_1(N_1)}}{N_1^{q_1}} \right) \left(\frac{e^{\theta_2(N_2)}}{N_2^{q_2}} \right) \cdots \left(\frac{e^{\theta_n(N_n)}}{N_n^{q_n}} \right) = C$$

in which C is a positive constant. We can then state that:

If the coexistence of the species results in their growth coefficients
being of the form

$$\sum_{1}^{n} {}_{s} \ F_{rs}(N_1,N_2,\ldots,N_n)(q_s - N_s \theta'_s(N_s))$$

$$(F_{rs} \ = \ - \ F_{sr}, F_{rr} = 0)$$

the differential equation for the variations of their numbers of individuals

$$\frac{dN_r}{dt} \ = \ N_r \sum_{1}^{n} {}_{s} \ F_{rs}(N_1,N_2,\ldots,N_n)(q_s \ - \ N_s \theta'_s(N_s))$$

will have the integral

$$\left(\frac{e^{\theta_1(N_1)}}{N_1^{q_1}} \right) \left(\frac{e^{\theta_2(N_2)}}{N_2^{q_2}} \right) \ \cdots \ \left(\frac{e^{\theta_n(N_n)}}{N_n^{q_n}} \right) \ = \ C$$

in which C *is a positive constant.*

In the case of only two species, the preceding integral becomes

$$\left(\frac{e^{\theta_1(N_1)}}{N_1^{q_1}} \right) \left(\frac{e^{\theta_2(N_2)}}{N_2^{q_2}} \right) \ = \ C$$

so that the time can be separated, and the problem is reduced to quadratures.
By conveniently limiting the form of the functions f_r , one can then generalize
the theories developed in Part I and Part II. Choosing both the F_{rs}'s and the
$\theta'(N_r)$'s as constants would be justified if the effects of the interactions among
individuals of different species were proportional to the number of their encounters
at that time. The linearity of the f_r's follows from the number of encounters
in a dt between individuals of species r and s being proportional to
$N_r N_s$ (Cf., Part II, §2, Number 1). This justification, however, should not cause
us to neglect the more general treatment.

Principles of Mathematical Biology: Part II

V. Volterra

Editors' note: We begin this contribution with the original English language summary for both parts of Volterra's paper. However most of Part I, "The founda-tions of the theory of the struggle for existence", summarizes material on con-servative systems which is also discussed at length in the 1927 monograph, Variazioni e fluttuazioni del numero d'individui in specie animali conviventi. In order to avoid redundancy we will report from Part I only those formulas and definitions which are necessary for the development of Part II, "The general properties of the struggle for existence".

SUMMARY

This memoir consists of two parts, of which the first deals with the foundations of the theory of the struggle for existence, and begins with the intro-duction of the important concept of quantity of life, besides that of population. The fundamental equations are then established for the case where the individuals of the biological association mutually devour each other, the reasoning being based on the principle of encounters and on the fundamental hypothesis of the existence of equivalents of the individuals constituting the different species which form the association.

Having now obtained the equations which determine the rates of change of the numbers of individuals and of the quantities of life of the species, the memoir proceeds to find the equations relating to the stationary (equilibrium) state, by establishing theorems which are afterwards applied for the purpose of obtaining the three general laws of the fluctuations. This investigation of the equilibrium state is then followed by the determination of the integrals of the fluctuation equations, and the deduction from these of their most important consequences.

The second part of the memoir contains the enumeration and demonstration of the general laws of the struggle for existence, which flow from the discussion of the principles and the integrals obtained in the first part. In this way there is established a new sort of dynamics, demographic dynamics, which, though substantially different from the dynamics of material systems, is developed from an analogous point of view.

The second part begins with the principle of the conservation of demographic energy, according to which there are two sorts of energy, one actual and one potential, which transform mutually the one into the other. The principle is the analogue of the principle of conservation of mechanical energy. It is followed by the enunciation of the three laws relating to the biological fluctuations, the experimental verification of which has been investigated by several naturalists. The success which has attended their efforts is well known.

Everybody knows the importance of Hamilton's principle in mechanics and in all the domains of physical science. An analogous variation principle can be found in biology, and from it one can deduce the fluctuation equations in the canonical Hamiltonian form and also in the form of a Jacobian partial differential equation. Their integrals in involution form the subject next studied and in this way are found cases where the integration is reducible to quadratures.

Hamilton's principle leads to the principle of least action (Maupertuis). There exists also in biology a closely related principle, which may be called the principle of least vital action. Its analytical form is such that it requires the existence of a true minimum, a state of affairs which does not always hold good in the analogous case in mechanics.

SELECTED FORMULAS AND DEFINITIONS FROM PART I.

The general equations for a conservative association can be written as

$$\frac{dN_r}{dt} = \left(\epsilon_r + \frac{1}{\beta_r} \sum_{1}^{n} s \; a_{sr}N_s\right)N_r \qquad (r=1,2,\ldots,n) \; ,$$

or as

(I)
$$\beta_r \frac{dN_r}{dt} = \left(\epsilon_r\beta_r + \sum_{1}^{n} s \; a_{sr}N_s\right)N_r$$

in which

$$a_{sr} = -a_{rs} \; , \qquad a_{rr} = 0 \; .$$

The ϵ_r will be called *coefficients of self-increase* and the

$$\epsilon_r + \frac{1}{\beta_r} \sum_{1}^{n} s \; a_{sr}N_s = \vartheta_r$$

effective coefficients of increase or *demographic coefficients*. The positive numbers $1/\beta_r$ are called *equivalents* of the species, in which β_r might, for example, represent the average weight of an individual of species r .

By introducing the *quantity of life* of a species in a time interval $(0,t)$,

$$X_r = \int_0^t N_r dt$$

equations (I) become

(II)
$$.\beta_r \frac{d^2X_r}{dt^2} = \left(\epsilon_r\beta_r + \sum_{1}^{n} s \; a_{sr}\frac{dX_s}{dt}\right)\frac{dX_r}{dt}$$

or, setting $dX_r/dt = X_r'$ and $d^2X_r/dt^2 = X_r''$,

(II')
$$\beta_r X_r'' = \left(\epsilon_r \beta_r + \sum_1^n{}_s a_{sr} X_s' \right) X_r'$$

having as the integral

(B)
$$\sum_1^n{}_r \beta_r \dot{X}_r' - \sum_1^n{}_r \epsilon_r \beta_r X_r + C = 0 \quad .$$

Rewriting equations (II') as

(II")
$$\beta_r \frac{d}{dt} \ln X_r' = \epsilon_r \beta_r + \sum_1^n{}_s a_{sr} X_s' \,,$$

and integrating, one gets

(C)
$$\Theta_r = \beta_r \ln X_r' + \sum_1^n{}_s a_{rs} X_s - \epsilon_r \beta_r t - C_r = 0$$

from which one derives the integral

(C')
$$\Theta = \sum_1^n{}_r X_r' \Theta_r = \chi + Z - \left(\sum_1^n{}_r \epsilon_r \beta_r t + C_r \right) X_r' = 0$$

where the C_r's are constants, and

(8)
$$Z = \sum_1^n{}_r \sum_1^n{}_s a_{rs} X_r' X_s \,,$$

(8')
$$\chi = \sum_1^n{}_r \beta_r X_r' \ln X_r' \quad .$$

Finally, combining integrals (B) and (C'), one has

(C")
$$\Omega = \Theta + \sum_1^n{}_r \epsilon_r \beta_r X_r - \sum_1^n{}_r \beta_r X_r' - C = \sum_1^n{}_r \beta_r X_r' \ln X_r' + \sum_1^n{}_r \sum_1^n{}_s a_{rs} X_s X_r' -$$

$$- \sum_1^n{}_r \epsilon_r \beta_r t X_r' - \sum_1^n{}_r C_r X_r' + \sum_1^n{}_r \epsilon_r \beta_r X_r - \sum_1^n{}_r \beta_r X_r' - C = 0 \quad .$$

The equilibrium conditions for system (I) are given by the linear system of equations

$$(III) \qquad \epsilon_r \beta_r + \sum_{1}^{n} {}_s a_{sr} N_r = 0$$

which is called the *fundamental system*. If it has only positive roots, q_1, q_2, ..., q_n, there *exists an equilibrium state* for the association. Notice that, since $a_{sr} = -a_{rs}$, the roots of III satisfy

$$(3) \qquad \sum_{1}^{n} {}_r \epsilon_r \beta_r q_r = 0 \quad .$$

Utilizing a remark by Levi (1931) equations (I) can be rewritten as

$$\sum_{1}^{n} {}_r \left(\beta_r \frac{dN_r}{dt} - q_r \beta_r \frac{1}{N_r} \frac{dN_r}{dt} \right) = 0$$

from which one gets the integral

$$(D) \qquad \sum_{1}^{n} {}_r (\beta_r N_r - q_r \beta_r \ln N_r) = C' \quad ,$$

and, passing from logarithms to numbers,

$$(E) \qquad \left(\frac{e^{N_1}}{N_1^{q_1}} \right)^{\beta_1} \left(\frac{e^{N_2}}{N_2^{q_2}} \right)^{\beta_2} \cdots \left(\frac{e^{N_n}}{N_n^{q_n}} \right)^{\beta_n} = c' \quad ,$$

in which c' is a constant.

PART II.   THE GENERAL PROPERTIES OF THE STRUGGLE FOR EXISTENCE

Table of Contents:

§1.   The principle of conservation of demographic energy

    1.   Let us set

$$\sum_{1}^{n} r \ \beta_r N_r \ = \ \sum_{1}^{n} r \ \beta_r X'_r \ = \ L$$

$$C_0 \ - \ \sum_{1}^{n} r \ \beta_r \epsilon_r X_r \ = \ M$$

in which C_0 is a constant which one might suppose to be the upper bound of $\sum_{1}^{n} r \ \beta_r \epsilon_r X_r$.   The integral (B) is then expressed as

$$L \ + \ M \ = \ \text{constant} \ .$$

From the biological point of view one can regard L as *actual demographic energy*, and M as *potential demographic energy*. These two forms of energy are then transformed one into the other, while their sum remains constant. This proposition is analogous to the theorem of kinetic energy in mechanics.

    2. Denote by ϑ_1, ϑ_2, ..., ϑ_n the demographic coefficients, and suppose that δX_1, δX_2, ..., δX_n **are some** *virtual changes in the quantities of life.*

One might regard

$$\sum_r \beta_r \vartheta_r \, \delta X_r$$

as the *work of growth* or *demographic work*. If we take

$$\vartheta_r = \varepsilon_r + \frac{1}{\beta_r} \sum_s^n a_{sr} N_s \ ,$$

and suppose that the δX_i are the natural increases taking place in the time dt , we will have

$$\delta X_r = N_r dt \ ,$$

and the demographic work will then be

$$\sum_r \beta_r \varepsilon_r N_r dt \ .$$

This means that the work which is derived from increases in the interactions between species is zero, and that the total work is simply that which is derived from the rates of increase of the individual species.

3. In many cases the past life of a species in a given habitat modifies its properties, affecting its vital coefficients. This happens, for example, when the individuals release catabolic products which intoxicate the habitat. At any one time the change in the habitat produced in this way by the different species is due to the actions of the individuals during previous times. If the action of such species remains constant, then at any one time it can be assumed to be proprotional to the quantity of life of the species. The change in the growth rate of species r can then be expressed by

$$\frac{1}{\beta_r} \sum_s^n c_{rs} X_s$$

in which the quantities c_{rs} are constant.

When different species have a symmetrical action, namely $c_{rs} = c_{sr}$, the preceding expression becomes

$$\frac{1}{\beta_r} \frac{\partial}{\partial X_r} \left[\frac{1}{2} \sum_1^n r \sum_1^n s \ c_{rs} X_r X_s \right] \ .$$

Setting

(13) $$P = \sum_1^n r \ \beta_r \epsilon_r X_r + \frac{1}{2} \sum_1^n r \sum_1^n s \ c_{rs} X_r X_s \quad ,$$

the fundamental equations become

(II) $$\beta_r X_r'' = \left(\frac{\partial P}{\partial X_r} + \sum_1^n s \ a_{sr} X_s' \right) X_r'$$

or

$$\sum_1^n r \ \beta_r X_r'' = \frac{dP}{dt} \ .$$

Their integral is

(B') $$\sum_1^n r \ \beta_r N_r = P - c_0$$

in which c_0 is a constant.

In order to use terminology which follows as closely as possible that of classical mechanics we will call P the *demographic potential*. The potential demographic energy will be

$$c_0 - P = N$$

and, as before, one will have

$$L + N = \text{constant} \ .$$

The demographic work produced by the system in the time dt will be dP .

In this section we have introduced a demographic potential made up of both first degree and second degree terms. In the following, unless otherwise stated, we will always deal with demographic potentials having only first degree terms.[1]

§2. The three properties of fluctuations in conservative associations

1. We have seen that in a conservative association with an odd number of species the fundamental determinant is zero. In general, equilibrium states would not be possible and the number of individuals in some of the species would have to increase without bounds or tend to zero. With an odd number of species, then, it is probable that the association would not be preserved, and it would end up as an association with a smaller, even number of species.

2. Consider now a conservative association with an even number of species and a non-zero fundamental determinant. We further suppose that the roots of the fundamental equations are all positive, so that there is an equilibrium state. In this case three fundamental properties of the fluctuations hold:

First property: THE CONSERVATION OF THE FLUCTUATIONS

The number of individuals of the different species are limited between two positive numbers, and there are always fluctuations which never dampen.

Second property: THE CONSERVATION OF THE AVERAGES

If one takes the limits of the averages over time spans approaching infinity (asymptotic averages) as the averages of the numbers of individuals in the different species, such averages will be constant, and they will not depend on the initial numbers of individuals in the species.

---

[1] By taking into account the degradation of catabolites in the medium one gets *bona fide* integro-differential equations, as in Volterra and D'Ancona (1935). Notice also that the symmetry of the coefficients c_{rs} , c_{sr} might not be verified in nature as would seem to occur, for instance, in the experiments by Régnier and Lambin (1934).

Third property: THE PERTURBATION OF THE AVERAGES

Suppose that one exploits all the species uniformly in proportion to their
numbers, and that such exploitation is small enough to allow for the possibility
of an equilibrium state.   There will always be some species which will benefit
from such an exploitation, that is to say their asymptotic averages will increase,
and there will always be some species which will suffer, that is to say their
averages will decrease.   Among the species which benefit at least one of them
will be preyed upon by another species, and among the species which suffer at
least one of them will prey upon another species.

To prove this property, recall that the asymptotic averages are the same
as the equilibrium numbers q_1, q_2, ..., q_n . It is clear that if the $\Delta\epsilon_1$,
$\Delta\epsilon_2$, ..., $\Delta\epsilon_n$ have all the same sign, among the Δq_1, Δq_2, ..., Δq_n there will
certainly be some which are not zero, and which will have different signs.

Suppose, for instance, that $\Delta q_1 > 0$. One can make two hypotheses: either
some of the a_{21}, a_{31}, ..., a_{n1} will be negative, so that species 1 is preyed
upon by others, or all such quantities will be positive or zero so that species
1 preys upon the others without itself being preyed upon by any other species.
In the second hypothesis, taking negative $\Delta\epsilon_1$, $\Delta\epsilon_2$, ..., $\Delta\epsilon_n$ one will find among
the species 2, 3, 4, ..., n some whose equilibrium populations will increase.
There will be then at least one species preyed upon by others whose population
will increase by decreasing the ϵ_1, ϵ_2, ..., ϵ_n .

We have supposed that Δq_1 is positive. It follows that among the
Δq_2, Δq_3, ..., Δq_n at least one of them must be negative. Suppose, then, that
$\Delta q_2 < 0$. Then, among the a_{12}, a_{32}, ..., a_{n2} some will be positive, meaning
that species 2 will prey upon others, or they will all be negative or zero, so
that species 2 is preyed upon by others without preying on any itself. It
follows that if the $\Delta\epsilon_1$, $\Delta\epsilon_2$, ..., $\Delta\epsilon_n$ are negative there will be species which
prey on others, and whose averages will decrease. This proves the third property.

3. Some further, detailed clarifications can help to remove any remaining

difficulty concerning the third property. One can in fact distinguish in a biological association among three categories of species: 1) Those which feed on
others without being fed upon by any; 2) those which do not feed on any, and are
feed upon by others; 3) those which are fed upon by other species, and, in turn
feed on some others.

It can happen that all three kinds of species are represented in an
association, or, alternatively, just two of them. If only one kind of species is
represented, it must necessarily be of the third kind.

In this light the final statement of the third property can be changed in
the following way: *Among the species which benefit at least one will belong either
to the second or to the third category; among the species which suffer at least
one will belong either to the first or to the third category.*

Suppose that only the first and the second category are represented in a
conservative association, in which case both categories must be represented by
the same number of species. Then, among the species which benefit by a decrease
in the coefficient of self increase, there is at least one which belongs to the
second category. Similarly, among the species which suffer, there will be at
least one which belongs to the first category.

In my previous work I had stated the third property in this last special
form, thus restricting the meaning of predator and prey species. The statement
immediately above removes any such restriction. It is clear that one can reason
in a wholly analogous manner in the case in which the ε_1, ε_2, ..., ε_n are
increased instead of being decreased.

4. Let us now study the special case of two species, the first feeding
on the second. The integral (E) becomes

$$\left(\frac{e^{\frac{N_1}{q_1}}}{N_1} \right)^{\beta_1} \left(\frac{e^{\frac{N_2}{q_2}}}{N_2} \right)^{\beta_2} = \text{constant}$$

in which q_1 and q_2 are positive. Considering N_1 and N_2 as the cartesian coordinates of a plane, one obtains a closed cycle. The phenomenon will then be periodic, and the asymptotic means become the means during one period.

The three properties of the fluctuations become:

First property: THE PERIODIC CYCLE.

The fluctuations of the two species are periodic.

Second property: THE CONSERVATION OF THE AVERAGES.

The average numbers of individuals in the two species during one period are constant, and they do not depend on the initial numbers.

Third property: THE PERTURBATION OF THE AVERAGES.

If one destorys the two species uniformly, and in proportion to their numbers, the average number of the prey species increases, and the average number of the predator species decreases.

§3. The variational principle.

1. We will show here that equations (II) can be reduced to a problem of the calculus of variations, and that the same applies to equations (II").

The tendency to reduce natural problems to minimal problems must have always been present since one believes that nature tends to operate, for different phenomena, in the most economic way. This idea was the starting poing for Maupertuis who, in a famous work, sought to establish one of the fundamental principles of nature, which he called the principle of least action. This principle was to become the foundation of classical mechanics.

Maupertuis started from the philosophical principle that nature always acts through the simplestmeans. He attempted to derive all the laws of motion and of statics out of this single metaphysical concept. Descartes had instead tried to begin with the principle of momentum, and Leibnitz with the principle of kinetic energy. Maupertuis defined, to begin with, the *quantity of action,* and tried to deduce the solutions of natural problems from the principle of least action. He applied it to the elastic collision of solid bodies. In this work he also related the famous principle by Fermat on the refraction of light to his principle.

It is clear that the principle of Maupertuis is much more comprehensive than those of Descartes and Leibnitz. The latter two, in fact, only give the integrals of the dynamic equations, whereas the principle of Maupertuis is equivalent to the equations themselves. Lagrange gave different bases to the dynamics, and showed that the principle of least action follows from his equations. In this way he reduced mechanics to a branch of that calculus of variations, which he contributed so much to create. Further progress was due to Hamilton who first established the principle of stationary action, and who then developed the principle of its variation. Finally Jacobi systematized the general theory, obtaining an equation which recently has been extended. It appears that day by day the implications of this equation become larger and larger.

We are going to see that one can follow the same path in biological problems. The equations of the struggle for existence can be posed in a canonical form and they can be reduced to a Jacobian equation.

2. In studying the integral of the fundamental conservative equations we found the expression

$$(8') \qquad\qquad \chi = \sum_{1}^{n} {}_{r} \ \beta_r X_r' \ \ln X_r' \ .$$

Being N_r the population of a species, $\dfrac{dN_r}{N_r}$ will be its *relative infinitesimal increase*, and $\displaystyle\int_1^{N_r} \dfrac{dN}{N}$ will be the total relative increase which is necessary to attain the present value, N_r, starting from a single individual. Now, since dX_r is the infinitesimal increase in the quantity of life of a species, one has

$$N_r = \frac{dX_r}{dt} = X_r' \ .$$

One can give the name *infinitesimal vital action* to the quantity

$$\beta_r \ln N_r dX_r \;=\; \beta_r N_r \ln N_r dt \;=\; \beta_r X_r' \ln X_r' dt \;\;,$$

adopting a terminology analogous to that of classical mechanics. Recall that $1/\beta_r$ is the *equivalent* of species r , in which β_r may be interpreted as the average weight of the individuals.

The total action in a time interval $(0,t)$ will then be

$$\int_0^t \beta_r N_r \ln N_r dt \;=\; \int_0^t \beta_r X_r' \ln X_r' dt \;\;,$$

and, if one refers to an association of n species, the *total vital action* will be

$$(F) \qquad A \;=\; \int_0^t \sum_1^n{}_r \beta_r X_r' \ln X_r' dt = \int_0^t \sum_1^n{}_r \beta_r N_r \ln N_r dt \;.$$

3. In terms of the *demographic potential* (§1, Number 3),

$$(13) \qquad P \;=\; \sum_1^n{}_r \beta_r \epsilon_r X_r \;+\; \frac{1}{2} \sum_1^n{}_r \sum_1^n{}_s c_{rs} X_r X_s \;\;,$$

and of the bilinear form (Cf. Editors' note),

$$(8) \qquad Z \;=\; \sum_1^n{}_r \sum_1^n{}_s a_{rs} X_r' X_s \;\;,$$

we can construct the expression

$$(14) \qquad \Phi = \chi + \frac{1}{2} Z + P \;=\; \sum_1^n{}_r \beta_r X_r' \ln X_r' + \frac{1}{2} \sum_1^n{}_r \sum_1^n{}_s a_{rs} X_r' X_s + P \;.$$

We will have

(14')
$$\frac{\partial \Phi}{\partial X'_r} = \beta_r \ln X'_r + \beta_r + \frac{1}{2} \Sigma_s a_{rs} X_s$$

(14")
$$\frac{\partial \Phi}{\partial X_r} = \frac{1}{2} \Sigma_s a_{sr} X'_s + \frac{\partial P}{\partial X_r} \ .$$

Now set

(G)
$$U = \int_{t_0}^{t} \Phi dt \ .$$

By writing

(G')
$$\delta U = 0$$

Euler's equations will give

(G")
$$\frac{d}{dt} \frac{\partial \Phi}{\partial X'_r} - \frac{\partial \Phi}{\partial X_r} = 0 \ ,$$

that is,

(G''')
$$\frac{d}{dt} (\beta_r \ln X'_r) = \frac{\partial P}{\partial X_r} + \sum_1^n {}_s a_{sr} X'_s \ .$$

These are the equations (II") which reduce to the equations (II) when the demographic potential has only first degree terms. In this way we have reduced the fundamental equations of the struggle for existence to a *problem of the calculus of variations*.

§4. The canonical equations.

1. Having set the fundamental equations in Lagrangian form, in which the demographic potential has second degree terms, we can now derive their canonical form. Set

$$(14''')\qquad p_r = \frac{\partial \Phi}{\partial X'_r} = \beta_r \ln X'_r + \beta_r + \frac{1}{2} \sum_{1}^{n} {}_s a_{rs} X_s ,$$

from which we get

$$(14^{iv})\qquad X'_r = e^{\frac{1}{\beta_r}\left(p_r - \beta_r - \frac{1}{2} \sum_{1}^{n} {}_s a_{rs} X_s\right)} .$$

Taking

$$(15)\qquad H = \Phi - \sum_{1}^{n} {}_r p_r X'_r = P - \sum_{1}^{n} {}_r \beta_r X'_r =$$

$$= P - \sum_{1}^{n} {}_r \beta_r e^{\frac{1}{\beta_r}\left(p_r - \beta_r - \frac{1}{2} \sum_{1}^{n} {}_s a_{rs} X_s\right)} ,$$

we have

$$(15')\qquad \frac{\partial H}{\partial p_r} = - e^{\frac{1}{\beta_r}\left(p_r - \beta_r - \frac{1}{2} \sum_{1}^{n} {}_s a_{rs} X_s\right)} = - X'_r = - \frac{dX_r}{dt} .$$

But, from (15), (14''), (G''), and (14'''), we have

$$(15'')\qquad \frac{\partial H}{\partial X_r} = \frac{\partial P}{\partial X_r} + \frac{1}{2} \sum_{1}^{n} {}_s a_{sr} X'_s = \frac{\partial \Phi}{\partial X_r} = \frac{d}{dt} \frac{\partial \Phi}{\partial X'_r} = \frac{dp_r}{dt} .$$

We obtain, then, the canonical equations

(IV)
$$\frac{dp_r}{dt} = \frac{\partial H}{\partial X_r} \quad , \quad \frac{dX_r}{dt} = - \frac{\partial H}{\partial p_r}$$

in which H is given by

(15)
$$H = P - \sum_{1}^{n} {}_r \beta_r e^{\frac{1}{\beta_r} \left(p_r - \beta_r - \frac{1}{2} \Sigma_s a_{rs} X_s \right)} .$$

2. The canonical equations have the integral

$$H = \text{constant} .$$

which is the conservation of demographic energy derived in §1, Number 3. We have
also found, when the demographic potential has only first degree terms, the
integral

(D)
$$\sum_{1}^{n} {}_r \beta_r (N_r - q_r \ln N_r) = \sum_{1}^{n} {}_r \beta_r (X'_r - q_r \ln X'_r) = \text{constant.}$$

We can write it as

$$K = \sum_{1}^{n} {}_r \left[\beta_r e^{\frac{1}{\beta_r} \left(p_r - \beta_r - \frac{1}{2} \sum_{1}^{n} {}_s a_{rs} X_s \right)} - q_r \left(p_r - \beta_r - \frac{1}{2} \sum_{1}^{n} {}_r a_{rs} X_s \right) \right]$$

$$= \text{constant} ,$$

and it is easy to verify that the Poisson bracket

$$(H, K)$$

is zero.

3. The Jacobian equation deduced from the canonical equations (IV) will be

$$\text{(IV')} \qquad C = \sum_{1}^{n} {}_{r} \ \beta_r \left[\varepsilon_r X_r - e^{\frac{1}{\beta_r}\left(\frac{\partial V}{\partial X_r} - \beta_r - \frac{1}{2}\Sigma_s a_{rs}X_s\right)} \right] \quad ,$$

and the integrals of the canonical equations will be

$$\frac{\partial V}{\partial a_r} = b_r \quad , \qquad \frac{\partial V}{\partial X_r} = p_r$$

in which $V(X_1, X_2, \ldots, X_n ; a_1, a_2, \ldots, a_n)$ is a complete integral of the partial derivative equation (IV'), and $a_1, a_2, \ldots, a_n ; b_1, b_2, \ldots, b_n$ are some constants.

§5. The involution integrals.

1. We have seen (Cf., Editors' note) that the integrals of the equations (II') are

$$\text{(C)} \qquad \beta_r \ \ell n \ \frac{dX_r}{dt} + \sum_{1}^{n} {}_{s}' \ a_{rs}X_s - \varepsilon_r \beta_r t = \text{constant.}$$

From Equations (14''') we can rewrite (C) as

$$\text{(16)} \qquad \frac{p_r + \frac{1}{2} \sum_{1}^{n} {}_{s} \ a_{rs}X_s}{\varepsilon_r \beta_r} - t = \text{constant} \ .$$

Setting

$$\frac{p_r + \frac{1}{2} \sum_{1}^{n} {}_{s} \ a_{rs}X_s}{\varepsilon_r \beta_r} = H_r \quad ,$$

and eliminating from them the time, t , one will find the integral of the canonical equations (III). When P has only first degree terms,

$$H_r - H_i = H_{ri} = \text{constant} .$$

It is easy to verify that

$$(17) \qquad\qquad (H, H_r) = 1 , \qquad (H_r, H_h) = \frac{a_{hr}}{\epsilon_h \beta_h \epsilon_r \beta_r}$$

from which

$$(18) \qquad\qquad\qquad (H, H_{ri}) = 0$$

$$(19) \qquad (H_{rh}, H_{g\ell}) = \frac{a_{gr}}{\epsilon_g \beta_g \epsilon_r \beta_r} + \frac{a_{\ell h}}{\epsilon_\ell \beta_\ell \epsilon_h \beta_h} + \frac{a_{r\ell}}{\epsilon_r \beta_r \epsilon_\ell \beta_\ell} + \frac{a_{hg}}{\epsilon_h \beta_h \epsilon_g \beta_g} .$$

Let us set

$$(20) \qquad\qquad\qquad - H - K = L - \sum_1^n i \; q_i \beta_i .$$

We will then have

$$L = \sum_1^n r \; \epsilon_r \beta_r q_r H_r = \sum_1^n r \; \epsilon_r \beta_r q_r H_{ri}$$

because of the relation (Cf. Editors' note)

$$(3) \qquad\qquad\qquad \sum_1^n r \; \epsilon_r \beta_r q_r = 0 .$$

As a consequence,

$$(L, H_h) = \sum_1^n r \epsilon_r \beta_r q_r (H_r H_h) = \frac{1}{\epsilon_h \beta_h} \sum_1^n r \; a_{rh} q_r = -1 ,$$

and

$$(L, H_{rh}) = 0 .$$

2. Therefore the integrals of the canonical equations,

$$H, L, H_{rh} ,$$

are independent, and "in involution". Each linear combination of the H_{12}, H_{13}, ..., H_{1n} is in involution with the functions H and L . If one could find n - 2 of such linear combinations which are independent, and also independent of L , the problem would be reduced to quadratures. One would have, in fact, n integrals independent and in involution.

This is not possible unless one imposes some constraints on the constant quantities a_{rs}, β_r, ε_r . One would obtain, in fact, some H_{12}, H_{13}, ..., H_{1n} which are linearly expressed through functions in involution, and which as a consequence are in involution themselves. Generally, this would contrast with equations (19). However, for special values of the constants $a_{rs} = - a_{sr}$, β_r , ε_r , the situation may be different.

3. Suppose there is any number of species, n , and notice that, from (19), a necessary and sufficient condition for the H_{1h}, H_{1g} to be in involution is that (Cf. 19)

$$\frac{a_{gh}}{\varepsilon_g \beta_g \varepsilon_h \beta_h} + \frac{a_{1g}}{\varepsilon_1 \beta_1 \varepsilon_g \beta_g} + \frac{a_{h1}}{\varepsilon_h \beta_h \varepsilon_1 \beta_1} = 0 .$$

Then, if

(21) $$a_{rs} = \varepsilon_r \beta_r \varepsilon_s \beta_s (m_s - m_r) ,$$

in which the m_1, m_2, ..., m_n are some constants, the H_{12}, H_{13}, ..., H_{1n} will be in involution. They will also be independent, since each contains a p_r

which the others do not contain. Therefore, *when the* a_{rs} *have the form (21) the problem is reduced to quadratures.*

4. This result can be directly verified in equations (I) (Cf. Editors' note). Setting, in fact,

$$\Sigma_s \epsilon_s \beta_s N_s = N \quad , \quad 1 - \Sigma_s \epsilon_s \beta_s m_s N_s = M$$

such equations become

$$\frac{1}{\epsilon_r} \frac{d}{dt} \ln N_r = m_r N + M ,$$

and, by eliminating M and N ,

$$\frac{\frac{1}{\epsilon_1} \cdot \frac{d}{dt} \ln N_1 - \frac{1}{\epsilon_2} \frac{d}{dt} \ln N_2}{m_1 - m_2} = \frac{\frac{1}{\epsilon_1} \frac{d}{dt} \ln N_1 - \frac{1}{\epsilon_3} \frac{d}{dt} \ln N_3}{m_1 - m_3} = \ldots =$$

$$= \frac{\frac{1}{\epsilon_1} \frac{d}{dt} \ln N_1 - \frac{1}{\epsilon_n} \frac{d}{dt} \ln N_n}{m_1 - m_n} .$$

This immediately gives n - 2 integrals which do not depend on time. But in the equations (I) we can eliminate dt , obtaining n - 1 equations with one multiplier, which proves that the integration is reduced to quadratures.

§6. The principle of least action in ecology.

1. We have seen that the equations (II)

$$\beta_r \frac{d^2 X_r}{dt^2} = \left(\epsilon_r \beta_r + \sum_1^n {}_s \ a_{sr} \frac{dX_s}{dt} \right) \frac{dX_r}{dt}$$

have the integral **(cf. Editors' note)**

(C") $\qquad \Omega = \sum\limits_{1}^{n} {}_{r} \ X'_r \Theta_r \ + \ \sum\limits_{1}^{n} {}_{r} \ \varepsilon_r \beta_r X_r \ - \ \sum\limits_{1}^{n} {}_{r} \ \beta_r X'_r \ - \ C \ = \ 0 \ .$

If we change the quantities X_r in a way such that $\delta t = 0$ (isochronic variation) we find

$$\delta \Omega \ = \ \sum\limits_{1}^{n} {}_{r} \ \delta X'_r \Theta_r \ + \ \sum\limits_{1}^{n} {}_{r} \ \delta X_r \left(\sum\limits_{1}^{n} {}_{s} \ a_{sr} X'_s \ + \ \varepsilon_r \beta_r \right) .$$

Supposing that equations (C) are satisfied by the X_1, X_2, ..., X_n we will have

$$\delta \Omega \ = \ \sum\limits_{1}^{n} {}_{r} \ \delta X_r \left(\sum\limits_{1}^{n} {}_{s} \ a_{sr} X'_s \ + \ \varepsilon_r \beta_r \right) .$$

Then, if the isochronic variation is such that

$$\sum\limits_{1}^{n} {}_{r} \ \delta X_r \left(\sum\limits_{1}^{n} {}_{s} \ a_{sr} X'_s \ + \ \varepsilon_r \beta_r \right) \ = \ 0 \ ,$$

we have

$$\delta \Omega \ = \ 0 \ ,$$

which is to say that the integral (C") is preserved by an isochronic variation.

It follows that the two conditions

(22) $\qquad \sum\limits_{1}^{n} {}_{r} \ \delta X_r \left(\sum\limits_{1}^{n} {}_{s} \ a_{sr} X'_s \ + \ \varepsilon_r \beta_r \right) \ = \ 0$

(22') $\qquad\qquad\qquad\qquad \delta \Omega \ = \ 0$

are equivalent with respect to any given isochronic variation. It is evident that we could arbitrarily choose $n - 1$ of the quantities δX_1, δX_2, ..., δX_n and the n-th would be determined by (22). Consequently, one can simultaneously set δX_1, δX_2, ..., δX_n to zero at the boundaries 0, t.

2. We have seen that the conditions (22) and (22') are equivalent. The latter means that integral (C") is invariant with respect to an isochronic variation. Let us now try to interpret the former condition.

To this end recall that the demographic coefficients of the different species are, at a certain instant, ϑ_1, ϑ_2, ..., ϑ_n , and that δX_1, δX_2, ..., δX_n are virtual variations in the quantities of life. One can consider (Cf. §1, Number 2).

$$\Sigma_r \beta_r \vartheta_r \delta X_r$$

as the *virtual demographic work*. Since the *demographic coefficients* or *effective coefficients of increase* can be equivalently expressed as

$$\varepsilon_r + \frac{1}{\beta_r} \sum_s^n {}_1 a_{sr} N_s = \varepsilon_r + \frac{1}{\beta_r} \sum_s^n {}_1 a_{sr} X'_s$$

if follows that

$$\sum_1^n r \left(\varepsilon_r \beta_r + \sum_1^n s\ a_{sr} X'_s \right) \delta X_r$$

will be the *virtual demographic work* for the variation δX_1, δX_2, ..., δX_n . It is for this reason that condition (22) means that *the virtual demographic work is zero*.

3. We have found that the equations of the struggle for existence can be reduced to a problem of the calculus of variation. That is to say, when they are satisfied, a certain expression is *stationary* for all infinitesimal variations of the parameters corresponding to successive states of a biological association (§3). We are now going to show that the natural change between states of the association corresponds effectively, *under certain conditions* to a *minimum* of the expression we have called *vital action* (§3, Number 2). The equations derive, then, from a principle which is analogous to the principle of least action in classical

mechanics.

To begin with let us write

$$2\Phi - \Omega = \sum_{1}^{n} {}_{r} \beta_r X'_r \ln X'_r + \frac{d}{dt} \left[\sum_{1}^{n} {}_{r} \varepsilon_r \beta_r t X_r + \sum_{1}^{n} {}_{r} (C_r + \beta_r) X_r \right] + C \quad .$$

One has

$$\chi = \sum_{1}^{n} {}_{r} \beta_r X'_r \ln X'_r = 2\Phi - \Omega - \frac{d}{dt} \left[\sum_{1}^{n} {}_{r} \varepsilon_r \beta_r t X_r + \sum_{1}^{n} {}_{r} (C_r + \beta_r) X_r \right] - C \quad ,$$

and, from this,

$$(F) \qquad A = \int_{0}^{t} \chi \, dt = \int_{0}^{t} (2\Phi - \Omega) \, dt - \left[\sum_{1}^{n} {}_{r} \varepsilon_r \beta_r t X_r + \sum_{1}^{n} {}_{r} (C_r + \beta_r) X_r \right]_{0}^{t} - Ct \quad .$$

Let us choose X_1, X_2, ..., X_n such that equations (II) are satisfied, namely that

$$\frac{d}{dt} \frac{\partial \Phi}{\partial X'_r} - \frac{\partial \Phi}{\partial X_r} = 0 \quad .$$

Suppose, furthermore, that the isochronic variations are zero at the boundaries 0, t, and that they satisfy conditions (22). From the equation (F) one has then

$$\delta A = 0 \quad .$$

4. This leads us to the final result. We have, in fact, from equation (F)

$$\delta^2 A = \delta^2 \int_{0}^{t} \chi \, dt = \int_{0}^{t} \sum_{1}^{n} {}_{r} \beta_r \frac{\delta X'^2_r}{X'_r} \, dt \quad .$$

Since the $X_r' = N_r$ are positive, it follows that

$$\delta^2 A > 0 \ .$$

Therefore, *any infinitesimal isochronic variation of the* $X_1, X_2, \ldots, X_n$ *which leaves the integral (C") invariant will determine an increase of the vital action. It is, therefore, the matter of a minimum, which proves the principle of least vital action.*

5. The same result could be obtained directly from the condition (22) rather than from the equivalent one (22'). It can also be obtained in a simpler and more direct way, following the procedure below.

In the expression

$$\chi = \sum_1^n {}_r \ \beta_r X_r' \ \ell n \ X_r' \ = \ \sum_1^n {}_r \ \beta_r N_r \ \ell n \ N_r$$

let us change the X_r to $X_r + \Delta X_r = X_r + \xi_r$, and correspondingly, the X_r' to $X_r' + \xi_r' = N_r + \gamma_r$, in which $\gamma_r = \xi_r'$. Supposing that $\gamma_r > -N_r$, $N_r \ \ell n \ N_r$ will become

$$(N_r + \gamma_r) \ \ell n \ (N_r + \gamma_r) \ = \ N_r \ \ell n \ N_r + \gamma_r (\ell n \ N_r + 1) + N_r f\left(\frac{\gamma_r}{N_r}\right) ,$$

in which we set

$$f(x) \ = \ (1+x) \ \ell n \ (1+x) - x \ \ .$$

Then, $A = \displaystyle\int_0^t \chi dt$ will become

$$\int_0^t \chi dt + \int_0^t \sum_1^n {}_r \ \beta_r (\ell n \ N_r + 1) \xi_r' dt + \int_0^t \sum_1^n {}_r \beta_r N_r f\left(\frac{\gamma_r}{N_r}\right) dt \ .$$

If the ξ_r are zero at the boundary 0, t, we have

$$(23) \qquad \int_0^t \sum_1^n r \; \beta_r (\ell n \; N_r + 1)\xi_r' dt \; = \; - \int_0^t \sum_1^n r \; \beta_r \frac{N_r'}{N_r} \; \xi_r dt \; .$$

If the equations

$$\sum_1^n r \; \xi_r \left(\sum_1^n s \; a_{sr} N_s + \varepsilon_r \beta_r \right) = 0$$

are satisfied, it follows from equations (I) that the expression (23) will be zero.

Therefore, $A = \int_0^t \chi \; dt$ will increase by

$$\int_0^t \sum_1^n r \; \beta_r N_r f\left(\frac{\gamma_r}{N_r}\right) \; dt \; ,$$

and this quantity will be positive for $\gamma_r > - N_r$, and it will only be zero when all the γ_r are zero.

In fact, $df(x)/dx = \ell n \; (1+x)$ and, therefore, $f(\gamma_r/N_r)$ will decrease as γ_r changes between $-N_r$ and zero. It will be zero for $\gamma_r = 0$, and it will increase for positive values of γ_r .

Taking into account that the populations cannot be negative, one then has:

Let us modify in an isochronic manner the natural change in a conservative association from one state to another by changing the populations of the different species. The vital action will increase if the quantities of life at the initial and at the final instants are unchanged, and if the virtual demographic work is zero at each instant (Cf. Number 4).

We are then dealing with an effective minimum of the vital action, and this constitutes the principle of least action in ecology.

6. In this way we have built up an *ecological dynamics* which is similar to the *dynamics of material systems*. In fact, the variational principle (§3) can be compared to Hamilton's principle, and we have also obtained in ecology a principle which corresponds to that of least action. In ecology, as in material systems, the passage from one principle to the other can take place in an analogous manner. In the mechanics of material systems this is accomplished by considering displacements for which no mechanical work is performed. In ecology this is done by considering variations in the vital action for which the demographic work is zero.

Notice that in ecology we are actually dealing with a minimum of action while this is not always true in the dynamics of material systems. We should not be surprised by this observation, however, since the general principles we have just compared have only an analogous appearance. The mechanical action and the vital action are, in fact, expressed by different functions.

References:

1. Levi, B., 1931. Boll. Un. Mat. It., Anno X, No. 4.

2. Régnier et Lambin, 1934. Étude d'un cas d'antagonisme microbien (B. Coli, Staphylococcus aureus), avec des remarques par Vito Volterra. Comptes rendus de l'Acad. des Sciences. Décembre 1934.

3. Volterra, Vito et Umberto D'Ancona, 1935. Les associations biologiques au point de vue mathématique. Paris, Hermann et Cie.

The general equations of biological strife in the case of

historical actions

Vito Volterra

I

1. I have begun to study the laws of the struggle for life by a group of species living in the same environment in such a way that some devour others. I have used for this purpose the so called *principle of encounter*, considering the encounters of the individuals of the various species, and what follows by reason of the actions that the indivudals exercise on one another.[1]

Making very simple and probable hypotheses, the equations to which it leads are the following:

$$\text{(A)} \qquad \beta_r \frac{dN_r}{dt} = \left(\varepsilon_r \beta_r + \sum_1^n {}_s\ a_{sr} N_s\right) N_r \ , \quad (a_{rs} = -a_{sr})$$

in which N_1, N_2, ... N_n denote the poulations of the various species, ε_1, ε_2, ... ε_n the coefficients of auto-increase of the various species and β_1, β_2, ... β_n the values of the individuals.

These equations are well known and their consequences can be summed up in the following general laws:

Law I. *Law of conservation of fluctuations.*

The populations of the different species are limited by positive numbers and there are always fluctuations that never die out.

Law II. *Law of conservation of means.*

If we take as means of the different species the asymptotic means they are constant and independent of the initial populations.

Law III. *Laws of perturbation of means.*

If all the species are destroyed uniformly and in proportion to the numbers of their individuals, there are always some species which profit and others which suffer. Among the first, there is at least one which is devoured by the others and among the second there is also at least one that devours others.

2. But in this treatment only immediate actions are considered. Now in all biological phenomena it is necessary to examine not only immediate actions but also those depending on the past, that is, on the changes which the species have undergone. These actions were first called *hereditary actions;* but this name was not well chosen, although it may have been useful to signify such phenomena in the inorganic world. It was found preferable to use the term *historical actions* or *actions belonging to memory*. We shall use the first of these two denominations.

To treat this case I have tried to extend the *method of encounter* which has already been happily employed in the case of two species, one of which devours the other. The same method was perfectly successful and the three general laws could be extended, introducing only a few modifications.

For this consult the work: *Lecons sur la théorie mathématique de la lutte pour la vie*. Paris, Gauthier-Villars, 1931.

But if we are determined to extend the same method to any number of species living together there arise difficulties which seem not easily resolvable.

Shall we then give up the study of a general case taking the historical phenomenon into account? Or will it be possible to change the method, no longer using the useful method of encounters?

II

3. Last year, at the Mathematical Society of France,[2] I gave a lecture in which I examined only the case of immediate actions, but did not follow the method of encounters. I used instead a more simple and intuitive, but perhaps not so rigorous, method. I believe that this method can be easily applied to the historical case not only of two species, but also to that of any number of species.

This method consists fundamentally in making all the actions exercised in the past, that is, all the historical actions, enter into the values of the coefficients of increase.

4. After these general considerations let us expound the matter in detail. Assume that we have n species whose populations are denoted by

$$N_1, N_2, \ldots N_n .$$

If their coefficients of increase are called $\varepsilon_1, \varepsilon_2, \ldots \varepsilon_n$, taking these constants as positive or negative according to whether the species tend to increase or to die out when not interfered with, we shall have the following equations expressing the variations of the populations:

$$\frac{dN_1}{dt} = \varepsilon_1 N_1 , \quad \frac{dN_2}{dt} = \varepsilon_2 N_2, \ldots \frac{dN_n}{dt} = \varepsilon_n N_n .$$

But we want to suppose that the various species act on one another dependently upon the number of these populations; and if we suppose this dependency to be of an immediate and linear nature, we shall have to substitute for $\varepsilon_1, \varepsilon_2, \ldots \varepsilon_n$

$$\varepsilon_1 + \sum_{1}^{n} {}_s A_{s1} N_s , \quad \varepsilon_2 + \sum_{1}^{n} {}_s A_{s2} N_s , \quad \ldots \varepsilon_n + \sum_{1}^{n} {}_s A_{sn} N_s .$$

The coefficient A_{sr} measures that unitary action (per individual) which the species s exercises upon the species r , while A_{rs} denotes the inverse action that species r exercises upon the species s ; and as it is supposed that these actions are such that, while one species injures the other, the latter profits from the first (for example, one species devours the other) the coefficients A_{sr}, A_{rs} may be assumed to have opposite signs. They will not however be of equal absolute value.

Without repeating the discussion made in last year's lecture, we shall make a simple and probable hypothesis; that their absolute values be in the inverse ratio of certain constant and positive coefficients, which we shall assume as values of the individuals of each species. In this way we can write:

(1)
$$A_{rs} = \frac{1}{\beta_r} a_{rs} \; , \quad A_{sr} = \frac{1}{\beta_s} a_{sr} \; , \quad a_{rs} = - a_{sr} \; ;$$

and we shall have as general equations of the struggle for life the equations
(A), that is, the very same equations which are found by the method of encounters.

5. We can generalise further and suppose that the equations (1) do not hold and
that the condition $A_{rr} \gtrless 0$ is satisfied, when instead of supposing that the
individuals of the various species devour each other, we assume that they act on
one another in such a way that they influence the generic coefficients of increase
of the populations. Then we find the following equations

$$\frac{dN_r}{dt} = \left(\epsilon_r + \sum_1^n s \; A_{sr} N_s \right) N_r \; ,$$

which have been classified and studied in paragraph 7, part II of *"Variazioni e
fluttuazioni del numero d'individui in specie animali conviventi."* R. Comit.
Talassografico Italiano, Mem. CXXXI, 1927.

III

6. Now we come to examine the historical case. By t we denote the actual
instant and by τ a preceding instant. The number of individuals of the species
s at time τ will be $N_s(\tau)$.

Let us suppose that the species s exercises over the coefficient of
increase of the species r an action which will be manifested in the future and
which varies with the distance in time. We shall denote such a (unitary) action
by $F_{sr}(t-\tau)$ when it is exercised by the species s in the infinitesimal interval
of time $(\tau, \tau + d\tau)$ and is manifested on the species r at time t . Then the
action corresponding to the population $N_s(\tau)$ will be

$$N_s(\tau)F_{sr}(t-\tau)d\tau \ .$$

If we take into account all these actions beginning from the origin of times at which they are supposed to have begun, up to the present moment t we shall have

$$\int_0^t N_s(\tau)F_{sr}(t-\tau)d\tau \ .$$

Considering historical actions for all n species on the coefficient of increase of the species r we shall have

$$\sum_1^n s \int_0^t N_s(\tau)F_{sr}(t-\tau)d\tau \ .$$

The coefficient of increase of the species r , taking into account all immediate and historical actions exercised upon it, will therefore become

$$\varepsilon_r + \sum_1^n s \left(A_{sr}N_s(t) + \int_0^t N_s(\tau)F_{sr}(t-\tau)d\tau \right) \ .$$

And as general equations of the struggle for life we shall assume

$$(B) \qquad \frac{dN_r}{dt} = \left\{ \varepsilon_r + \sum_1^n s \left[A_{sr}N_s(t) + \int_0^t N_s(\tau)F_{sr}(t-\tau)d\tau \right] \right\} N_r(t) \ .$$

In this way we shall have n integro-differential equations.

It is not impossible that some of the A_{sr} , F_{sr} may be zero, and up to now we shall assume nothing about their signs.

7. We may suppose that the historical actions may be prolonged indefinitely in the past, and then the equations (B) must be replaced by the following:

(B')
$$\frac{dN_r}{dt} = \left\{ \varepsilon_r + \sum_1^n {}_s \left[A_{sr}N_s(t) + \int_{-\infty}^t N_s(\tau)F_{sr}(t-\tau)d\tau \right] \right\} N_r(t) \ .$$

Of course we must suppose that the necessary conditions for convergence of these integrals are satisfied. We may also suppose that the historical actions cease after a certain interval of time, and then it is sufficient to change the lower limit of the preceding integral into $t - T_0$, denoting by $T_0 > 0$ this interval of time. Then (B') becomes

(B")
$$\frac{dN_r}{dt} = \left\{ \varepsilon_r + \sum_1^n {}_s \left[A_{sr}N_s(t) + \int_{t-T_0}^t N_s(\tau)F_{sr}(t-\tau)d\tau \right] \right\} N_r(t) \ ,$$

or

(B"')
$$\frac{dN_r}{dt} = \left\{ \varepsilon_r + \sum_1^n {}_s \left[A_{sr}N_s(t) + \int_0^{T_0} N_s(t-\tau)F_{sr}(\tau)d\tau \right] \right\} N_r(t) \ .$$

8. We easily see that there may be stationary states. For that purpose we shall substitute in the equations (B"') for $N_1, N_2, \ldots N_n$ the constants $K_1, K_2, \ldots K_n > 0$.

The equations become

$$\varepsilon_r + \sum_1^n {}_s \left(A_{sr} + \int_0^{T_0} F_{sr}(\tau)d\tau \right) K_s = 0 \ ;$$

and if we write

$$A_{sr} + \int_0^{T_0} F_{sr}(\tau)d\tau = C_{sr} \quad,$$

we shall have

(C) $$\varepsilon_r + \sum_1^n {}_s C_{sr}K_s = 0 \;.$$

If the determinant of the C_{sr} is different from 0 and the roots positive, these will represent the populations of equilibrium (stationary states).

9. Let us take again the equation (B), multiply both sides by $\dfrac{dt}{N_r(t)}$ and then integrate between 0 and θ . We find

$$\ln \frac{N_r(\theta)}{N_r(0)} = \varepsilon_r\theta + \sum_1^n {}_s \int_0^\theta \left[A_{sr}N_s(t) + \int_0^t N_s(\tau)F_{sr}(t-\tau)d\tau \right] dt \;.$$

Applying a well known Dirichlet's transformation, we obtain

$$\ln \frac{N_r(\theta)}{N_r(0)} = \varepsilon_r\theta + \sum_1^n {}_s \int_0^\theta \left[A_{sr} + \int_t^\theta F_{sr}(\tau-t)d\tau \right] N_s(t)dt \;,$$

from which it follows that

$$N_r(\theta) = N_r(0)e^P$$

where

$$P = \varepsilon_r\theta + \sum_1^n {}_s \int_0^\theta \left[A_{sr} + \int_t^\theta F_{sr}(\tau-t)d\tau \right] N_s(t)dt \;.$$

Now we may apply a method of successive approximations and integrate the integro-differential equations. I have already applied the method of successive approximations to the resoltuion of the integro-differential equations relative to the historical case for two species; these equations are a particular case   (n=2) of those now found.

The procedure for the application of the method of successive approximations to the case when   n = 2   has been exposed in detail in the work "*Lecons sur la théorie mathématique de la lutte pour la vie*" (see page 192 and the following). I think it unnecessary to explain here the easy extension of this procedure to the general case.

10.   There are many interesting particular cases to examine. For instance if the actions are reciprocal (as in the case in which the individuals of the various species devour each other) it will be convenient to assume A_{sr}, A_{rs}, with different signs, and so F_{sr}, F_{rs} . Some of these may be zero.

If we admit both for the A_{sr}, A_{rs} and for the F_{sr}, F_{rs}, that they will be in certain cases inversely proportional to the values

$$\beta_1, \; \beta_2, \; \cdots \; \beta_n$$

of the individuals, we shall put

$$A_{sr} = \frac{a_{sr}}{\beta_r} \; , \quad A_{rs} = \frac{a_{rs}}{\beta_s} \; , \quad a_{sr} = - a_{rs} \; ,$$

$$F_{sr}(t-\tau) = \frac{f_{sr}(t-\tau)}{\beta_r} \; , \quad F_{rs}(t-\tau) = \frac{f_{rs}(t-\tau)}{\beta_s} \; , \quad f_{sr} = - f_{rs} \; ,$$

so that the equations (B''') become

$$(B^{iv}) \qquad \beta_r \frac{dN_r}{dt} = \left\{ \beta_r \epsilon_r + \sum_1^n s \left[a_{sr} N_s(t) + \int_0^{T_0} N_s(t-\tau) f_{sr}(\tau) d\tau \right] \right\} N_r(t) \; .$$

The equations which give the stationary states (conditions of equilibrium) will then be

(C')
$$\beta_r \epsilon_r + \sum_{1}^{n} s \; c_{sr} K_s = 0$$

when we put

$$C_{sr} = \frac{c_{sr}}{\beta_r}, \quad c_{sr} = - c_{rs} \; .$$

From (C') we deduce

$$\sum_{1}^{n} r \; \beta_r \epsilon_r K_r = 0 \; ,$$

from which we may deduce, for the historical case, the same laws of reciprocity that we have enunciated in previous works in the case of immediate actions. (See the conference mentioned in §2.)

It is however necessary to observe that while in the last case the populations of equilibrium coincide with the averages of the populations, we cannot say the same, at least in general, in the historical case.

11. We confine ourselves, in the present short essay, to the preceding general considerations and leave aside any further developments.

Finally we consider it convenient to summarise the following conclusions:

I.   That with conveniently appropriate hypotheses the general case of historical actions, already studied for two species, can also be examined for any number of species.

II. That in this case we can obtain some general integro-differential equations which include all those previously given for the problems of biological struggle.

III. That we can completely and easily integrate these equations with simple methods of successive approximations.

IV. That we can undertake to study the stationary case.

Footnotes

1. *Variazioni e fluttuazioni del numero di individui in specie animali conviventi.* Memorie della R. Accademia dei Lincei S. 6, 2 (1926), 31-113.

2. *Fluctuations dans la lutte pour la vie, leurs lois fondamentales et de reciprocite.* Conférence de la réunion internationale des mathématicians. Paris 1938.

The growth of mixed populations: Two species competing for a common food supply

Alfred J. Lotka

The general analysis of the growth of mixed populations of any number of species in mutual interdependence of any kind which has been given by the writer in prior publications[1] covers many special cases. It is of interest to note how it applies to and readily furnishes the solution of a special case that has since been separately discussed by Volterra,[2] namely that of two species competing for a common food supply.

Volterra, following in this respect well-established lines familiar from prior literature,[3] starts out from the supposition that, in the absence of restraining influences, the rate of growth of a population would be proportional to the existing population thus

$$\frac{dN}{dt} = rN \tag{1}$$

resulting in an exponential (Malthusian) law of population growth; but that the natrual limitations of the food supply convert the coefficient r into a diminishing function of N. In the simplest case this would be a linear function, so that we should have

$$\frac{dN}{dt} = r_0 N (1-phN) \tag{2}$$

where ph is a constant which I have written as the product of two constants for reasons that will appear presently. Equation (2) is simply the Verhulst-Pearl law of population growth, which, as we know, has been found to fit very acceptably a number of observed examples of population growth.

Now, when two populations compete for a common food supply, Volterra writes, essentially,

$$\frac{dN_1}{dt} = r_1 N_1 \left\{ 1 - p_1(hN_1 + kN_2) \right\}$$

$$\frac{dN_2}{dt} = r_2 N_2 \left\{ 1 - p_2(hN_1 + kN_2) \right\}$$

(3)

a system of equations that may be regarded as an almost self-evident extension of the equation (2), except that one may question why the same constants h, k appear in the two equations. We shall take up this question later. For the present we shall accept Volterra's original setting. He does not solve his equations, but discusses certain fundamental properties of the functions defined by them. As a matter of fact, by the general method set forth in my prior publications, a solution is readily obtained in series form, and at the same time the conclusions reached by Volterra drop out very readily, together with further information which is not found in his discussion.

We will proceed as follows. Volterra's equations are of the form

$$\frac{dN_1}{dt} = a_1 N_1 \qquad + a_{11} N_1^2 \qquad + a_{12} N_1 N_2$$

$$\frac{dN_2}{dt} = \qquad a_2 N_2 \qquad + a_{22} N_2^2 + a_{21} N_1 N_2$$

(4)

Equilibria. A stationary state occurs whenever $\frac{dN_1}{dt}$ and $\frac{dN_2}{dt}$ both vanish together. This defines three possible equilibria (to be more exact, stationary states) as follows:

a) $N_1 = 0$ $N_2 = 0$ (5)

b) $N_1 = 0$ $N_2 = -\frac{a_2}{a_{22}} = \frac{1}{p_2 k}$ (6)

$$\text{c)} \quad N_2 = 0 \qquad N_1 = -\frac{a_1}{a_{11}} = \frac{1}{p_1 h} \tag{7}$$

If the coefficients a are constants, there are no other equilibria within real finite values of N_1, N_2 .

Stability of Equilibrium: 1. *At origin*. To determine the nature of the equilibrium at the origin ($N_1 = 0$, $N_2 = 0$) we form the characteristic equation of the linear terms in equations (4), that is,

$$\begin{vmatrix} a_1 - \lambda & 0 \\ 0 & a_2 - \lambda \end{vmatrix} = 0 \tag{8}$$

which gives

$$\begin{aligned} \lambda_1 &= a_1 \\ \lambda_2 &= a_2 \end{aligned} \tag{9}$$

Now both a_1 and a_2 , from the nature of things, are positive quantities, since the case of real interest is that in which each species is viable separately under the prevailing conditions. Hence both roots of the characteristic equation are positive, and the equilibrium at the origin is unstable.

2. *At Second Equilibrium*. To examine the character of the second equilibrium, we transform the equations (4) to a new origin by writing

$$N_1 = N_1 \tag{10}$$

$$n_2 = N_2 + \frac{a_2}{a_{22}} \tag{11}$$

we thus obtain

$$\frac{dN_1}{dt} = \left(a_1 - \frac{a_{12}a_2}{a_{22}}\right) N_1 + a_{11}N_1^2 + a_{12}N_1n_2$$

$$\frac{dn_2}{dt} = -a_2n_2 - \frac{a_{21}a_2}{a_{22}} N_1 + a_{22}n_2^2 + a_{21}N_1n_2 \qquad (12)$$

and, forming the characteristic equation, we find here

$$\begin{vmatrix} \left(a_1 - \dfrac{a_{12}a_2}{a_{22}}\right) - \lambda & 0 \\[2em] -\dfrac{a_{21}a_2}{a_{22}} & -a_2 - \lambda \end{vmatrix} = 0 \qquad (13)$$

or, in our original notation

$$\begin{vmatrix} r_1\left(1 - \dfrac{p_1}{p_2}\right) - \lambda & 0 \\[2em] -r_2\dfrac{h}{k} & -r_2 - \lambda \end{vmatrix} = 0 \qquad (14)$$

from which it is seen that the equilibrium is stable if, and only if

$$r_1\left(1 - \frac{p_1}{p_2}\right) < 0 \qquad (15)$$

i.e., if

$$p_2 - p_1 < 0 \qquad (16)$$

3. *At Third Equilibrium.* By the same reasoning we find that the third equilibrium is stable if, and only if

$$p_1 - p_2 < 0 \qquad\qquad (17)$$

It will be seen that, except in the special[4] case that $p_2 = p_1$, one of the two equilibria must be stable, the other unstable.

When p_2 is not equal to p_1 , it is, for reasons of symmetry, immaterial to which of the two coefficients we ascribe the greater value. Let us, then, write

$$p_2 < p_1 \qquad\qquad (18)$$

so that the second equilibrium is the stable one,

i.e. $\qquad\qquad N_1 = 0 \qquad N_2 = \dfrac{1}{p_2 k}$

The general solution of the system of equations (12) can be written in the form of exponential series

$$N_1 = \Sigma P_{rs}\, e^{(r\lambda_1 + s\lambda_2)} \qquad\qquad N_2 = \Sigma Q_{rs}\, e^{(r\lambda_1 + s\lambda_2)} \qquad (19)$$

Numerical example. For the sake of obtaining a visual presentation of the form of the functions defined by the differential equations (3), (4), (12), and their series solution (19), several numerical examples were worked, of which the following is here selected for reproduction in the accompanying graph Figure 1. The values given to the various constants in this example were arbibtrary, except that in order to establish some contact with a concrete case, the value of the exponent λ_1 and the ultimate population N_∞ of the one species were those actually observed in the human population in the United States.

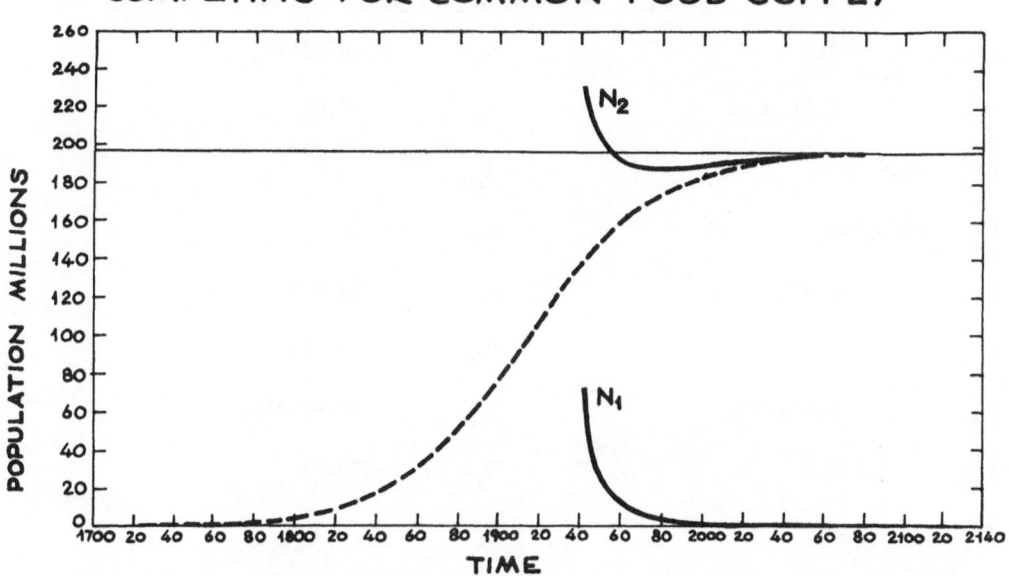

FIGURE 1.

The following is a table of the numerical values[5] of the several constants in this example:

h	20	p_1	1/1,315,153,333
k	10	p_2	1/10 × 197,273,000
r_{01}	0.10	λ_1	-0.03134
r_{02}	0.03134	λ_2	-0.05
P_{10}	-986,365	Q_{10}	0.0
P_{01}	+100,000.	Q_{01}	29,770
P_{02}	+102.85	Q_{02}	+72.228
P_{03}	+0.13084	Q_{03}	+0.14360
P_{04}	+0.00018325	Q_{04}	+0.00026977
P_{11}	1,706.5	Q_{11}	-712.43
P_{21}	+19.762	Q_{21}	+10.306
P_{12}	-3.1083	Q_{12}	-2.5999
P_{31}	-0.19113	Q_{31}	-0.11656
P_{22}	+0.056626	Q_{22}	+0.053969
P_{13}	-0.0056759	Q_{13}	-0.0069826

P_{20}	+491.83	P_{30}	-24.66	P_{40}	0.1233

It will be seen that one of the two populations, in the circumstances to which the graph in Figure 1 relates, at first diminishes, presently turns the corner, and then increases, approaching a certain straight line asymptotically; the other population diminshes continually and approaches zero. Thus one competitor drives the other out completely. The last point is one of the results given by Volterra[6] who, however, does not give any method for tracing the actual integral curve in detail.

Another graph, which is particularly instructive, is prepared by eliminating the independent variable t from the two equations (4), and writing

$$\frac{dN_1}{dN_2} = \frac{f_1(N_1,N_2)}{f_2(N_1,N_2)} \tag{20}$$

where f_1 and f_2 are quadratic functions of N_1 and N_2. The locus of all points at which the integral curves of (20) have a slope s is given by

$$\frac{dN_1}{dN_2} = \frac{f_1}{f_2} = s \tag{21}$$

or

$$f_1 - sf_2 = 0 \tag{22}$$

This defines the *isoclines* as a family of conics which in point of fact are, in the present case, hyperbolas. The construction has been carried out with the results shown in Figure 2, which exhibits a number of properties of the isoclines. By their aid a map of the family of integral curves has been drawn which is reproduced here in Figure 3. A part of the negative field has been included merely for its geometrical interest; it has, of course, no concrete meaning for our present problem.

The following characteristics of this map are particularly noteworthy:

At the origin of N_1, N_2 there is an unstable equilibrium characterized by a stream of integral curves all directed away from the origin (when the time is taken into consideration). This corresponds to the two positive roots of the characteristic equation.

There is a second unstable equilibrium at $N_1 = 1/p_1 h$, $N_2 = 0$. Here the integral curves approach, then turn away, *avoiding* the equilibrium point. This corresponds to two roots λ of opposite sign, of the characteristic equation.

The third equilibrium at $N_1 = 0$, $N_2 = 1/p_2 k$, is stable, the integral curves streaming in from all sides. This corresponds to the two negative roots λ of the characteristic equation.

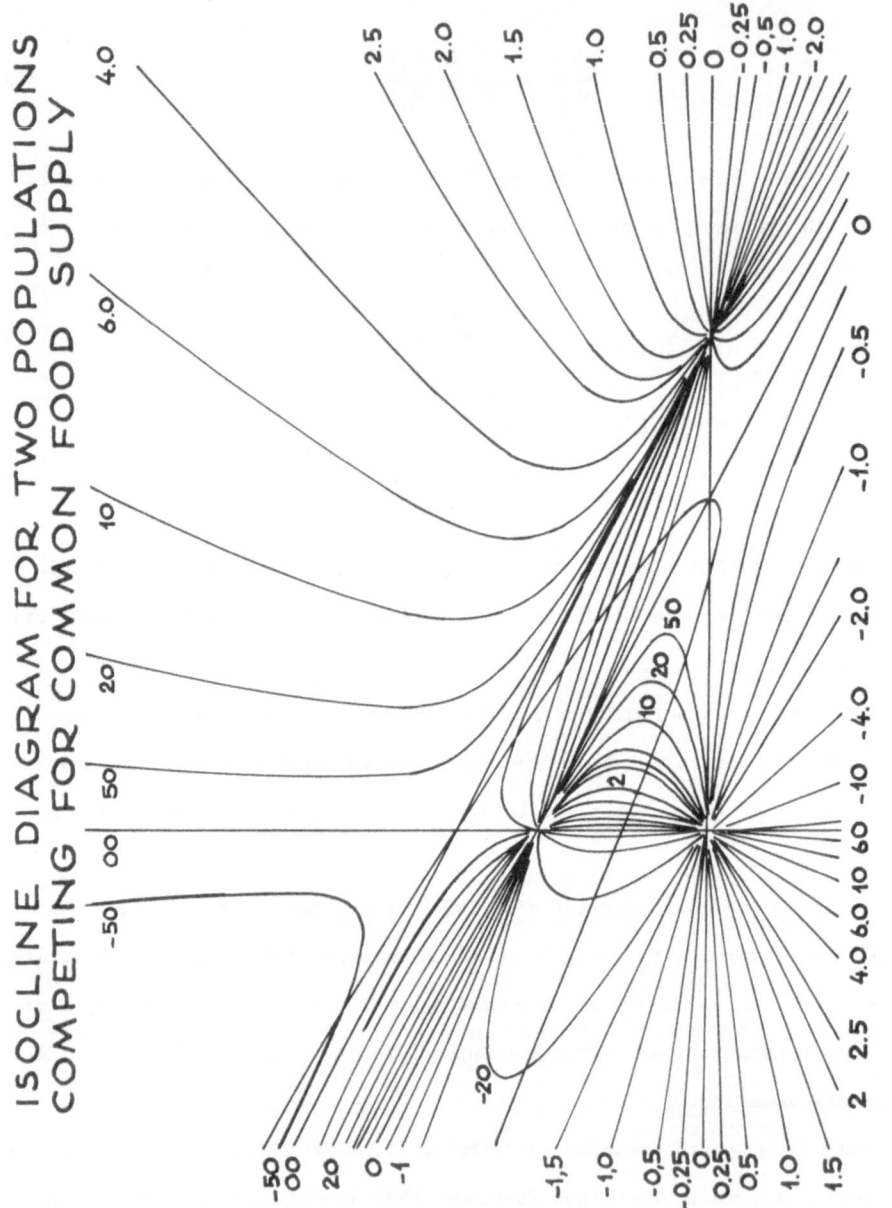

ISOCLINE DIAGRAM FOR TWO POPULATIONS
COMPETING FOR COMMON FOOD SUPPLY

FIGURE 2.

283

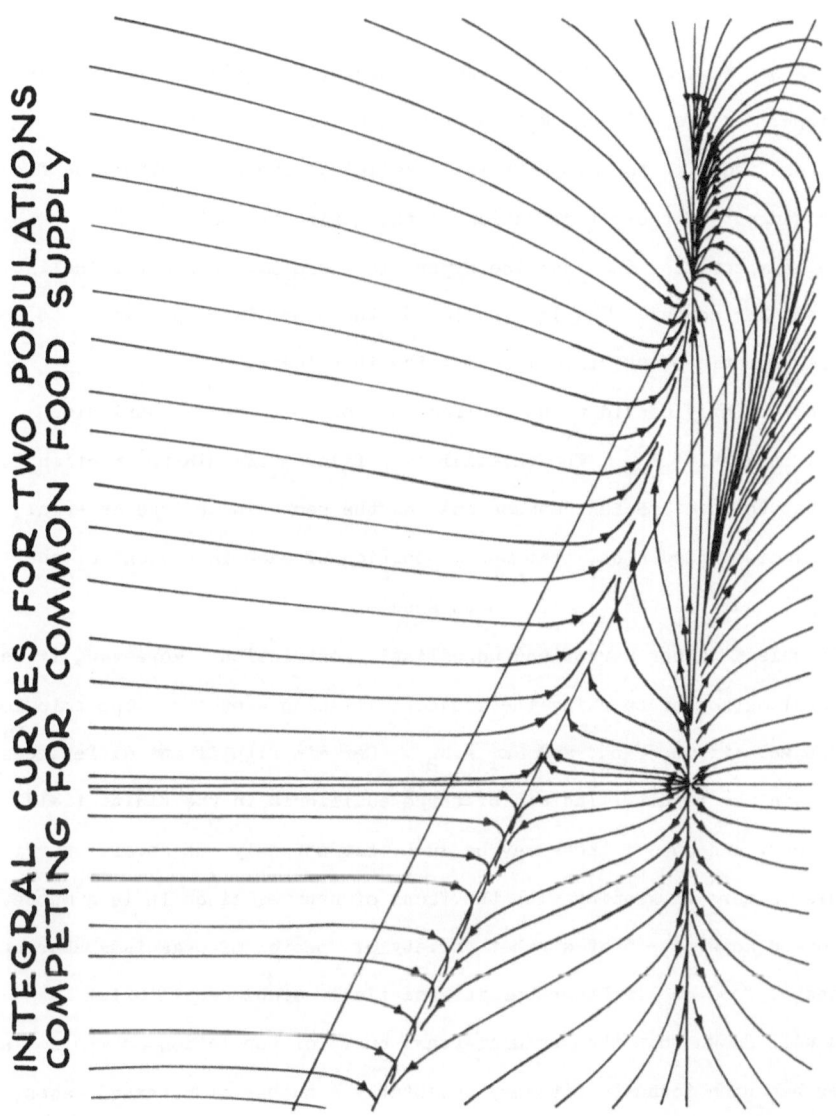

INTEGRAL CURVES FOR TWO POPULATIONS
COMPETING FOR COMMON FOOD SUPPLY

FIGURE 3.

The locus of the centers of the isocline hyperbolas is a parabola. In particular, the center of the isocline for slope ∞ lies at the intersection of the parabola with the axis of N_1 ; the center of the isocline for slope zero lies at the intersection of the parabola with the axis of N_2 .

The axes of N_1 , N_2 themselves are isoclines, the axis of N_1 corresponding to slope ∞ , the axis of N_2 corresponding to the slope zero.

The second isocline for slope ∞ is parallel to the second isocline for slope 0 , the tangent of their inclination to the horizontal being $-\frac{k}{h}$.

Of the asymptotes of the isocline hyperbolas, one always has the inclination $-\frac{k}{h}$ to the horizontal. The inclination of the other is proportional to the slope s characteristic of the isocline to which it belongs.

Let us now briefly consider the implications of Volterra's restriction that $h_1 = h_2$, and $k_1 = k_2$. The physical significance of this restriction is, essentially, that the two species consume one and the same single food material, or, if they consume a mixed diet, that the proportion of each ingredient of the diet which they consume is the same for both species.

Now this is a rather narrow and unrealistic restriction. Moreover, if we adopt the general method of treating the subject, it is unnecessary. The solution applies just as well if $h_1 \neq h_2$ and $k_1 \neq k_2$. Certain significant differences, however, appear in the result. Instead of three equilibria in the finite field, there are now four, and one of these may be such that not only one species survives, but both. This is more in keeping with the facts of nature, since it is a matter of the most common knowledge that a great variety of species of organisms sharing certain sources of food do live together in essentially stable equilibrium.

It is well known that the Verhulst-Pearl curve of population growth for a single species has been found to fit very acceptably a number of observed cases, among them the growth of the human population of the United States, and also certain laboratory populations of fruit flies and other organisms.

It is perhaps hardly to be expected that concrete examples of the law of growth for two populations here discussed shall be found in nature. There is better prospect of realizing it in a laboratory population, though the difficulties of establishing the requisite conditions will here be considerably greater than in the case of a single population. It would be interesting to see the experiment actually made.

But it is possible that the treatment which has here been developed in the analysis of the growth of multiple populations, may find more immediate application in the field of economics. For our variables N_1 and N_2 may be conceived as denoting the size or extent of two (or more) commercial enterprises competing for common sources of supply and for a common market. It will be recalled that Cournot's treatment of the problem of competition has been criticized on the ground that under the conditions of the problem, as analyzed by him, any one competitor who should possess the slightest advantage over the others, would ultimately displace them entirely, and hold the field in absolute monopoly. This criticism, however, is justified only on the assumption that the sources of supply and the markets are equally accessible, in their entirety, to all the competitors. In actual fact, with competitors scattered over an area, each has a certain surrounding territory in which he has an advantage over his competitors. In these circumstances the criticism levelled at Cournot falls to the ground.[7] These observations are strongly reminiscent of the facts we observed in the analysis of competition among growing populations, regarding the effect of varying in *some* degree at least the composition of the diet of the competing populations. In the same way two competing commercial firms, though they may sell to the same set of people, will not sell to their several local zones in identical proportions. That an application of an analysis similar to that here set forth should present itself as a possibility in dealing with economic systems is only natural, since economic competition is, after all, only a special form of the more general phenomenon of biological competition.

Footnotes:

1. Among these may be mentioned Physical Review, 1912, 34: 235; Proc. American
   Academy Arts and Sciences 55: 137; 1920; American Journal of Hygiene 3:
   January Supplement; 1923; Elements of Physical Biology, Baltimore, 1925.
   This last contains references to the author's other publications relative
   to this subject.

2. Memorie della R. Acad. Nazion. dei Lincei 1926 ser 6, vol. 2, part 3, page 5;
   Lecons sur la théorie mathématique de la lutte pour la vie, Paris,
   Gauthiers-Villars, 1931, page 9.

3. See, for example, Lotka, A. J., Elements of Physical Biology, 1925, page 64.

4. This special case, as Volterra has shown, can be integrated in finite terms.
   It is, however, of minor interest, since such an exact relation between
   the coefficients p_1 and p_2 represents, in concrete cases, an infinitely
   improbable condition.

5. A considerable number of significant figures has been retained in these
   constants and throughout the computations, in order to furnish an arithmetical check on the series solution (19). This check was obtained by
   substituting the solution (19) separately in the left hand member and the
   right hand member of equations (4) or (12). In the absence of a special
   investigation of the conditions of convergence of the series (19) this
   arithmetical check is necessary, and was found to be well satisfied within
   the limits of the curves shown in Figure 1.

6. A similar conclusion had been previously reached by J.B.S. Haldane regarding
   the competition between two Mendelian phenotypes. Trans. Cambridge
   Philos. Soc. 23: 39; 1924.

7. Compare H. Hotelling, Economic Journal (London) 41: 41; 1929.

On Volterra's theory of the struggle for existence

A. N. Kolmogoroff

Summary - The author investigates differential equations with reference to the struggle for existence. They are analogous to those considered by Volterra, but their form is based on purely qualitative hypotheses.

1. Vito Volterra [1] has described the interaction between a predator species and its prey through the differential equations

$$\frac{dN_1}{dt} = (\epsilon_1 - \gamma_1 N_2)N_1 \ ,$$

(Ia)

$$\frac{dN_2}{dt} = (-\epsilon_2 + \gamma_2 N_1)N_2 \ .$$

N_1 and N_2 are, respectively, the quantities of individuals in the prey and predator species and they depend on time, t. ϵ_1 and $-\epsilon_2$ are their rates of increase, and γ_1, γ_2 interaction constants. Obviously the analytical expression chosen by Volterra for the second term of equations (Ia) can only be viewed as a first approximation to reality. Various authors have proposed other relationships to express the dependence of dN_1/dt and dN_2/dt on N_1 and N_2. We shall disregard these special hypotheses, chosen in a wholly arbitrary way, and instead write the interaction equations in the following form:

$$\frac{dN_1}{dt} = K_1(N_1,N_2) \cdot N_1 \ ,$$

(I)

$$\frac{dN_2}{dt} = K_2(N_1,N_2) \cdot N_2 \ .$$

We shall assume that the functions $K_1(N_1,N_2)$ and $K_2(N_1,N_2)$ and their first derivatives are continuous for any $N_1 \geq 0$, $N_2 \geq 0$. Moreover, we shall impose on these functions certain qualitative conditions which seem reasonable on biological grounds. To this end we shall denote by S the direction of the vector $\overrightarrow{OP}$ which connects the origin, 0, with any point on the (N_1,N_2) plane, and by dK_1/dS and dK_2/dS the derivatives of the functions K_1 and K_2 along this direction.

We shall impose on the function $K_1(N_1,N_2)$ the following conditions:

(I$_1$) $\dfrac{\partial K_1}{\partial N_2} < 0$; that is, for any given number of individuals in the prey species, their rate of multiplication is a decreasing function of the number in the predator species.

(II$_1$) $\dfrac{dK_1}{dS} < 0$; that is, for any given ratio of the numbers in the prey and the predator species, the rate of multiplication in the prey species decreases as their numbers increase (2).

(III$_1$) $K_1(0,0) > 0$; that is, the prey species has a positive rate of growth when the numbers of both species are sufficiently low.

(IV$_1$)   There exists an A > 0 such that $K_1(0,A) = 0$; that is, when the predator species is sufficiently numerous, the prey species cannot increase.

(V$_1$)   There exists a B > 0 such that $K_1(B,0) = 0$; that is, when the prey species is sufficiently numerous, it cannot increase even in the absence of predators.

We shall impose on the function $K_2(N_1,N_2)$ the following conditions:

(I$_2$) $\dfrac{\partial K_2}{\partial N_2} < 0$; that is, the rate of increase of the predator species is a decreasing function of its own numbers.

(II$_2$) $\dfrac{dK_2}{dS} > 0$; that is, for any given ratio of the numbers of the prey and the predator species, an increase in the numbers of both species favors the predator species.

(III$_2$)   There exists a C > 0 such that $K_2(C,0) = 0$; that is, no matter how few predators there are, they are unable to increase unless $N_1 > C$.

Finally, we impose the condition

(I$_{1,2}$)   C < B .

It is easy to verify that without this last condition the predator species will inevitably disappear.

Figure 1 presents an arrangement of the lines $K_1 = 0$ and $K_2 = 0$ which meets the conditions I$_1$ to V$_1$, I$_2$ to III$_2$, and I$_{1,2}$. It follows from our conditions that the two lines cross at a unique point Z, and they divide the positive quadrant of the (N_1,N_2) plane into four domains denoted by I, II, III and IV.

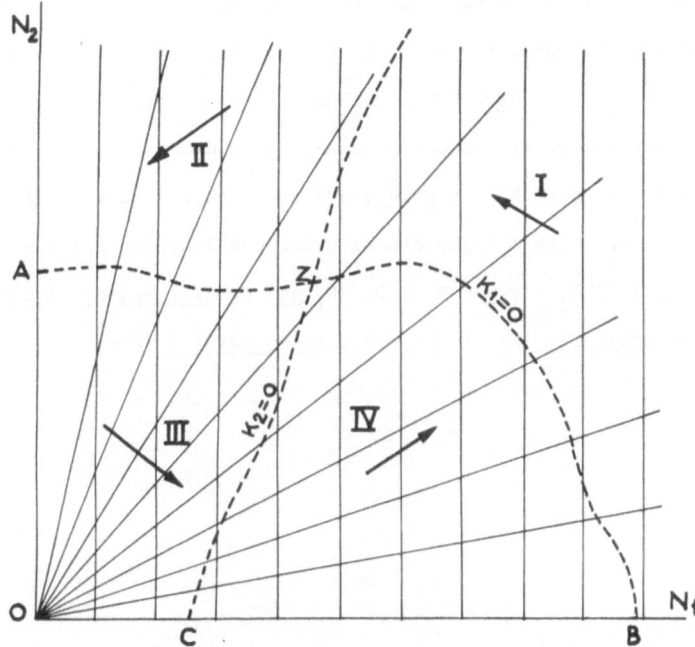

Fig. 1.

Let us now turn to the behavior of the integral curves of system (I), as resulting from the conditions imposed on K_1 and K_2. To begin with, it is easily recognized that the coordinate axes are integral curves.

In the domain of interest, $N_1 \geq 0$, $N_2 \geq 0$, there are only three singular points: the origin, 0, and the points B and Z. A more detailed analysis shows that, except for the coordinate axes, no integral curve can start or end at the origin. Besides the axis $N_2 = 0$, there is a unique integral curve, L, which originates at B and moves toward the domain I. From such considerations it is already possible to derive a qualitative result of biological interest: no integral curve starting in the domain $N_1 > 0$, $N_2 > 0$ can move asymptotically toward the coordinate axes. In other words, if initially both $N_1 > 0$ and $N_2 > 0$, neither species can completely disappear (3).

The following analysis is based primarily on the different possible behaviors of the integral line L. Only three qualitatively distinct cases are possible.

Case (a) The integral curve L approaches the point Z along a spiral path, passing an infinite number of times from domain I to domains II, III and IV.

In this case the point Z is a <u>focus</u>, and the integral curves originating at any $N_1 > 0$, $N_2 > 0$ approach Z asymptotically through infinite spirals. In other words, <u>for any initial</u> $N_1 > 0$ <u>and</u> $N_2 > 0$. N_1 <u>and</u> N_2 <u>will undergo damped oscillations</u> <u>approaching asymptotically the stable equilibrium point Z</u> (Fig. 2).

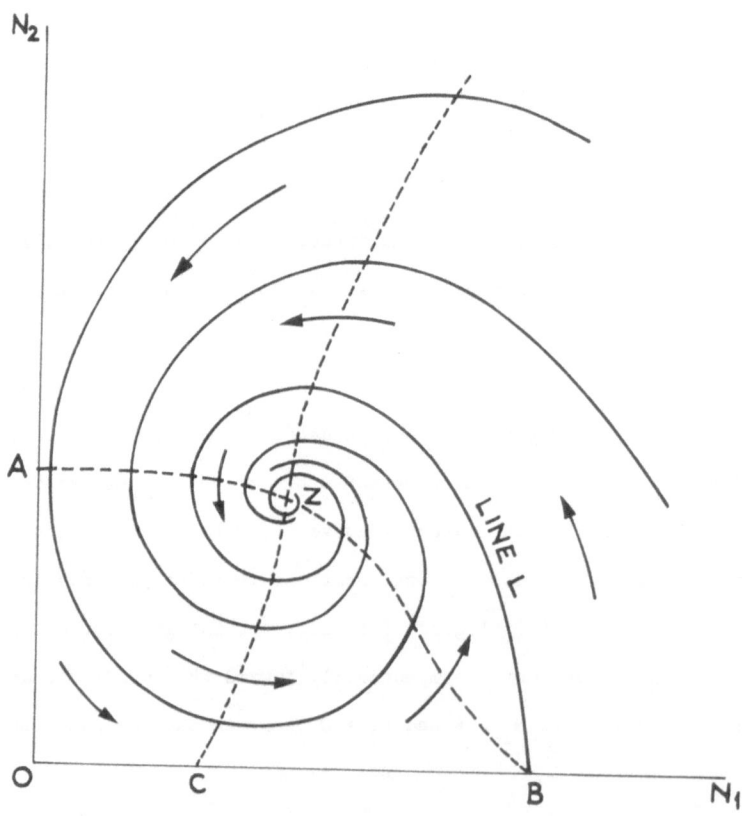

Fig. 2.

Case (b) The integral curve L approaches the point Z along a specific direction. In this case the point Z is a <u>node, and for any initial</u> $N_1 > 0$, $N_2 > 0$ <u>the system</u> <u>converges toward the stable equilibrium state Z, without undergoing damped</u> <u>oscillations</u>.

Case (c) The integral curve L approaches a closed cycle F, passing an infinite number of times from domain I to domains II, III and IV.

Every integral curve originating from a point $N_1 > 0$, $N_2 > 0$ outside the <u>limit</u> <u>cycle</u> F approaches F <u>asymptotically</u>. In other words, <u>there can occur periodic</u> <u>oscillations of</u> N_1 <u>and</u> N_2 <u>with well defined period and amplitude, corresponding to</u> <u>the cycle F.</u> <u>If the initial point</u> $N_1 > 0$, $N_2 > 0$ <u>lies in the domain outside the</u> <u>cycle F, there will be oscillations which approach asymptotically the oscillations</u> <u>along the cycle F</u> (Fig. 3, continuous line). It is not possible to determine what

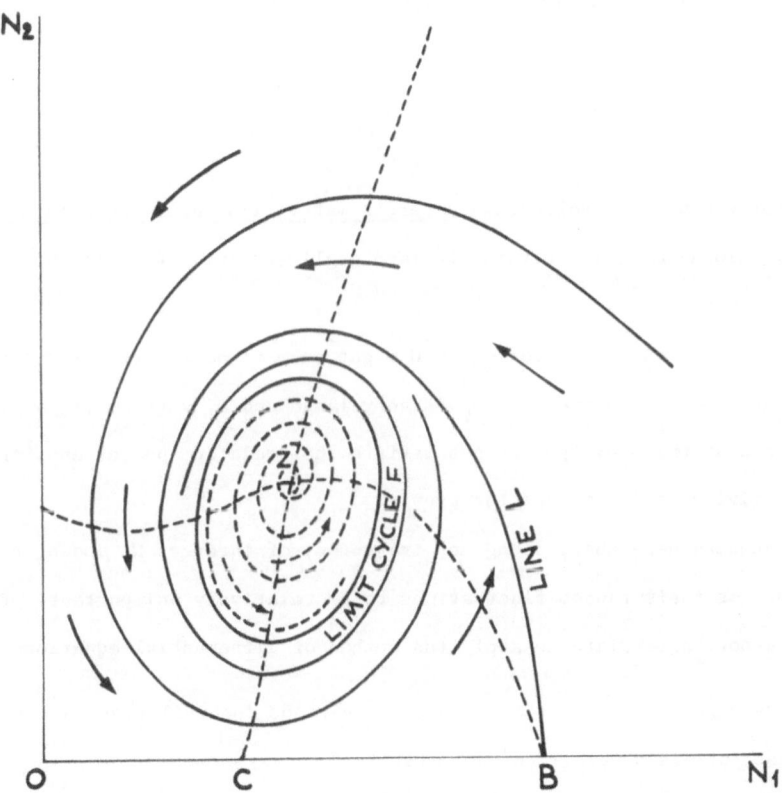

Fig. 3.

will happen inside the cycle F without introducing further hypotheses. In general there can be an arbitrary number of stable or unstable cycles, each with the same center. The point Z can be a center, a stable or unstable node, or a stable or unstable focus.

The simplest case is the following:

Case (c_1) The cycle F is unique and simple (and, therefore, it is bilaterally stable).

In this case the point Z is either an unstable focus or an unstable node. <u>Any initial non-equilibrium state inside the cycle F leads to oscillations of increasing amplitude which approach asymptotically the oscillations of cycle F.</u> <u>This case is represented in Fig. 3 by dotted lines (in the case where Z is a focus)</u> (4).

The case studied by Volterra, equations (Ia), can be obtained only as a limit case of case (a) or of case (c).

FOOTNOTES

(1) See, for example, V. Volterra. <u>Ricerche matematiche nelle associazioni biologiche</u>. Giornale dell'Istituto Italiano degli Attuari. Vol. II, no. 3. July, 1931.

(2) At first sight the condition $\frac{\partial K_1}{\partial N_1} < 0$ might seem to be a more natural one, but it is not sufficient. In fact, if N_2 is very large and N_1 very small, an increase in the numbers of the prey species can satiate the predator species and, therefore, improve the living conditions of the prey.

(3) It is assumed here that, along any trajectory of interest, N_1 and N_2 remain large enough for their random fluctuations to be relatively unimportant. For small N_1, N_2 it is not appropriate to apply the method of differential equations.

(4) It seems highly probable that, among the various possibilities in case (c), only the case (c_1) is of practical interest.

On asymptotically stable periodic solutions in biological differential equations

V. A. Kostitzin

In (1) A. Kolmogoroff draws attention to the possible existence of asymptotically periodic solutions of the system of differential equations

$$(1) \qquad x' = \phi(x,y) , \qquad y' = \psi(x,y) .$$

This possibility already had been pointed out by A. J. Lotka (2) who first studied system (1) in a biological context. This possibility also follows immediately from the well known investigations by H. Poincare and E. Picard. In this note I will show that this possibility can occur in relatively simple cases.

Consider in this respect the following system:

$$(2) \qquad \begin{cases} x' = \epsilon x - \lambda y - x(\alpha^2 x^2 + \beta^2 y^2) , \\[2ex] y' = \epsilon y + \lambda x - y(\alpha^2 y^2 + \beta^2 x^2) . \end{cases}$$

This system only differs from the more familiar biological equations by the order of the limiting terms, which is three rather than two. The interactions among species are not always of the most elementary form, so that third order terms are highly plausible.

By changing to the polar coordinates (r,θ), system (2) becomes

$$(3) \qquad \begin{cases} \dfrac{r'}{r} = \epsilon - r^2 \left[\alpha^2 \cos^2 2\theta + \dfrac{1}{2} (\beta^2 + \alpha^2) \sin^2 2\theta \right] \\[3ex] \theta' = \lambda - \dfrac{r^2}{2} (\beta^2 - \alpha^2) \sin 2\theta \cos 2\theta . \end{cases}$$

Suppose that the multiplication coefficient, ϵ, is positive. The curve

$$(4) \qquad r' = 0 , \qquad r^2 = \dfrac{\epsilon}{\alpha^2 \cos^2 2\theta + \dfrac{1}{2} (\alpha^2 + \beta^2) \sin^2 2\theta}$$

is a closed one, and the vector-radius oscillates between $\sqrt{2\varepsilon/\beta^2+\alpha^2}$ and

$\sqrt{\varepsilon/\alpha}$. On the inside of this curve r' is positive, and on the outside it is negative. The curve

$$(5) \qquad\qquad \theta' = 0 , \quad r^2 = \frac{2\lambda}{(\beta^2-\alpha^2) \sin 2\theta \cos 2\theta}$$

decomposes the (r,θ) plane into two regions, in which θ' has different signs. One of these regions is connected and has the shape of a cross. The other is made up of the four separate domains on the outside of this cross.

In the case

$$(6) \qquad\qquad 8\lambda^2\alpha^2 (\beta^2 + \alpha^2) > \varepsilon^2 (\beta^2 - \alpha^2)^2 ,$$

the region of positive r' completely lies within the cross-shaped region

$$r^2 < \frac{2\lambda}{(\beta^2-\alpha^2) \sin 2\theta \cos 2\theta} .$$

Under this condition a point (x,y) within the interior of the closed curve (4) moves, in a finite time, toward the outside of it. Conversely, a point outside this curve moves, in a finite time, toward the inside. It is easy to show that as time, t , grows indefinitely, any integral curve of system (2) tends closer and closer to a closed limit curve. That is, the process tends toward a specified periodic limit.

Equations (2) can be changed substantially without losing the asymptotic nature of this solution. A similar case can occur when the vital coefficients of the biological equations

$$x' = x(\varepsilon_1-h_{11}x-h_{12}y) , \qquad y' = y(\varepsilon_2-h_{21}x-h_{22}y)$$

change in different·regions of the (x,y) plane. In such a case the (x,y) plane is subdivided into domains in each of which the relationships between the two (x,y) groups are expressed by equations of the same type, but with different vital co-efficients. As a result the system might tend, asymptotically, towards a periodic regime. In each special case, the probability of the establishment of such a regime can be numerically evaluated. One has then, together with damped oscillations, a type of quasi-periodic behavior which must be very common in nature. Quasi-periodic behaviors can help us perceive order in the apparent chaos of nature. They also permit us to comprehend the local realization of rare states.

References:

1. Kolmogoroff, A. N. Sulla teoria di Volterra della lotta per l'esistenza. Giorn. Instituto Ital. Attuari, 7, 74-80, 1936.

2. Lotka, A. J. Elements of physical biology. Baltimore: Williams and Witkins, 1925, pp. 147-148.

PART III

PARASITISM, EPIDEMICS AND SYMBIOSIS

INTRODUCTION

It is often difficult to draw a line between parasitism and symbiosis on the one hand, and predation and competition on the other. Also, the general methodology employed to study each of these cases is much the same. What sets apart the treatment of parasitism and symbiosis is that the study of biologically meaningful cases requires a variety of special assumptions, posing more specific mathematical problems. Among the many works in this area we have chosen to present two papers (Numbers I and IV) in a series of five by Lotka (1923b, c, d, and e) and Sharpe and Lotka (1923) on malaria epidemics. Of the other three papers in this series by Lotka, II deals with an approximation, III with numerical results, and V is a non-technical synopsis directed toward epidemiologists. We have also included Kostitzin's short monograph "Symbiose, parasitisme et évolution" (1934).

The series of papers by Lotka and the monograph by Kostitzin seem to epitomize both the highest levels of sophistication attained in this area before the war, and the variety of biological and mathematical problems posed by parasitism and symbiosis. It might be useful to sketch the development of the thoughts from which Lotka's contributions to epidemiology matured. Detailed, quantitative information on the time course of a variety of epidemics had already begun to accumulate in the second half of the 19th century. They had displayed a number of regularities which soon attracted the attention of mathematicians. Brownlee (1906, 1910) and Ross (1911) were among the first to model such regularities in terms of differential equations. Ross concentrated on the steady state behavior of malaria epidemics. Almost simultaneously, and independently, Lotka and Martini had also begun to study epidemiological equations analogous to those proposed by Ross. Martini, and even more so Lotka, placed in their analyses much greater emphasis than Ross on the time course of epidemics. We present Lotka's papers primarily because they provide a more thorough analysis than those by Martini (1921, 1941). Also, the series by Lotka had a much greater effect on further developments of the field, and especially on the works of Kostitzin. These papers are virtually unknown to modern theoretical ecologists undoubtedly because they were not discussed at any length in Lotka's "Elements of physical biology" (1924). The reader is referred for

further developments of the deterministic theory of epidemics to the papers by
Kermack and McKendrick (e.g., 1927-1937), Puma (1939), and to the discussion of
D'Ancona (1954). Kermack and McKendrick particularly emphasized the effect of
changing population densities, and explicitly accounted for the immigration of
infected individuals. For a general discussion of parasitology one might also refer
to Lotka (1923a).

Papers I, II and III in Lotka's series analyzed Ross' equations and emphasized
the dramatic effects on the time course of the epidemic of the initial rates of
contagion in the human and mosquito populations. The figure reported here (Fig. 1)

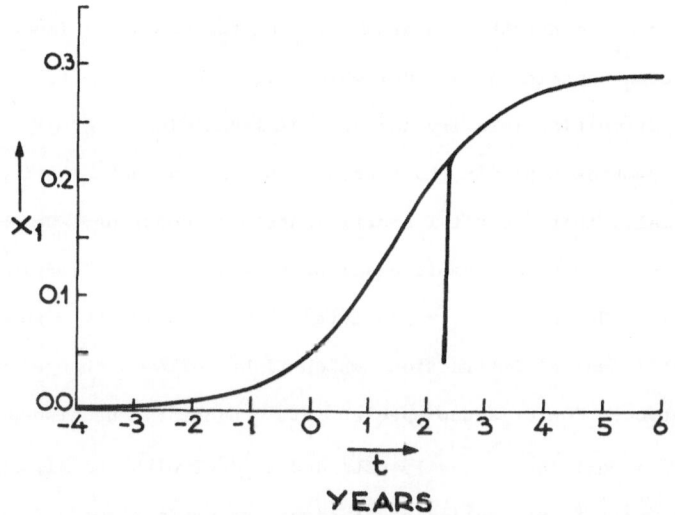

Fig. 1. Curve of growth of endemic malaria. The upper (S-shaped)
curve relates to the particular case in which the malaria rates in the
human and in the mosquito population stand (nearly) in the ratio which
they have at equilibrium. The lower curve represents the course of events
when the initial malaria rate is 4.2 percent in the human population, and
1.4 percent in the mosquito population. (The zero of the time scale has
arbitrarily been placed in the same position as in Fig. 9 of Part III.
The lower curve, accordingly, starts at about 2.3 years.) (From Part V.)

(from V) gives a compelling example of the usefulness of detailed numerical investigations in dealing with these problems. In the other paper we report, IV, a constant lag representing the incubation time of the parasite is introduced in much the same system of differential equations as that of Ross. The quantitative effects of this lag are studied in detail, again with special reference to epidemiological data. As can easily be expected, the lag has no effect on either the position or the stability of the final equilibrium. In a seasonal epidemic, however, the time course is far more interesting, and in particular its initial stages are slightly slowed down by such lags. This result led Lotka to emphasize the general importance of taking into account the effects of delays in population interactions.

Kostitzin's short monograph starts, as he often did, with a general philosophical discussion of the merits and limitations of mathematical models in ecology and evolution. He was especially critical of the widespread assumption that the vital coefficients are constant. He also stressed the potential importance of symbiosis among different species as a model which might lead to a better understanding of the evolution of multicellularity and of social behavior.

In dealing with symbiosis proper, Kostitzin considered a number of special cases, assuming initially that the vital coefficients were constant. He emphasized the frequent occurrence of several stable equilibria, and their implications for the evolution of symbiosis. He explicitly represented this evolution by a sudden variation of the coefficient of interaction, which might follow a change in behavior, and/or the sudden introduction of a new species into a habitat. He also pointed out that much the same treatment can be easily extended to deal with the almost continuous variations in the interaction coefficient, resulting from selection at the genetic level. He then dealt with a number of mathematically more sophisticated examples in which the vital coefficients depend on the life stages of both the host and the parasite.

Parasitism is introduced by Kostitzin as being, mathematically, a special case of symbiosis in which the vital coefficients are modified so that only one of the species gains from the association. This approach is particularly interesting in the light of theories which attribute the origin of many cases of symbiosis to non-lethal forms of predation, parasitism, and competition. The special cases presented by parasitism are mainly those which arise from the distribution of the number of parasites in individual hosts, and the related case of parasitic castration. He dealt with both at length, pointing out the usefulness of their predictions in interpreting natural observations, such as the infestation of hermit crabs by the parasitic barnacle <u>Chlorogaster</u>.

In the final section of this work Kostitzin returns to a general discussion of the evolutionary implications of models, both at the ecological and at the genetic levels. A short appendix on the qualitative study of differential equations helps to make this short monograph particularly self-contained.

Contribution to the Analysis of Malaria

Epidemiology. I. General Part

by

Alfred J. Lotka

Contents

Introduction

In his work *The Prevention of Malaria** Sir Ronald Ross has developed a

---

[*] Second (English) edition, John Murray, London, 1911. The first edition does not contain the Appendix on the mathematical theory of happenings, pp. 651 et seq., section 66 of the second edition.

system of differential equations to represent, in certain circumstances, the course
of events in a community that has become infected with malaria. The author does
not, in the work cited, give the general solution of the differential equations
developed by him, but restricts his discussion to the final state of equilibrium to
which they lead.

Sir Ronald Ross has also set up, by another method, another type of
equation relating to the same subject.* This he discusses in greater detail, and
a further contribution to their study has been made by H. Waite[†] in a paper cited
and discussed by Sir Ronald in his book.[§]

The writer has, in several previous publications,[¶] had occasion to refer
to the system of differential equations given by Ross, and has indicated the form
of their general solution, and certain peculiarities of this solution. In these
previous publications the case was treated incidentally, as an example to illustrate
certain related matters which were the main topic. No numerical applications were
made.

The purpose of the present paper is to make a more exhaustive study of
the case, with numerical applications based on Ross's estimates of the approximate
values of the constants entering into his equations. While the subject will here
be treated as a chapter in quantitative epidemiology, it also possesses a certain
interest as an illustration of certain principles relating to the more general field
of the evolution of organic systems. It is partly for this reason that the analysis
here given is carried out in greater detail and with perhaps more refinement than
the numerical data available at present would warrant. This attitude finds its

---

*Ibid., pp. 155 et seq.; sections 27 et seq.; also, R. Ross, Report on the Preven-
tion of Malaria in Mauritius, 1908, pp. 30-40.

†Biometrika, vol. 7, 1910, p. 421.

§"Prevention of Malaria," p. 652.

¶Nature, Feb., 1912, p. 497; Science Progress, 1920, vol. 14, p. 413.

justification in the fact that we are here concerned more with the exhibition and exemplification of a method, than with concrete numerical results, though such results will be shown.

A further purpose of this paper is to examine somewhat in detail the relation between the system of differential equations set up by Ross in Section 66 of his book, and his alternative treatment, in Section 27, of the case under consideration. This relation has never been fully laid bare. That the two methods can not be wholly equivalent is seen from the fact that the method of Section 27 presents the course of events in a single equation which, when carried to the limit of infinitesimals, appears as one differential equation of the first order; whereas the second method, that of Section 66, leads to a system of two simultaneous differential equations of the first order, such as are equivalent to one differential equation of the second order for each of the two dependent variables.

Consideration will also be given to Mr. Waite's paper.

Establishment of the Ross Equations

It will be well to begin by reproducing, in brief recapitulation, the argument by which Sir Ronald Ross develops his system of differential equations relating to malaria. We shall here adopt his notation, as it appears in Section 66 of his book, according to which

p = number of total human population at a given instant, (1)

z = number of human population affected with malaria, (2)

fz = number of human population so affected, and infective, (3)

r = recovery rate, i.e., fraction of affected population that reverts

to the non-affected (healthy) state per unit of time, (4)

M = human mortality, i.e., deathrate per head per unit of time, (5)

N = human natality, i.e., birthrate per head per unit of time, (6)

t = time, (7)

p', z', f', r', M', N', denote corresponding quantitites relative to the mosquito population.

If a mosquito bites a human being, on an average, b' times per unit of time, then the $f'z'$ infective mosquitoes will place $b'f'z'$ infective bites (on human beings) per unit of time, and a fraction $(p - z)/p$ of these will fall on unaffected ("healthy") persons. If we assume[*] that every person thus bitten becomes affected, then the number of new infections per unit of time in the human population will be $b'f'z'(p - z)/p$.

Similarly, if a human individual is bitten, on an average, b times per unit of time, the number of new infections among mosquitoes will be $bfz(p' - z')/p'$.

Now, evidently, the total number of times, $b'p'$ per unit of time, that mosquitoes bite human beings, is equal to the total number of times, bp, that human beings are bitten, per unit of time, by mosquitoes. We have therefore

$$b = \frac{b'p'}{p} \tag{8}$$

and finally, for the number of new infections, per unit ot time, among mosquitoes, we have the expression $b'fz(p' - z')/p$.

If emigration and immigration in the area under consideration is negligible for both men and mosquitoes, then evidently

$$\left.\begin{array}{l}\text{Rate of increase} \\ \text{of affected indi-} \\ \text{viduals among} \\ \text{the human popu-} \\ \text{lation}\end{array}\right\} = \left\{\begin{array}{l}\text{New infec-} \\ \text{tions per} \\ \text{unit of time}\end{array}\right\} - \left\{\begin{array}{l}\text{deaths} \\ \text{per unit} \\ \text{of time}\end{array}\right\} - \left\{\begin{array}{l}\text{recoveries} \\ \text{per unit} \\ \text{of time}\end{array}\right\} \tag{9}$$

that is to say,

$$\frac{dz}{dt} = \frac{b'f'z'}{p}(p - z) - Mz - rz . \tag{10}$$

Similarly, for the mosquito population

$$\frac{dz'}{dt} = \frac{b'fz}{p}(p' - z') - M'z' - r'z' . \tag{11}$$

---

[*] This assumption is not necessary. We may introduce a suitable factor g giving the proportion, of those bitten by infective mosquitoes, which becomes infected. Little is known regarding this factor, and it is here put equal to unity. But the argument remains unaltered if g has some other (constant, fractional) value. See "Prevention of Malaria," p. 667.

An analysis of the probable numerical values of M, M', r, r', leads Sir Ronald Ross to the conclusion that M is negligible as compared with r , while on the contrary r' is negligible as compared with M'; furthermore, if we consider the special case of a population in which birthrate and deathrate are just balanced, we may write N' in place of M'. We thus obtain, finally,

$$\frac{dz}{dt} = \frac{b'f'z'}{p} (p - z) - rz , \tag{12}$$

$$\frac{dz'}{dt} = \frac{b'fz}{p} (p' - z') - N'z' , \tag{13}$$

which is the system of equations given by Ross, as it appears when emigration and immigration are negligible. We may simplify these equations somewhat by putting

$$\frac{p'}{p} = A, \tag{14}$$

$$\frac{N'}{b'f} = B,$$

$$\frac{r}{b'f'} = C , \tag{15}$$

and, further, we may, without sacrificing generality, refer our reflections to the unit of population, so that

$$p = 1, \tag{16}$$

$$p' = A. \tag{17}$$

Noting that in this case the "malaria rate" m, defined as z/p, is simply given by z, the equations (12), (13) appear in the form

$$\frac{dm}{dt} = b'f'\{z'(1 - m) - Cm\} = - b'f'Cm + b'f'z' - b'f'mz' , \tag{18}$$

$$\frac{dz'}{dt} = b'f\{m(A - z') - Bz'\} = b'fAm - b'fBz' - b'fmz' . \tag{19}$$

Equations (18), (19) are of the general form

$$\frac{dx_1}{dt} = f_1(x_1, x_2) = a_{11}x_1 + a_{12}x_2 + a_{112}x_1x_2 , \tag{20}$$

$$\frac{dx_2}{dt} = f_2(x_1, x_2) = a_{21}x_1 + a_{22}x_2 + a_{212}x_1x_2 , \tag{21}$$

where we have written, for the sake of symmetry in notation,

$$m = x_1 , \tag{22}$$

$$z' = x_2 . \tag{23}$$

It is sometimes convenient to consider equations (20), (21) in the form

$$\frac{dx_1}{f_1} = \frac{dx_2}{f_2} = dt \tag{24}$$

or

$$\frac{dx_1}{dx_2} = \frac{f_1}{f_2} , \tag{25}$$

in which last form the variable t no longer appears, having been eliminated by division.

Equilibrium; Singular Points. A state of equilibrium ensues when

$$\frac{dx_1}{dt} = \frac{dx_2}{dt} = 0 \tag{26}$$

i.e., when

$$f_1 = f_2 = 0 \tag{27}$$

$$a_{11}x_1 + a_{12}x_2 + a_{112}x_1x_2 = a_{21}x_1 + a_{22}x_2 + a_{212}x_1x_2 = 0 . \tag{28}$$

At the same time

$$\frac{dx_1}{dx_2} = \frac{f_1}{f_2} = \frac{0}{0} \tag{29}$$

becomes indeterminate, so that (28) defines singular points in the integral curves of (25). It is seen by inspection of (28) that one such singular point is situated at

$$x_1 = x_2 = 0 . \tag{30}$$

If another such singular point lies at

$$x_1 = p , \tag{31}$$

$$x_2 = q . \tag{32}$$

We must have, by (28)

$$a_{11}p + a_{12}q + a_{112}pq = 0 , \tag{33}$$

$$a_{21}p + a_{22}q + a_{212}pq = 0 . \tag{34}$$

Dividing by pq we find

$$\left.\begin{array}{c} \dfrac{a_{11}}{q} + \dfrac{a_{12}}{p} + a_{112} = 0 , \\[2em] \dfrac{a_{21}}{q} + \dfrac{a_{22}}{p} + a_{212} = 0 , \end{array}\right\} \tag{35}$$

a system of linear equations for $1/p, 1/q$. Solving, we find

$$p = \frac{a_{21}a_{12} - a_{11}a_{22}}{a_{11}a_{212} - a_{21}a_{112}} , \tag{36}$$

$$q = \frac{a_{21}a_{12} - a_{11}a_{22}}{a_{22}a_{112} - a_{12}a_{212}} . \tag{37}$$

We shall have occasion further on to note a simple graphic construction for the singular point p, q when the numerical values of certain of the malaria constants are given.

Transformation of Origin to Second Singular Point

A greater interest attaches, in some ways, to the second equilibrium, $p, q,$ than to the first, $0, 0$. For this reason it is desirable to be prepared to operate, in place of the original system of equations (20), (21), with a transformed system obtained by the substitution

$$\xi_1 = x_1 - p \ , \tag{38}$$

$$\xi_2 = x_2 - q \ , \tag{39}$$

which gives

$$\frac{d\xi_1}{dt} = (a_{11} + a_{112}q)\xi_1 + (a_{12} + a_{112}p)\xi_2 + a_{112}\xi_1\xi_2$$

$$+ a_{11}p + a_{12}q + a_{112}pq \tag{40}$$

and a similar equation for $d\xi_2/dt$. Now the sum of the last three terms in (40) vanishes, by (33), so that we have

$$\frac{d\xi_1}{dt} = (a_{11} + a_{112}q)\xi_1 + (a_{12} + a_{112}p)\xi_2 + a_{112}\xi_1\xi_2 \ , \tag{41}$$

$$\frac{d\xi_2}{dt} = (a_{21} + a_{212}q)\xi_1 + (a_{22} + a_{212}p)\xi_2 + a_{212}\xi_1\xi_2 \ . \tag{42}$$

These equations may also be written in another form. By virtue of the relation (36) (37) we obtain from (41) (42)

$$\frac{d\xi_1}{dt} = - a_{12}\frac{q}{p}\xi_1 - a_{11}\frac{p}{q}\xi_2 + a_{112}\xi_1\zeta_2 \ , \tag{43}$$

$$\frac{d\xi_2}{dt} = - a_{22}\frac{q}{p}\xi_1 - a_{21}\frac{p}{q}\xi_2 + a_{212}\xi_1\xi_2 \ . \tag{44}$$

General Solution. We are now prepared to consider the general solution of the system of differential equations (20), (21), or the equivalent, and, in form, precisely similar system (41), (42); (43), (44).

This solution can not, in general, be expressed in the form of a finite expression in terms of any of the standard algebraic or transcendental functions. It can, however, be given in the form of infinite series.

Solution Near Ordinary Point. Near an ordinary point a series solution can be developed for x_1 and x_2 in terms of t, or for x_2 in terms of x_1, by Taylor's theorem.

Expansion in Powers of t. The work of determining successive higher derivatives of x_1 and x_2 with regard to t is rendered more symmetrical by writing (20) in the form

$$\frac{dx_1}{dt} = x_1(a_{11} + a_{112}x_2) + x_2(a_{12} + a_{112}x_1) - a_{112}x_1x_2 \ . \tag{45}$$

It is then found, by successive differentiation, that

$$\frac{d^2x_1}{dt^2} = \frac{dx_1}{dt}(a_{11} + a_{112}x_2) + \frac{dx_2}{dt}(a_{12} + a_{112}x_1) \ , \tag{46}$$

$$\frac{d^3x_1}{dt^3} = \frac{d^2x_1}{dt^2}(a_{11} + a_{112}x_2) + \frac{d^2x_2}{dt^2}(a_{12} + a_{112}x_1) - 2a_{112}\frac{dx_1}{dt} \cdot \frac{dx_2}{dt} \ , \tag{47}$$

$$\frac{d^4x_1}{dt^4} = \frac{d^3x_1}{dt^3}(a_{11} + a_{112}x_2) + \frac{d^3x_2}{dt^3}(a_{12} + a_{112}x_1)$$
$$- 3a_{112}\left[\frac{d^2x_1}{dt^2} \cdot \frac{dx_2}{dt} + \frac{dx_1}{dt} \cdot \frac{d^2x_2}{dt^2}\right] \ , \tag{48}$$

$$\frac{d^5x_1}{dt^5} = \frac{d^4x_1}{dt^4}(a_{11} + a_{112}x_2) + \frac{d^4x_2}{dt^4}(a_{12} + a_{112}x_1)$$
$$- a_{112}\left[\frac{4d^3x_1}{dt^3} \cdot \frac{dx_2}{dt} + \frac{6d^2x_1}{dt^2} \cdot \frac{d^2x_2}{dt^2} + \frac{4dx_1}{dt} \cdot \frac{d^3x_2}{dt^3}\right] \ . \tag{49}$$

The law of formation of the successive derivatives is clear by comparison of the numerical coefficients in the right-hand member of (49) with those of the terms in the right-hand member of

$$(a + b)^4 = a^4 + b^4 + 4a^3b + 6a^2b^2 + 4ab^3 . \tag{50}$$

A precisely similar treatment applies to equation (21) written in the form

$$\frac{dx_2}{dt} = x_1(a_{21} + a_{212}x_2) + x_2(a_{22} + a_{212}x_1) - a_{212}x_1x_2 . \tag{51}$$

It is not necessary to carry this out in detail here.

Expansion of x_1 *in Terms of* x_2 .   A power series for x_1 in terms of x_2 is readily obtained, near an ordinary point, by Taylor's theorem, since dx_1/dx_2 is given by (20), (21), (25), and the higher derivatives are obtainable by successive differentiation. It is unnecessary to carry this out here. It may be remarked, however, that the expansion assumes a particularly simple form for point at which either x_1 or x_2 vanishes.

Solution near Singular Point. Near a singular point x_1, $x_2 = 0$, 0 or x_1, $x_2 = p,q$ the development by Taylor's theorem fails because the first and all the higher derivatives with regard to t vanish, while the derivative dx_2/dx_1 becomes indeterminate. A general solution can, however, be written in the form of exponential series.[*]

$$x_1 = P_1e^{\lambda_1 t} + P_2e^{\lambda_2 t} + P_{11}e^{2\lambda_1 t} + P_{12}e^{(\lambda_1+\lambda_2)t} + P_{22}e^{2\lambda_2 t} + \ldots , \tag{52}$$

$$x_2 = Q_1e^{\lambda_1 t} + Q_2e^{\lambda_2 t} + \ldots . \tag{53}$$

The Exponential Coefficients λ_1, λ_2 . At the singular point $0, 0$ the values of λ_1, λ_2 are given by the two roots of the quadratic equation

---

[*] See, for example, Picard, Traité d'Analyse, vol. 3 (1896), p. 14

$$\begin{vmatrix} a_{11} - \lambda & a_{12} \\ a_{21} & a_{22} - \lambda \end{vmatrix} = 0 , \tag{54}$$

i.e.,

$$\lambda^2 - \lambda(a_{11} + a_{22}) + (a_{11}a_{22} - a_{21}a_{12}) = 0 \tag{55}$$

which gives

$$\lambda = \frac{1}{2} \{(a_{11} + a_{22}) \pm \sqrt{(a_{11} + a_{22})^2 - 4(a_{11}a_{22} - a_{21}a_{12})}\} \tag{56}$$

or, by a simple rearrangement,

$$\lambda = \frac{1}{2} \{(a_{11} + a_{22}) \pm \sqrt{(a_{11} - a_{22})^2 + 4a_{21}a_{12}}\} . \tag{57}$$

Similarly, at the singular point p,q, the values of λ_1', λ_2' are given, in accordance with (41), (42) by

$$\begin{vmatrix} (a_{11} + a_{112}q) - \lambda' & (a_{12} + a_{112}p) \\ (a_{21} + a_{212}q) & (a_{22} + a_{212}p) - \lambda' \end{vmatrix} = 0 , \tag{58}$$

i.e.,

$$\lambda'^2 - \{(a_{11} + a_{22}) + (a_{212}p + a_{112}q)\}\lambda - (a_{11}a_{22} - a_{21}a_{12}) = 0 \tag{59}$$

or, according to (43), (44), by

$$\begin{vmatrix} - a_{12} \frac{q}{p} - \lambda' & - a_{11} \frac{p}{q} \\ - a_{22} \frac{q}{p} & - a_{21} \frac{p}{q} - \lambda' \end{vmatrix} = 0 , \tag{60}$$

i.e.,

$$\lambda'^2 + \left(a_{21}\frac{p}{q} + a_{12}\frac{q}{p} \right)\lambda' - (a_{11}a_{22} - a_{21}a_{12}) = 0 , \qquad (61)$$

$$\lambda' = -\frac{1}{2}\left\{ \left(a_{21}\frac{p}{q} + a_{12}\frac{q}{p} \right) \right.$$

$$\left. \pm \sqrt{ \left(a_{21}\frac{p}{q} + a_{12}\frac{q}{p} \right)^2 + 4(a_{11}a_{22} - a_{21}a_{12}) } \right\} \qquad (62)$$

$$= -\frac{1}{2}\left\{ \left(a_{21}\frac{p}{q} + a_{12}\frac{q}{p} \right) \pm \sqrt{ \left(a_{21}\frac{p}{q} - a_{12}\frac{q}{p} \right)^2 + 4a_{11}a_{22} } \right\} . \qquad (63)$$

Stability of Equilibrium. It is readily seen by inspection that the series solution (52), (53) converges for large values of t, provided that the real parts of λ_1 and λ_2 are both negative. In that case we have, for $t = \infty$

$$x_1 = x_2 = 0 , \qquad (64)$$

so that the equilibrium at 0, 0 is stable. Similarly if λ_1', λ_2' are both negative the equilibrium at p, q is stable. On the contrary, if one or both of the roots for λ in (54), or for λ' in (58) or (60) have their real part positive, then the corresponding equilibrium is unstable.

Now it should be noted that according to (55) we have

$$\lambda_1\lambda_2 = (a_{11}a_{22} - a_{21}a_{12}) , \qquad (65)$$

whereas according to (59) or (61)

$$\lambda_1'\lambda_2' = -(a_{11}a_{22} - a_{21}a_{12}) \qquad (66)$$

$$= -\lambda_1\lambda_2 . \qquad (67)$$

From (67) it follows that at least one of the four quantities λ_1, λ_2, λ_1', λ_2' must be positive. Hence it is impossible for both equilibria, the one at 0,0

and the one at p,q to be stable; at least one of them must be unstable, and
both may be, so far as these considerations alone show.

   Oscillatory Solutions. If the roots λ are complex, the solution (52),
(53) represents a damped oscillatory process, the exponential function assuming in
this case the trigonometric form. The condition for oscillations near the
equilibrium 0, 0 is that the discriminant of (57) be less than zero, i.e.,

$$(a_{11} - a_{22})^2 + 4a_{21}a_{12} < 0 . \qquad (68)$$

From (68) it is clear that there can be no oscillation near the equilibrium 0,0
if a_{21} and a_{12} are of the same sign. Similarly, in the neighborhood of the
equilibrium p,q, the condition for oscillation is, according to (63),

$$\left(a_{21}\frac{p}{q} - a_{12}\frac{q}{p}\right)^2 + 4a_{11}a_{22} < 0 . \qquad (69)$$

From (69) it is seen that there can be no oscillations near the equilibrium p,q
if a_{11} and a_{22} are of the same sign.* In point of fact, in the malaria

---

* It should be noted that this conclusion is general, since it depends only on the
linear terms of $f_1(x_1,x_2)$, $f_2(x_1,x_2)$, and is therefore independent of special
form which these functions happen to assume in the malaria equations. Furthermore,
the observation here made has the following general significance:

   Inasmuch as the coefficients a in the Taylor expansions of the functions
f_1, f_2, are

$$a_{ij} = \frac{\partial}{\partial x_j} \cdot \frac{dx_i}{dt} ,$$

the conclusion noted in the text may be expressed as follows:

   If, near equilibrium, the two species of matter to which x_1, x_2 refer
mutually promote each other's growth; or if, on the contrary, they are mutually
antagonistic; in either such case the approach to that equilibrium can not be of
the oscillatory type. Such oscillatory approach can occur only if one species
promotes the growth of the second, while the second exerts an unfavorable influence
upon the growth of the first -- or vice versa.

   This may also be expressed by saying that if, near equilibrium, the two
species have a positive value, each for the other, or a negative value, each for the
other, then there can be no oscillatory approach to equilibrium.

   (For the concept of value here implied, see Lotka, A.J., "An Objective
Standard of Value Derived from the Principle of Evolution," Jl. Washington Acad.
Sci., 1914, vol. 4, pp. 409, 447, 499; "Efficiency as a Factor in Organic Evolution,"
ibid., 1915, vol. 5, pp. 360, 397. "The Economic Conversion Factors of Energy,"
Proc. Natl. Acad. Sci., 1921, vol. 7, p. 192).

equations (18), (19), a_{11} and a_{22} are both negative, while a_{21} and a_{12} are both positive, so that the course of events in the establishment of endemic malaria in accordance with these equations can not be oscillatory. For the sake of generality, however, the following point may here be noted:

Oscillation at both Singular Points Impossible. Since $\lambda_1' \lambda_2' = -\lambda_1 \lambda_2$ it is impossible for λ and λ' to be both complex. For the product of two conjugate complex quantities is always positive. Hence if one singular point is of the oscillatory type, the other must be non-oscillatory.

The Two Singular Points Coincide. If one of the roots λ is zero, it follows that one of the roots λ' also is zero. As a matter of fact, in this case the two singular points coincide, and a special case arises, in which the solution (52), (53) fails.

The Multiplicative Constants P, Q *in the Series Solution (52), (53).* When $t = 0$, we have as initial condition

$$x_{1,0} = P_1 + P_2 + P_{11} + \dots = \Sigma P ,\qquad(70)$$

$$x_{2,0} = Q_1 + Q_2 + Q_{11} + \dots = \Sigma Q .\qquad(71)$$

Two of the constants P_1, P_2 $\dots$; Q_1, Q_2 are therefore arbitrary, being determined by initial conditions. The remaining constants are not independent. Substituting the solution (52), (53) in the left- and the right-hand member of (20), (21) and equating coefficients of homologous terms we find:

$$\frac{Q_1}{P_1} = \frac{\lambda_1 - a_{11}}{a_{12}} = \frac{a_{21}}{\lambda_1 - a_{22}} ,\qquad(72)$$

$$\frac{Q_2}{P_2} = \frac{\lambda_2 - a_{11}}{a_{12}} = \frac{a_{21}}{\lambda_2 - a_{22}} ,\qquad(73)$$

In the case of malaria, the infected human population promotes the growth of malaria among mosquitoes, and vice versa.

$$P_{11} = P_1 Q_1 \left\{ \frac{(2\lambda_1 - a_{22})a_{112} + a_{12}a_{212}}{(2\lambda_1 - a_{11})(2\lambda_1 - a_{22}) - a_{21}a_{12}} \right\} , \qquad (74)$$

$$P_{22} = P_2 Q_2 \left\{ \frac{(2\lambda_2 - a_{22})a_{112} + a_{12}a_{212}}{(2\lambda_2 - a_{11})(2\lambda_2 - a_{22}) - a_{21}a_{12}} \right\} , \qquad (75)$$

$$P_{12} = (P_1 Q_2 + P_2 Q_1) \left\{ \frac{(\lambda_1 + \lambda_2 - a_{22})a_{112} + a_{12}a_{212}}{(\lambda_1 + \lambda_2 - a_{11})(\lambda_1 + \lambda_2 - a_{22}) - a_{21}a_{12}} \right\} . \qquad (76)$$

Similar expressions are obtained for Q_{11}, Q_{22} and Q_{12}. A general formula for these constants may be written in the following form:

$$P_{mn} = \frac{\Sigma(PQ)_{mn} \{(m\lambda_1 + n\lambda_2 - a_{22})a_{112} + a_{12}a_{212}\}}{(m\lambda_1 + n\lambda_2 - a_{11})(m\lambda_1 + n\lambda_2 - a_{22}) - a_{21}a_{12}} , \qquad (77)$$

$$Q_{mn} = \frac{\Sigma(PQ)_{mn} \{(m\lambda_1 + n\lambda_2 - a_{11})a_{212} + a_{21}a_{112}\}}{(m\lambda_1 + n\lambda_2 - a_{11})(m\lambda_1 + n\lambda_2 - a_{22}) - a_{21}a_{12}} , \qquad (78)$$

where the symbols P_{mn}, Q_{mn}, and $\Sigma(PQ)_{mn}$ are to be understood as follows:

$$\left. \begin{array}{l} P_{mn} = P_{\underbrace{111 \ldots}_{m \text{ times}} \underbrace{222 \ldots}_{n \text{ times}}} \\[20pt] Q_{mn} = Q_{\underbrace{111 \ldots}_{m \text{ times}} \underbrace{222 \ldots}_{n \text{ times}}} \end{array} \right\} , \qquad (79)$$

$$\Sigma(PQ)_{mn} = \Sigma P_{ij} Q_{kr} , \qquad (80)$$

the summation being extended over all such (positive) values of the indices i, j, k, r, as make

$$\left. \begin{array}{l} i + k = m \\ j + r = n \end{array} \right\} . \qquad (81)$$

So, for example,

$$\Sigma(PQ)_{1,3} = P_1 Q_{222} + P_{12} Q_{22} + P_{122} Q_2 + P_2 Q_{122} + P_{22} Q_{12} + P_{222} Q_1. \qquad (81a)$$

Certain of the coefficients P, Q, admit of simplification. Thus, for example, in view of (55), that is to say, since

$$\lambda_1 + \lambda_2 = a_{11} + a_{22}, \qquad (82)$$

(76) becomes

$$P_{12} = \frac{a_{22} - a_{11}}{a_{12}} \cdot \frac{a_{11} a_{112} + a_{12} a_{212}}{a_{11} a_{22} - a_{21} a_{12}} \qquad (83)$$

and

$$Q_{12} = \frac{a_{22} - a_{11}}{a_{12}} \cdot \frac{a_{22} a_{212} + a_{21} a_{112}}{a_{11} a_{22} - a_{21} a_{12}}. \qquad (84)$$

Pure Coefficients. Special interest attaches to pure coefficients, that is, those containing only one kind of index, either 1 or 2. These may be reduced from the form (74), (75), in view of (55), to the form

$$P_{m,0} = \frac{\Sigma(PQ)_{m,0}}{(m-1)} \frac{m a_{112} \lambda_1 + (a_{12} a_{212} - a_{22} a_{112})}{(m+1)\lambda_1^2 - (a_{11} + a_{22})\lambda_1}, \qquad (85)$$

$$Q_{m,0} = \frac{\Sigma(PQ)_{m,0}}{(m-1)} \frac{m a_{212} \lambda_1 + (a_{21} a_{112} - a_{11} a_{212})}{(m+1)\lambda_1^2 - (a_{11} + a_{22})\lambda_1}, \qquad (86)$$

$$P_{0,n} = \frac{\Sigma(PQ)_{0,n}}{(n-1)} \frac{n a_{112} \lambda_2 + (a_{12} a_{212} - a_{22} a_{112})}{(n+1)\lambda_2^2 - (a_{11} + a_{22})\lambda_2}, \qquad (87)$$

$$Q_{0,n} = \frac{\Sigma(PQ)_{0,n}}{(n-1)} \frac{n a_{212} \lambda_2 + (a_{21} a_{112} - a_{11} a_{212})}{(n+1)\lambda_2^2 - (a_{11} + a_{22})\lambda_2}. \qquad (88)$$

Particular Solutions. We may arbitrarily put

$$P_1 = 0. \qquad (89)$$

It then follows by (72) that

$$Q_1 = 0 \tag{90}$$

and hence by (74)

$$P_{11} = Q_{11} = 0$$

and by (76)

$$P_{12} = Q_{12} = 0 \tag{91}$$

and so on, all coefficients P and Q having an index 1 vanish. Hence if, in the general solution, we strike out all terms containing the index 1 (and the exponential coefficient λ_1), we obtain a particular solution. Another particular solution is, of course, obtained by putting $P_2 = Q_2 = 0$ and striking out all terms containing the index 2.

These particular solutions are of interest, especially in certain numerical cases, and will therefore be considered a little more closely here.

We have

$$P_2 = P_2 , \tag{92}$$

$$P_{22} = \frac{P_2 Q_2}{1} \frac{2a_{112}\lambda_2 + (a_{12}a_{212} - a_{22}a_{112})}{3\lambda_2^2 - (a_{11} + a_{22})\lambda_2} \tag{93}$$

$$= P_2^2 \frac{\rho_1}{1} \varphi_2 , \text{ say,} \tag{94}$$

$$P_{222} = \frac{P_2 Q_{22} + P_{22}Q_2}{2} \frac{3a_{112}\lambda_2 + (a_{12}a_{212} - a_{22}a_{112})}{4\lambda_2^2 - (a_{11} + a_{22})\lambda_2} \tag{95}$$

$$= \frac{P_2 P_{22}}{2} \left(\frac{Q_{22}}{P_{22}} + \frac{Q_2}{P_2} \right) \varphi_3 , \text{ say,} \tag{96}$$

$$= P_2 P_{22} \frac{\rho_1 + \rho_2}{2} \varphi_3 \tag{97}$$

$$= P_2^3 \frac{\rho_1}{1} \varphi_2 \frac{\rho_1 + \rho_2}{2} \varphi_3 \tag{98}$$

Similarly, for the higher coefficients, it is found

$$P_{2222} = P_2^4 \left\{ \frac{\rho_1}{1} \varphi_2 \frac{\rho_1 + \rho_2}{2} \varphi_3 \frac{\rho_1 + \rho_3}{3} \varphi_4 + \frac{\rho_1^2}{1} \varphi_2^2 \frac{\rho^2}{3} \varphi_4 \right\}, \tag{99}$$

$$P_{22222} = P_2^5 \left\{ \frac{\rho_1}{1} \varphi_2 \frac{\rho_1 + \rho_2}{2} \varphi_3 \frac{\rho_1 + \rho_3}{3} \varphi_4 \frac{\rho_1 + \rho_4}{4} \varphi_5 \right.$$

$$\left. + \frac{\rho_1^2}{1} \varphi_2^2 \frac{\rho_2}{3} \varphi_4 \frac{\rho_1 + \rho_4}{4} \varphi_5 + \frac{\rho_1^2}{1} \dot{\varphi}_2^2 \frac{\rho_1 + \rho_2}{2} \varphi_3 \frac{\rho_2 + \rho_3}{4} \varphi_5 \right\}. \tag{100}$$

The higher coefficients become very complicated, and the law of their formation is not very obvious. When the coefficients P have been determined, the coefficients Q are most readily obtained by the relation

$$Q_n = \rho_n P_n, \tag{101}$$

where the subscript n is written in place of the subscript 2 repeated n times.

The factors ρ are given by

$$\rho_n = \frac{(n\lambda_2 - a_{11})a_{212} + a_{21}a_{112}}{(n\lambda_2 - a_{22})a_{112} + a_{12}a_{212}} \tag{102}$$

and the functions φ_n by

$$\varphi_n = \frac{(n\lambda_2 - a_{22})a_{112} + a_{12}a_{212}}{(n-1)\lambda_2^2 - (a_{11} + a_{22})}. \tag{103}$$

The relation between ρ_n and n, and between φ_n and n is, in each case, hyperbolic. For (102), (103) may be written

$$\left(\rho_n - \frac{a_{212}}{a_{112}} \right) \left(n + \frac{\lambda_1}{a_{112}q} \right) = \frac{\lambda_1}{a_{112}q} \left(\frac{q}{p} - \frac{a_{212}}{a_{112}} \right) \tag{104}$$

$$\left(\varphi_n - \frac{a_{112}}{\lambda_2} \right) \left(n - \frac{\lambda_1}{\lambda_2} \right) = \frac{\lambda_1}{\lambda_2} \left(\frac{1}{q} + \frac{a_{112}}{\lambda_2} \right). \tag{105}$$

Special Case. Consider the case in which $|\lambda_2|$ is small as compared with $|a_{11}|$ and also $|a_{22}|$. Then, by (73) direct, or by (102) in view of (73)

$$\frac{Q_2}{P_2} = \rho_1 = \frac{a_{21}}{\lambda_2 - a_{22}} = \frac{\lambda_2 - a_{11}}{a_{12}} \tag{106}$$

$$= -\frac{a_{21}}{a_{22}} = -\frac{a_{11}}{a_{12}} \text{ approx.} \tag{107}$$

At the same time, for values of n not too large, (102) gives, in view of (36), (37),

$$\rho_1 = \rho_n = \frac{a_{21}a_{112} - a_{11}a_{212}}{a_{12}a_{212} - a_{22}a_{112}} = \frac{q}{p} \text{ (approx.)} \tag{108}$$

Consider further the condition

$$\frac{a_{212}}{a_{112}} = \frac{q}{p} \text{ approx.,} \tag{109}$$

which, in virtue of (108), (107) here becomes

$$\frac{a_{212}}{a_{112}} = \rho_1 = -\frac{a_{11}}{a_{12}} \text{ approx.,} \tag{110}$$

$$a_{11}a_{112} = -a_{12}a_{212} \text{ approx.} \tag{111}$$

Let a_{11} and a_{22} be of the same sign. Then $|\lambda_2|$, being small in comparison with $|a_{11}|$ and $|a_{22}|$ separately, is necessarily small as compared with $|a_{11} + a_{22}|$, i.e., as compared with $|\lambda_1 + \lambda_2|$. Thus $\dfrac{|\lambda_2|}{|\lambda_1 + \lambda_2|}$ is a small fraction, and so is, necessarily, $\left|\dfrac{\lambda_2}{\lambda_1}\right|$.

It follows that in the equation for λ

$$\lambda^2 - (a_{11} + a_{22})\lambda + a_{11}a_{22} - a_{21}a_{12} = 0 , \tag{56}$$

we may, in first approximation, neglect the quadratic term, and write

$$\lambda = \frac{a_{11}a_{22} - a_{21}a_{12}}{a_{11} + a_{22}} \quad \text{approx.} \tag{112}$$

Hence

$$\frac{\lambda_2}{a_{112}} = \frac{a_{11}a_{22} - a_{21}a_{12}}{a_{22}a_{112} + a_{11}a_{112}} \quad \text{approx.} \tag{113}$$

Hence, by (111)

$$\frac{\lambda_2}{a_{112}} = \frac{a_{11}a_{22} - a_{21}a_{12}}{a_{22}a_{112} - a_{12}a_{212}} \quad \text{approx.} \tag{114}$$

$$= -q \quad \text{approx.} \tag{115}$$

Recapitulation. Given the following conditions:

1. $|\lambda_2|$ small as compared with $|a_{11}|$ and $|a_{22}|$; \hfill (116)

2. a_{11} and a_{22} of the same sign; \hfill (117)

3. $\dfrac{a_{212}}{a_{112}} = \dfrac{q}{p}$ (approx.) \hfill (118)

Then

1. $\left| \dfrac{\lambda_1}{\lambda_2} \right|$ is large as compared with unity; \hfill (119)

2. $\dfrac{a_{112}}{\lambda_2} = -\dfrac{1}{q}$ (approx.). \hfill (120)

Application. These results are now to be applied to equations (104) (105). Substituting (120) in (104), (105) we obtain

$$\left(\rho_n - \frac{a_{212}}{a_{112}} \right) \left(n - \frac{\lambda_1}{\lambda_2} \right) = -\frac{\lambda_1}{\lambda_2} \left(\frac{q}{p} - \frac{a_{212}}{a_{112}} \right) \quad \text{approx.} \tag{121}$$

$$\left(\varphi_n - \frac{a_{112}}{\lambda_2} \right) \left(n - \frac{\lambda_1}{\lambda_2} \right) = \frac{\lambda_1}{\lambda_2} \left(\frac{1}{q} + \frac{a_{112}}{\lambda_2} \right) \quad \text{approx.} \tag{122}$$

In this form, it must be understood, the equations hold (approximately) only when the conditions (116), (117), (118) are satisfied, that is to say, when the bracket on the right is small, and when $|\lambda_1/\lambda_2|$ is large. The significance of this is best seen by a graphic representation which will be taken up presently. Let it be noted, first, however, that, according to (121) we have, for values of n not greatly in excess of unity, and therefore negligible in the sum

$$\left[n - \frac{\lambda_1}{\lambda_2} \right] ,$$

$$\rho_n = \ldots = \rho_1 = \rho_0 \tag{123}$$

$$= \frac{Q_2}{P_2} = -\frac{q}{p} \quad \text{approx.} \tag{124}$$

Similarly (122) leads to

$$\varphi_n = \ldots = \varphi_2 = \varphi_0 \tag{125}$$

$$= -\frac{1}{q} \tag{126}$$

These relations are brought out very clearly in a graphic representation. The relation between φ_n and n, for example, when plotted in a system of rectangular coordinates, appears as a hyperbola (Fig. 1), whose center has the coordinates

$$\varphi = \frac{a_{112}}{\lambda_2} \tag{127}$$

$$n = \frac{\lambda_1}{\lambda_2} . \tag{128}$$

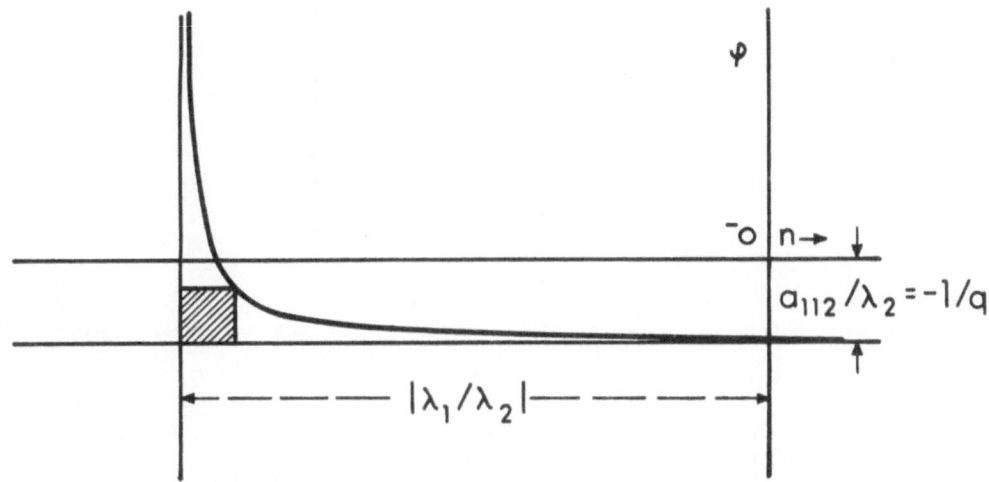

Fig. 1. Plot of φ against n, showing that for values of n near unity,
φ is sensibly constant, as it lies on the very flat portion of
a hyperbola.

The condition (120), that $1/q + a_{112}/\lambda_2$ is small, means, on the diagram,
that the shaded square shall not be very great. On the other hand the condition
that $|\lambda_1/\lambda_2|$ shall be large means that the center of the hyperbola shall be at a
considerable distance to the left of the origin. It is quite plain from the figure
that in these circumstances the values φ_0, φ_1, φ_2 ..., for moderate values of
n , lie on a very flat part of the hyperbola, and are therefore nearly equal. They
are also nearly equal to the constant ordinate of the asymptote, namely
$a_{112}/\lambda_2 = - 1/q$ (approx.).

Solution of Simple Form: Geometric Series. In the case in which the
approximations (123), (125) are justified, the particular solution defined by
(92) to (100) assumes a very simple form; for then the coefficients P are

$$P_2 = P_2 \, ,$$

$$P_{22} = P_2^2 \rho_1 \varphi_2 \, ,$$

$$P_{222} = P_2^3 \rho_1^2 \varphi_2^2 \, ,$$

$$P_{(2)n} = P_2^n \rho_1^{n-1} \varphi_2^{n-1}$$

$$(129)$$

That is to say, the first n terms of the series are simply a geometric progression, so that we may write:

$$x_1 = P_2 e^{\lambda_2 t} \left\{ 1 + P_2 \rho_1 \varphi_2 e^{\lambda_2 t} + P_2^2 \rho_1^2 \varphi_2^2 e^{2\lambda_2 t} + \dots \right\} + U \, , \qquad (130)$$

where U is a remainder term representing that part of the series to which it is not justifiable to apply the approximation (123), (125). The series (130) is readily summed. It gives

$$x_1 = \frac{P_2 e^{\lambda_2 t}}{1 - P_2 \rho_1 \varphi_2 e^{\lambda_2 t}} + U \, . \qquad (131)$$

If U is small, this reduces to

$$x_1 = \frac{P_2}{e^{-\lambda_2 t} - P_2 \rho_1 \varphi_2} \qquad (132)$$

or, in view of (124), (126)

$$x_1 = \frac{P_2}{e^{-\lambda_2 t} + \dfrac{P_2}{p}} \, . \qquad (133)$$

In particular, when $t = 0$,

$$x_0 = \frac{P_2}{1 + \frac{P_2}{p}} \quad , \tag{134}$$

$$P_2 = -\frac{x_0 p}{x_0 - p} \tag{135}$$

$$= -\frac{x_0 p}{\xi_0} \tag{136}$$

Introducing this value of P_2 in (134) we have

$$x_1 = \frac{p}{1 - \frac{\xi_0}{x_0} e^{-\lambda_2 t}} \tag{137}$$

or

$$\xi_1 = x_1 - p = \frac{p}{\frac{x_0}{\xi_0} e^{\lambda_2 t} - 1} \tag{138}$$

In these formulae the origin of time is, so far, arbitrary. If it is so chosen that

$$\frac{\xi_0}{x_0} = -1 \quad , \tag{139}$$

then formula (137) assumes the particularly simple form

$$x_1 = \frac{p}{1 + e^{-\lambda_2 t'}} \tag{140}$$

so that

$$x_0 = \frac{p}{2} \quad , \tag{141}$$

$$x_\infty = p \quad , \tag{142}$$

$$x_{-\infty} = 0 \quad . \tag{143}$$

At the same time (138) becomes

$$\xi_1 = - \frac{p}{1 + e^{\lambda_2 t'}} \tag{144}$$

so that

$$\xi_0 = - \frac{p}{2} , \tag{145}$$

$$\xi_\infty = 0 , \tag{146}$$

$$\xi_{-\infty} = - p \tag{147}$$

Particular Solution at Second Singular Point. The discussion of the particular solution (130), (140), (144) has so far been conducted with reference to the singular point 0, 0. Precisely analogous conclusions are, of course, reached with regard to the singular point p,q. If we denote by primed letters the constants relating to this singular point, we find

$$\left(\rho_n' - \frac{a_{212}}{a_{112}} \right) \left(n - \frac{\lambda_1'}{a_{112} q} \right) = - \frac{\lambda_1'}{a_{112} q} \left(\frac{q}{p} \frac{a_{212}}{a_{112}} \right) , \tag{148}$$

$$\left(\varphi_n' - \frac{a_{112}}{\lambda_2'} \right) \left(n - \frac{\lambda_1'}{\lambda_2'} \right) = - \frac{\lambda_1'}{\lambda_2'} \left(\frac{1}{q} - \frac{a_{112}}{\lambda_2'} \right) , \tag{149}$$

$$\rho_1' = \frac{Q_2'}{P_2'} = \frac{q}{p} \text{ approx.} \tag{150}$$

$$\varphi_2' = \frac{1}{q} \text{ approx.} \tag{151}$$

$$\xi_1 = \frac{p}{\frac{x_0}{\xi_0} e^{-\lambda_2' t} - 1} \tag{152}$$

$$= - \frac{p}{1 + e^{-\lambda_2' t'}} \tag{153}$$

or

$$x_1 = \frac{p}{1 - \dfrac{\xi_0}{x_0} e^{\lambda'_2 t}} \qquad (153a)$$

$$= \frac{p}{1 + e^{\lambda'_2 t'}} \quad . \qquad (154)$$

It will be observed that these equations are closely analogous to those derived with reference to the origin 0, 0. It may be well, however, to draw attention to certain changes in sign which occur.

Now the solutions (144), (154) are, strictly, incompatible; this is not surprising, inasmuch as each represents an approximation. Nevertheless, this point requires a little attention.

According to (107)

$$a_{11}a_{22} - a_{21}a_{12} = 0 \text{ approx.} \qquad (155)$$

Hence

$$\lambda_1\lambda_2 = 0 \text{ approx.} \qquad (156)$$

and also

$$\lambda'_1\lambda'_2 = 0 \text{ approx.} \qquad (157)$$

Therefore, if the characteristic equation (55), relating to the origin, has a small root λ , then the characteristic equation (60), relating to the singular point p, q also has such a small root λ' . Hence in (60) also we may neglect, in first approximation, the quadratic term, so that

$$\lambda'_2 = - \frac{a_{11}a_{22} - a_{21}a_{12}}{(a_{11} + a_{22}) + (a_{112}q + a_{212}p)} , \qquad (158)$$

which, in view of (118) is equivalent to

$$\lambda_2' = -\frac{a_{11}a_{22} - a_{21}a_{12}}{a_{11} + a_{22}} \quad \text{approx.,} \tag{159}$$

$$\lambda_2' = -\lambda_2 \quad \text{approx.} \tag{160}$$

It will now be seen that the substitution of (160) in (153) brings it into harmony with (144), so that the two solutions agree to this degree of approximation. The significance of these points will be seen in dealing with numerical examples. It may here be noted that, if desired, (144) and (153) may be combined to give a somewhat closer approximation, by writing

$$\frac{\lambda_2 - \lambda_2'}{2} = L \tag{161}$$

and

$$\xi_1 = -\frac{p}{1 + e^{Lt'}} \tag{162}$$

The particular solution (144) or (153) or one of its equivalents will be recognized as representing the typical curve of autocatalytic growth under the simplest conditions.

It will thus be seen that under the special conditions for which this solution was developed, and only under those conditions, the growth of endemic malaria, according to Ross's equations, follows the law of simple autocatalytic growth; the same law that T. Brailsford Robertson[*] has exhibited in growth curves of individual organisms, and R. Pearl and L. J. Reed[+] in growth curves of populations. This matter will be further discussed in a later section dealing with a

---

[*] Archiv. f. d. Entwickelungsmechanik der Organismen, 1907, vol. 25, part 4; 1908, vol. 26, p. 108; also, Wo. Ostwald, Die Zeitlichen Eigenschaften der Entwickelungsvorgänge, Leipzig, 1908.

[+] Proc. Natl. Acad. Sci., vol. 6, 1920, p. 275.

numerical example, and in comparing Ross's second malaria equation as here discussed, with his first equation, as given in section (27) of his "Prevention of Malaria," and as integrated by Waite.

From symmetry it is obvious that results precisely analogous to those developed above for x_1 can be derived also for x_2 . If, in order to avoid double subscripts, we write, as before, x_0 and ξ_0 for the initial values of x_1 and ξ_1 , and if we denote by y_0 and η_0 the initial values of x_2 and ξ_2 , we have, for example,

$$x_2 = \frac{q}{1 - \dfrac{\eta_0}{y_0}\, e^{-\lambda_2 t}} \tag{163}$$

$$= -\frac{q}{1 + e^{-\lambda_2 t'}} \quad . \tag{164}$$

Human and Mosquito Malaria Rates in Constant Ratio. Dividing (140) by (164) we find

$$\frac{x_2}{x_1} = \frac{q}{p} = \text{const} = \rho_1 \ . \tag{165}$$

It is thus seen that in the circumstances here considered, and to the degree of approximation taken in view, the malaria rate among men and among mosquitoes stands in constant proportion. It must not be supposed that this result is in any sense general. The matter will be further elucidated farther on, especially in dealing with numerical examples. It is worth noting here, however, that if we substitute in the original equation

$$\frac{dx_1}{dt} = a_{11}x_1 + a_{12}x_2 + a_{112}x_1 x_2 \tag{20}$$

the relation (165), we find

$$\frac{dx_1}{dt} = (a_{11} + \rho_1 a_{12})x_1 + a_{112}\rho_1 x_1^2 \,. \tag{166}$$

But, according to (106),

$$\rho_1 a_{12} = \lambda_2 - a_{11} \,, \tag{167}$$

while, according to (108), (115)

$$a_{112}\rho_1 = a_{112}\frac{q}{p} = -\frac{\lambda_2}{p} \,. \tag{168}$$

Hence

$$\frac{dx_1}{dt} = \lambda_2 x_1 \left(1 - \frac{x_1}{p}\right). \tag{169}$$

Integrating,

$$x_1 = \frac{p}{1 - \frac{\xi_0}{x_0} e^{-\lambda_2 t}} \tag{170}$$

a result identical with (137).

In general the supposition that $x_2/x_1 =$ const. is, of course, doubly unwarranted. For in the first place the initial values of x_1 and x_2 are wholly arbitrary, and are not in any way connected. Secondly, even if the initial values are arbitrarily established in the ratio ρ_1 , they will not, in general, retain that ratio, though they will, in the special circumstances here taken in view, do so approximately. This will also appear clearly in the numerical example to be considered later.

Particualr Solution in Reciprocal Form. The approximate solution (131) not only possesses a certain direct interest in itself, but it also points a way to express the exact (particular) series solution in a form that will in some cases be more rapidly convergent than the series in its original form. For, if the particular solution

$$x_1 = P_2 e^{\lambda_2} + P_2^2 \rho_1 \varphi_2 e^{2\lambda_2 t} + P_2^3 \rho_1 \varphi_2 \frac{\rho_1 + \rho_2}{2} \varphi_3 e^{3\lambda_2 t} + \ldots$$

be written in the form

$$x_1 = \frac{P_2 e^{\lambda_2 t}}{1 + \alpha e^{\lambda_2 t} + \beta e^{2\lambda_2 t} + \gamma e^{3\lambda_2 t} + \ldots} \tag{171}$$

$$= \frac{P_2}{e^{-\lambda_2 t} + \alpha + \beta e^{2\lambda_2 t} + \ldots} \tag{172}$$

it is found that

$$\alpha = - \rho_1 \varphi_2 , \tag{173}$$

$$\beta = \rho_1^2 \varphi_2^2 - \rho_1 \varphi_2 \frac{\rho_1 + \rho_2}{2} \varphi_3 , \tag{174}$$

etc.,

and, in the case that ρ_n and φ_n are slowly changing functions of n (as in the case considered above), the coefficients $\beta, \gamma, \ldots$ are evidently small.

This fact also will be clearly exemplified in a numerical example.

For purposes of computation of the series appearing in (171) it may be convenient here to indicate the law of the reciprocal of a power series. We have

$$1/(1 + ax + bx^2 + cx^3 + dx^4 + \ldots) = 1 + \alpha x + \beta x^2 + \gamma x^3 + \delta x^4 + \ldots, \tag{175}$$

where

$$\alpha = - a ,$$
$$\beta = (a^2 - b) ,$$
$$\gamma = - (a^3 - 2ab + c) , \tag{175a}$$
$$\delta = (a^4 - 3a^2 b + 2ac - d + b^2) ,$$

etc.

Convergence of the Series Solution. No attempt will be made to establish here in a general way the conditions for the convergence of the series (52), (53). The correct fit of the solution will, in the numerical computations that follow, be tested by direct substitution.

Only two points will here be noted with regard to the conditions of convergence of the series: first, that in the particular solutions whose coefficients are given by (92) to (100), both ρ and φ are *in general* finite for all values of n . For, in order to make ρ_n for example, infinite, we should have to have

$$n = - \frac{\lambda_1}{a_{112}} q .$$ (176)

This condition can not, in general, be satisfied, since n is, in the nature of things, restricted to integral values.

Secondly, for large values of n , both ρ and φ approach finite limiting values, namely

$$\rho_\infty = \frac{a_{212}}{a_{112}} ,$$ (177)

$$\varphi_\infty = \frac{a_{112}}{\lambda_2} .$$ (178)

Topography of Integral Curves

$x_1 x_2$ *Diagram.* It will add materially to our appreciation of the relations involved in the type of systems and processes here considered, if we study by graphic methods the general "lay of the land" in a plot of the integral curves defined by the differential equation

$$\frac{dx_1}{dx_2} = \frac{f_1}{f_2} = \frac{a_{11} x_1 + a_{12} x_2 + a_{112} x_1 x_2}{a_{21} x_1 + a_{22} x_2 + a_{212} x_1 x_2} .$$ (25)

Certain facts are obvious at sight, such as, for example the one already noted, that at the origin $0, 0$, as well as at the singular point $p, q,$ the derivative

$$\frac{dx_1}{dx_2} = \frac{f_1}{f_2} = \frac{0}{0}$$ (29)

is indeterminate.

On the other hand, if $x_1 x_2$ are both very large, the derivative approaches the value

$$\frac{dx_1}{dx_2} = \frac{a_{112}}{a_{212}} = \text{const.} \tag{179}$$

Along the axis of x_1, where $x_2 = 0$, we have

$$\frac{dx_1}{dx_2} = \frac{a_{11}}{a_{21}} \tag{180}$$

and along the axis of x_2, where $x_1 = 0$,

$$\frac{dx_1}{dx_2} = \frac{a_{12}}{a_{22}} . \tag{181}$$

Direction of Entry into Origin. While (25) does not directly give us any information regarding the slope of the integral curves at the origin, since the right-hand member there reduces to the indeterminate form $0/0$, we may in the following manner find the value of this slope, or in other words, of the direction of entry of the integral curves into the origin.

In order that an integral curve may enter the origin we must have, in the immediate neighborhood of the origin (where the terms of second degree, $a_{112}x_1 x_2$, $a_{212}x_1 x_2$ are negligible),

$$\frac{dx_1}{dx_2} = \frac{x_1}{x_2} , \tag{182}$$

$$= K , \text{ say,} \tag{183}$$

$$x_1 = K x_2 . \tag{184}$$

Substitute this in (25), omitting terms of second degree,

$$K = \frac{a_{11}K + a_{12}}{a_{21}K + a_{22}} , \tag{185}$$

$$K^2 + \frac{(a_{22} - a_{11})K}{a_{21}} - \frac{a_{12}}{a_{21}} = 0 , \tag{186}$$

a quadratic for K , so that, in general, there will be two and only two directions of entry of the integral curves into the origin, namely

$$K = -\frac{1}{2a_{21}} \{(a_{22}-a_{11}) \pm \sqrt{(a_{22} - a_{11})^2 + 4a_{21}a_{12}}\} . \qquad (187)$$

Comparing with this equation (57) for λ

$$\lambda = +\frac{1}{2} \{(a_{11}+a_{22}) \pm \sqrt{(a_{11} - a_{22})^2 + 4a_{21}a_{12}}\} , \qquad (57)$$

it is seen that between K and λ there is the relation

$$\lambda - a_{21}K = a_{22} \qquad (188)$$

or

$$K_1 = \frac{\lambda_1 - a_{22}}{a_{21}} \qquad (189)$$

$$= \frac{P_1}{Q_1} \text{ by (72) .} \qquad (190)$$

Similarly

$$K_2 = \frac{\lambda_2 - a_{22}}{a_{21}} \qquad (191)$$

$$= \frac{P_2}{Q_2} \text{ by (73) .} \qquad (192)$$

We have thus found, in (190), (192), a geometrical interpretation of the
constants P_1/Q_1 , P_2/Q_2 . These quotients give the tangents of the two angles
of entry of the integral curves into the origin. The same conclusion is also
reached in another way, which at the same time throws further light on the
situation. Judging from the seemingly symmetrical way in which the two roots K_1,
K_2 are obtained, it might appear that the two directions of entry were essentially
equivalent. As a matter of fact this is not the case. We may distinguish several
cases, as follows:

Case 1. λ_1, λ_2 both real and negative. From the general solution (52), (53), in the case of stable equilibrium, when λ_1 and λ_2 are both real and negative, we have, for very large values of t (when terms of higher than second degree are negligible)

$$\frac{x_1}{x_2} = \frac{P_1 e^{\lambda_1 t} + P_2 e^{\lambda_2 t}}{Q_1 e^{\lambda_1 t} + Q_2 e^{\lambda_2 t}} \quad . \tag{193}$$

If $|\lambda_1| > |\lambda_2|$ this becomes, ultimately,

$$\frac{x_1}{x_2} = \frac{P_2 e^{\lambda_2 t}}{Q_2 e^{\lambda_2 t}} = \frac{P_2}{Q_2} \quad , \tag{194}$$

that is to say, *in general* the actual direction of entry is given by P_2/Q_2 , the ratio of the constants associated with the particular λ which is of lesser absolute value.

What then is the meaning, in this case, of the direction of entry P_1/Q_1? The answer is not far to seek. P_1/Q_1 is the direction of entry of *one* individual integral curve, namely that corresponding to the particular solution

$$x_1 = P_1 e^{\lambda_2 t} + P_{11} e^{2\lambda_2 t} + P_{111} e^{3\lambda_2 t} + \ldots ,$$

$$x_2 = Q_1 e^{\lambda_2 t} + Q_{11} e^{2\lambda_2 t} + \ldots .$$

In other words, one and only one integral curve enters in the direction $\tan \theta = P_1/Q_1$, all others enter along $\tan \theta = P_2/Q_2$.

Case 2. λ_1, λ_2 both positive. Reflections precisely analogous to those set forth regarding case 1 lead here, also, to similar conclusions.

Case 3. λ_1, λ_2 real and of opposite sign. In this case only one integral curve enters along P_1/Q_1 and only one along P_2/Q_2 ; these correspond to the particular solutions containing respectively only P's and Q's with subscripts 1 and those containing only P's and Q's with subscript 2. The integral curves do not *in general* enter the origin at all.

Case 4. If K_1, K_2 are complex, the integral curves "enter the origin at an imaginary angle." Expressing the facts in the language of real geometry, we would say that the integral curves, near the origin, are spirals winding around the origin indefinitely, getting closer and closer to it without ever entering it. From (189), (191) it is at once seen that in this case λ_1, λ_2 also are complex, or, in other words, the solution (52), (53) is oscillatory. This case is excluded by the nature of the malaria constants, as has already been pointed out.

Case 5. A special case arises when the determinant

$$\begin{vmatrix} a_{11} & a_{12} \\ a_{21} & a_{22} \end{vmatrix} = D$$

vanishes. For then (186) may be written

$$\left. \begin{aligned} K^2 + \frac{a_{22} - a_{11}}{a_{21}} \; K - \frac{a_{11}a_{22}}{a_{21}^2} &= 0 \\[2em] \left(K + \frac{a_{22}}{a_{21}} \right) \left(K - \frac{a_{11}}{a_{21}} \right) &= 0 \end{aligned} \right\} \tag{194}$$

$$K_1 = -\frac{a_{22}}{a_{21}} , \tag{195}$$

$$K_2 = \frac{a_{11}}{a_{21}} , \tag{196}$$

$$= \frac{a_{12}}{a_{22}} . \tag{197}$$

The vanishing of the determinant D brings with it that one of the roots of the characteristic equation for λ is zero. This is an exceptional case in which the singular point p, q coincides with the origin, and the series solution (52), (53) fails.

Mapping the Field by Isoclines. Turning now from the special consideration of what happens near the singular points 0, 0 and p,q, and taking in view the entire field of the plane x_1, x_2, we may proceed as follows:

If we select some particular value s of the slope of the integral curves of (25), we have:

$$s = \frac{dx_1}{dx_2} = \frac{f_1}{f_2} = \frac{a_{11}x_1 + a_{12}x_2 + a_{112}x_1x_2}{a_{21}x_1 + a_{22}x_2 + a_{212}x_1x_2} \tag{198}$$

or

$$f_1 - sf_2 = \varphi(x_1, x_2) = 0 . \tag{199}$$

Equation (198) or (199) defines the locus of all those points in the x_1, x_2 plane, at which the slope of the integral curves is s . (Such a locus of line elements of equal inclination is termed an *isocline.)* Multiplying equation (198) across gives

$$x_1(a_{11} - sa_{21}) + x_2(a_{12} - sa_{22}) + x_1x_2(a_{112} - sa_{212}) = 0 , \tag{200}$$

which we may write

$$y_2x_1 + y_1x_2 - x_1x_2 = 0 \tag{201}$$

or

$$(x_1 - y_1)(x_2 - y_2) = y_1y_2 , \tag{202}$$

where

$$y_1 = -\frac{a_{12} - sa_{22}}{a_{112} - sa_{212}} \tag{203}$$

$$y_2 = -\frac{a_{11} - sa_{21}}{a_{112} - sa_{212}} . \tag{204}$$

It is thus seen from (202) that the locus of the points at which the integral curves have slope s (the "isocline s") is a hyperbola passing through the origin, having its asymptotes parallel to the axes of x_1, x_2, and having its center at

$$x_1 = y_1 = - \frac{a_{12} - sa_{22}}{a_{112} - sa_{212}} \; , \tag{205}$$

$$x_2 = y_2 = - \frac{a_{11} - sa_{21}}{a_{112} - sa_{212}} \; . \tag{206}$$

In the particular case that $s = 0$ (locus of maxima or minima of the integral curves) we have, according to (205), (206), for the center of this particular isocline

$$x_1' = - \frac{a_{12}}{a_{112}} \; , \tag{207}$$

$$x_2' = - \frac{a_{11}}{a_{112}} \; . \tag{208}$$

On the other hand, for $s = \infty$, we have

$$x_1'' = - \frac{a_{22}}{a_{212}} \; , \tag{209}$$

$$x_2'' = - \frac{a_{21}}{a_{212}} \; . \tag{210}$$

Putting

$$h = - \frac{sa_{212}}{a_{112}} \; , \tag{211}$$

(203), (204) may be written

$$y_1 = \frac{x_1' + hx_1''}{1 + h} \; , \tag{212}$$

$$y_2 = \frac{x_2' + hx_2''}{1 + h} \; , \tag{213}$$

from which it is seen that the centers of the isoclines (hyperbolas) lie on a straight line passing through the two fixed points x_1', x_2' and x_1'', x_2''. Furthermore, the location of the center of any isocline upon this straight line immediately tells us the slope s characteristic of this isocline; for it is seen from (212), (213) that h is the ratio (internal or external) in which the point y_1, y_2 divides the join of the points x_1', x_2' and x_1'', x_2''. And from this ratio h (easily read from a drawing), the slope s immediately follows by (211):

$$s = - \frac{ha_{112}}{a_{212}} . \qquad (214)$$

From (205), (206), the parametric representation of the locus of the centers of isoclines, we obtain its direct representation by eliminating the parameter s

$$y_1(a_{11}a_{212} - a_{21}a_{112}) + y_2(a_{22}a_{112} - a_{12}a_{212}) + (a_{22}a_{11} - a_{21}a_{12}) = 0 . \qquad (215)$$

Multiplying through by

$$\frac{a_{21}a_{12} - a_{11}a_{22}}{(a_{11}a_{212} - a_{21}a_{112})(a_{22}a_{112} - a_{12}a_{212})}$$

and introducing p, q, according to (36), (37), we find

$$y_2 p + y_1 q - pq = 0 . \qquad (216)$$

This relation we might also have derived directly from (201) by the substitution

$$\left.\begin{array}{l} x_1 = p , \\ x_2 = q , \end{array}\right\}$$

which must satisfy (201) in view of (199). For from (199) it is evident that φ vanishes whenever f_1 and f_2 both vanish separately, as they do at a singular point (equation 29). In other words, all isoclines pass through all singular points.

The form of the equation (216) leads to several interesting conclusions. In the first place we note that if p and q are given, $y_1 y_2$ lie on a straight line, the equation being linear in these quantities. This is only a confirmation

of what (212), (213) have already taught us. It is further to be noted that this

locus of the centers of the isoclines cuts off from the axes of x_1, x_2, the

intercepts p, q. This suggests a simple graphic construction for finding the

second singular point p, q. Plot the points P, Q, having respectively the

coordinates x_1', x_2' and x_1'', x_2'' (equations (207) to (210)); join PQ. Let it

cut the axes of x_1, x_2 at R, S. Complete the rectangle ROST. Then T is the

singular point p, q. (Fig. 2.)

   Method of Plotting Isoclines. The close parallel in the form of equation

(216) for the singular point p, q, and the equation (201) for any point x_1,

x_2 of an isocline suggests that a similar method may be employed to plot the

isoclines.

   We select, accordingly, any point P' on the locus PQ of centers of

isoclines. Through P' we draw any straight line P'R'S', cutting off the

intercepts R'O, OS' from the axes of x_1, x_2. We complete the rectangle R'O'S'T'.

Then T' is a point on the isocline having its center at P'. Any number of such

points may be further obtained by rotating P'Q' about P' and repeating the

construction of the point T'. (Fig. 2)

   In the particular, if the point P' is chosen so as to coincide with P ,

and if the diameter PQ is rotated about P, then T describes the isocline

s = 0, the locus of the maxima (or minima) of the integral curves.

   General Application of Results to Malaria. For the sake of generality

results have so far been developed on the basis of general form of equations (20),

(21), without paying much attention to the special character of the constants

appearing in the malaria equations (18), (19). Before passing on to the study of

numerical examples it will be well to make an intermediate step in the process of

specialization, and to apply the results to the malaria equations in their form

$$\frac{dm}{dt} = - b'f'Cm + b'f'z' - b'f'mz' , \tag{18}$$

$$\frac{dz'}{dt} = b'fAm - b'fBz' - b'fmz' . \tag{19}$$

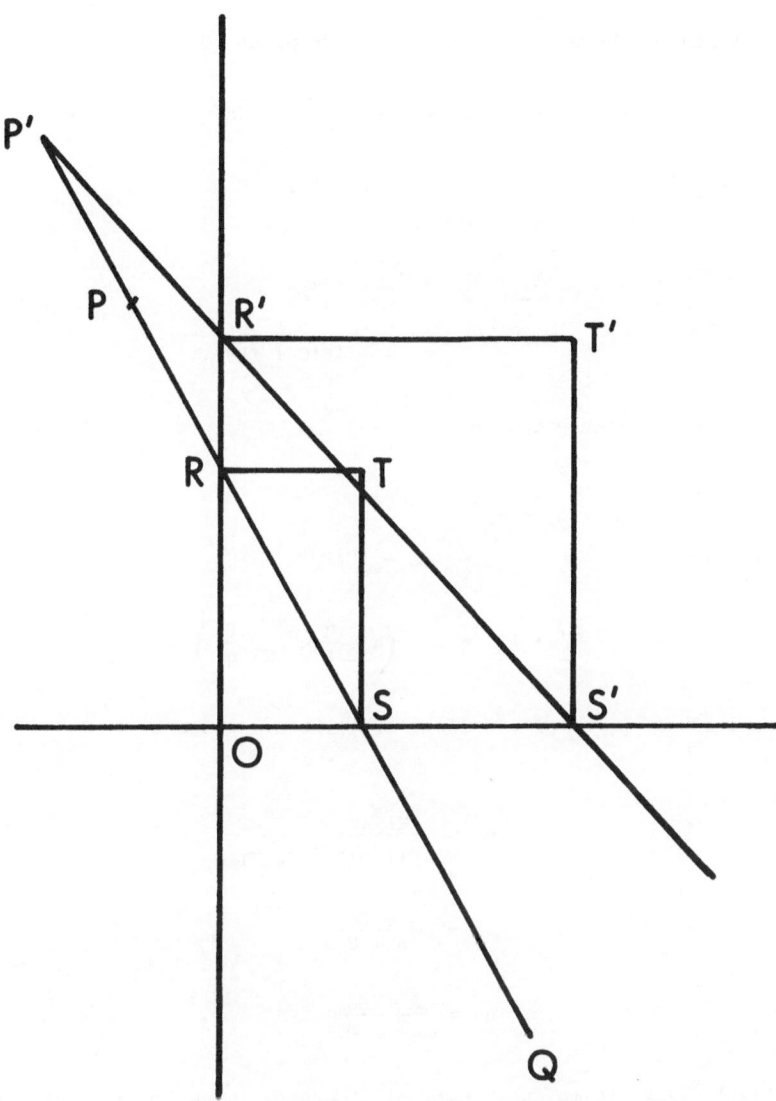

Fig. 2. Graphic construction of equilibrium point T, and of other points T' on an isocline.

It is to be noted that in these equations all constants are essentially positive quantities, by definitions (1) to (6).

We have, therefore, at the singular point 0, 0

$$\lambda_1 \lambda_2 \;=\; (a_{11}a_{22} - a_{21}a_{12}) \tag{65}$$

$$\;=\; (BC - A)ff' \tag{218}$$

and

$$\lambda_1 + \lambda_2 \;=\; a_{11} + a_{22}$$

$$\;=\; -b'(f'C + fB) \,. \tag{219}$$

While at the singular point p, q

$$\lambda_1' \lambda_2' \;=\; -\lambda_1 \lambda_2$$

$$\;=\; (A - BC)ff' \tag{220}$$

$$\lambda_1' + \lambda_2' \;=\; -\left(a_{12}\frac{q}{p} + a_{21}\frac{p}{q}\right).$$

The coordinates p, q of the singular point are given by

$$p \;=\; \frac{a_{21}a_{12} - a_{11}a_{22}}{a_{11}a_{212} - a_{21}a_{112}}$$

$$\;=\; \frac{A - BC}{A + C} \tag{221}$$

$$q \;=\; \frac{A - BC}{1 + B} \,. \tag{222}$$

From (221), (222) it is seen that p, q are positive if, and only if,

$$A \;-\; BC > 0 \,. \tag{223}$$

But in that case, by (220), λ_1' and λ_2' are both of the same sign, and by (221) this sign is negative, so that the equilibrium is stable.

Thus it is seen that whenever the singular point p, q lies in the positive quadrant it is a position of stable equilibrium; and by similar reasoning it is easily shown that the origin is then a point of unstable equilibrium, according to equations (218), (219).

If, on the contrary,

$$A - BC < 0 \, , \tag{224}$$

then the singular point p, q lies in the negative quadrant, and does not, therefore, correspond to any real equilibrium at all, since malaria rates can not assume negative values; the origin, on the contrary, is seen, by reasoning similar to that followed above, to be in this case a point of stable equilibrium.

The limiting case

$$A - BC = 0 \, , \tag{225}$$

to which attention has already been drawn by Sir Ronald Ross, makes the two singular points coincide at the origin, which is then a point of stable equilibrium.

The practical significance of these observations is obvious: Whenever the origin is a point of stable equilibrium, that is to say, whenever

$$A - BC \leqq 0 \, , \tag{226}$$

malaria can not sustain itself in the system, but will diminish and ultimately disappear entirely. This also is one of the important conclusions originally pointed out by Ross.

Graphic Construction for Malaria Equilibrium.   In drawing, for the malaria equations in their form (18), (19), the equilibrium diagram, we have, for the coordinates of the point P , according to (209), (210)

$$x_1' = + 1 \tag{227}$$

$$x_2' = - C \, , \tag{228}$$

and for those of the point   Q

$$x''_1 = - B ,$$ (229)

$$x''_2 = + A .$$ (230)

The construction is carried out in Fig. 3.   From a consideration of the
similar triangles   PWS, PMQ   it is easily seen that the intercepts   RO, OS,   and
therefore the coordinates   p, q   of the point   T   are

$$p = \frac{A - BC}{A + C} ,$$

$$q = \frac{A - BC}{1 + B} ,$$

as required according to (221), (222).

The numerical values assumed as an example for purposes of this construction
are:

$$b' = 0.237 \text{ (per month)},$$
$$A = 3.541 \text{ i.e., } a = 11.671$$
$$B = 5.556,$$
$$C = 0.292 .$$

These numbers were arbitrarily so chosen as to give an equilibrium malaria rate of
50 per cent. in the human population.

It is of particular interest to know how the malaria equilibrium is
affected when   A , the ratio of the mosquito population   p'   to the human population p, is varied, all other constants remaining unchanged.   If   A   only is
changed, this means that the point   Q   of the diagram moves horizontally while the
point   P   remains fixed.   In other words,   PQ   rotates about   P .   The point   T
then describes a hyperbola, which, in these circumstances, is the locus of the
singular point   p, q,   and is also at the same time the isocline   s = 0 , since
rotation is taking place about the point   P, for which   h = 0.   All other isoclines
must change as   T   moves, since they must bend around to pass through   T .   Thus
the isocline   s = 0 , and it only, is common to all systems of isoclines drawn for

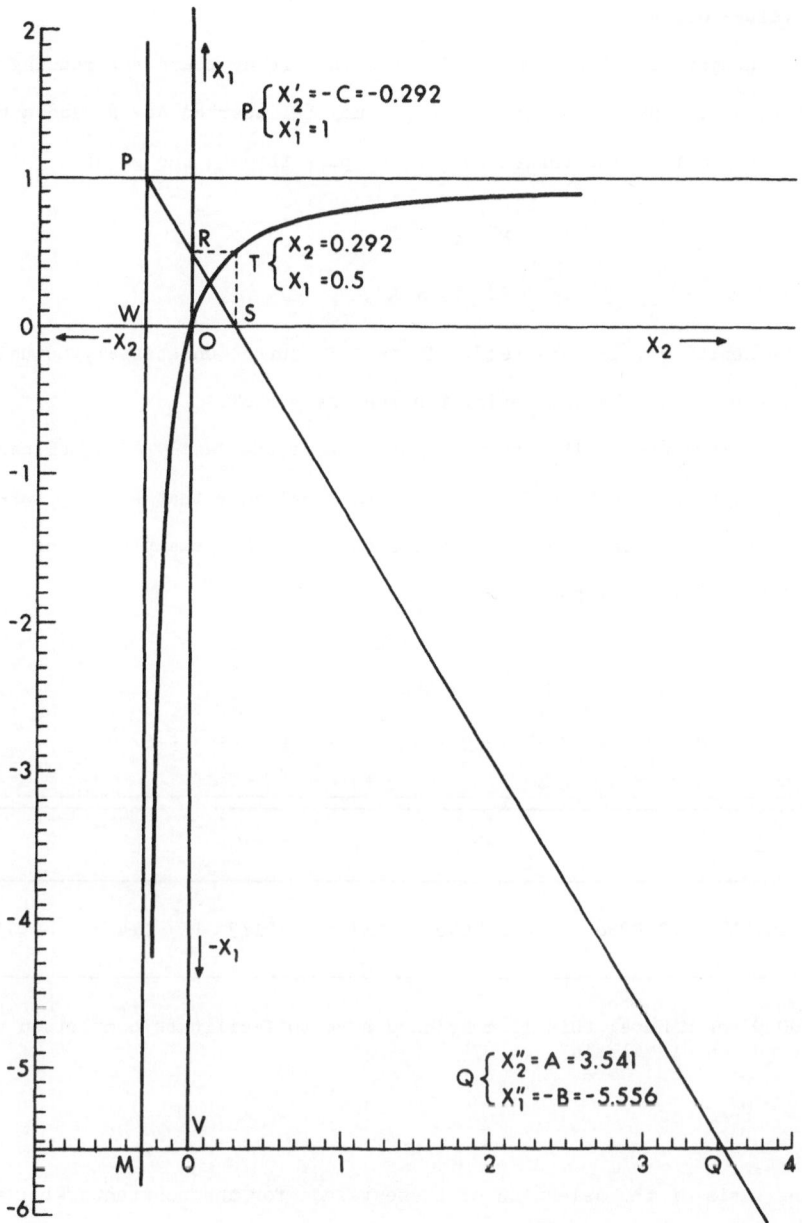

Fig. 3. Equilibrium diagram drawn for the case b' = 0.237; other constants as indicated on the drawing and in Table 1.

different values of A .

This graphic construction furnishes a very ready means for reading off the values of p, q for any value of A. A pin may be inserted at P and a black thread attached to it. The thread is made to pass through the point

$$x_1'' = - B ,$$

$$x_2'' = + A ,$$

and the intercepts OR, OS are read. It is, of course, unnecessary actually to construct the point T by completing the rectangle ROST.

Map of Isoclines. The general character of the family of isoclines is seen in Fig. 4, which has been drawn for a numerical case that will be taken up in detail in a later section. The constants chosen for this example are as follows, the unit of time being one month:

TABLE 1

Constants used in construction of isoclines, Fig. 4.

a	A	N'	r	f	f'	β	b'
64	19.418	3.2958	0.231046	1/4	1/3	1/4	0.82396[*]

[*] b' = .000019 per minute; this is mentioned here to facilitate comparison with Ross, "Prevention of Malaria", p. 674.

The basis of the selection of these values for the constants will be set forth in the numerical part of this series of papers.[*]

---

[*] *Editors' note:* Parts II, III, and V are not included in the present volume. The reader is referred to the original papers for a numerical treatment of the general equations.

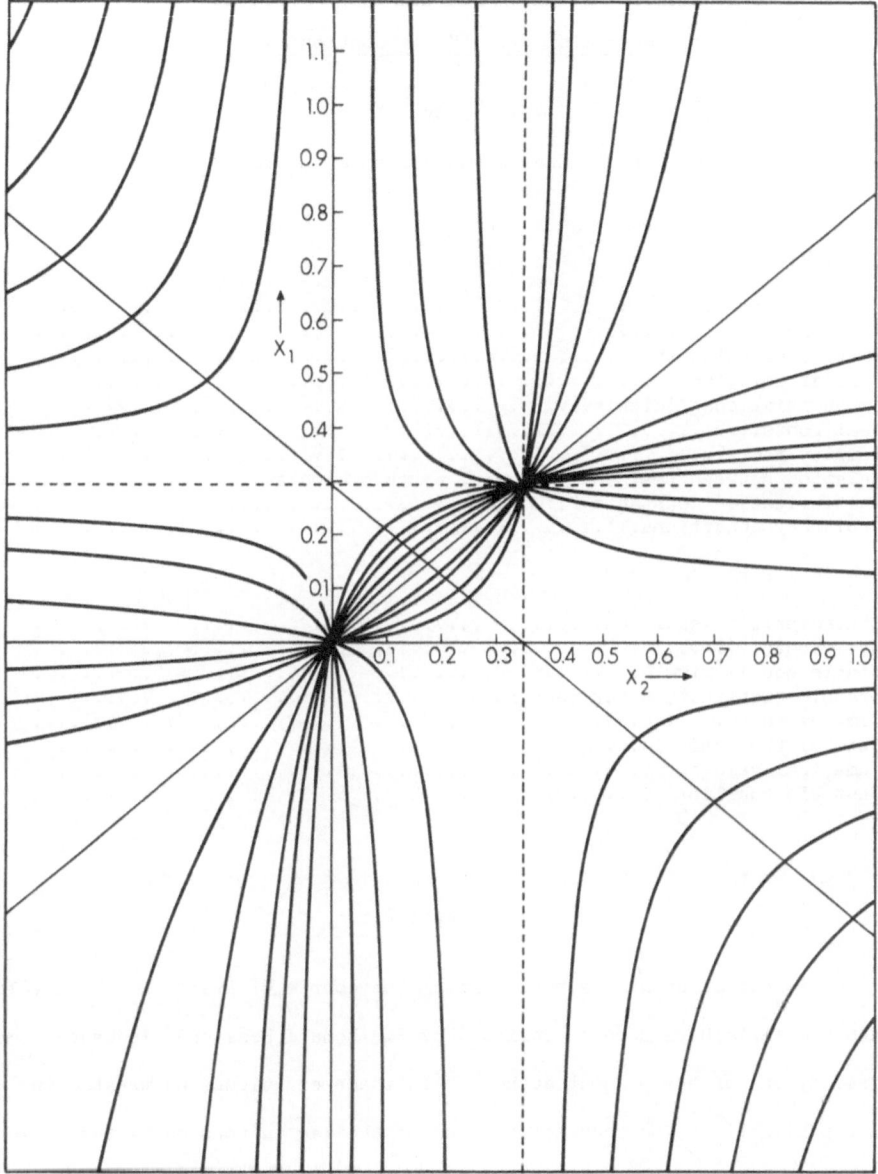

Fig. 4. Map of isoclines defined by the malaria equations.
Drawn for values of the constants shown in Table 1.
Curves in negative portions of the field, and where
$x_1 > 1$, have only geometrical significance.

Contribution to the analysis of malaria

epidemiology. IV. Incubation lag

By

F. R. Sharpe and Alfred J. Lotka

A. General Part

Differential equations representing the course of events in the development
of endemic malaria have been established by Sir Ronald Ross and discussed somewhat
in detail by one of the present authors. In these equations, as treated in
anterior publications, no account is taken of the lag introduced by the period of
incubation of the malaria parasite. If we take into account the periods of incubation u, v, respectively, of this parasite in man and in the mosquito, the basic
equations, though otherwise unchanged in form, now contain the dependent variables
taken, not only at time t (the independent variable), but also at times (t-u),
(t-v).

The mode of development of the basic equations remains essentially

unchanged when account is thus taken of the incubation lag, and it will therefore
be unnnecessary to repeat here the presentation of that development. It will
suffice to recall the basic relation

$$\left\{ \begin{array}{l} \text{Rate of} \\ \text{increase in} \\ \text{number of} \\ \text{affected} \\ \text{individuals} \end{array} \right\} \quad = \quad (\text{Infection rate}) - (\text{Death rate}) - (\text{Recovery rate})$$

which, in analytical form, appears as

$$\frac{dz}{dt} \;=\; \frac{b'f'z'}{p}\,(p-z) - Mz - rz \quad . \tag{1}$$

The symbols z, z' here denote, respectively, the number of affected individuals
in the human population p and in the mosquito population p' . It should be
noted, however, that in the first term in the right-hand member z, z' are to be
taken at time (t-u), because infections taking place at that time actually become
effective only at time t .

Similarly in the first term on the right of the corresponding equation

$$\frac{dz'}{dt} \;=\; \frac{bfz}{p'}\,(p'-z') - M'z' - r'z' \quad , \tag{2}$$

the variables z, z' are to be taken at time (t-v) .

Taking due account of this we may proceed as follows: To simplify our
notation we write

$$\frac{z}{p} \;=\; x \quad , \tag{3}$$

$$\frac{z'}{p} \;=\; y \quad . \tag{4}$$

For x(t) we shall simply write x , similarly for y(t) we shall write y .
But we shall denote x(t-u) by x_{-u}, x(t-v) by x_{-v}, y(t-u) by y_{-u}, and
y(t-v) by y_{-v} . Following the same steps as set forth in the first paper of
this series we obtain, in place of the equations here designated as (20), (21),

$$\frac{dx}{dt} = Ax + By_{-u} + Cx_{-u}y_{-u} ,$$
(5)

$$\frac{dy}{dt} = ax_{-v} + by + cx_{-v}y_{-v} ,$$
(6)

the coefficients A, B, etc., being identical with a_{11}, a_{12}, etc., of the former
notation.

Equilibria. It should be noted that the equilibria defined by

$$\frac{dx}{dt} = \frac{dy}{dt} = 0$$
(7)

are independent of the lags u,v. For when x,y are constant,

$$x_{-u} = x$$
(8)

and similarly for y . At points of equilibrium the equations (5), (6) simply
reduce to the form discussed in the first paper of this series. It was there
observed that there are two positions of equilibrium, namely one at the point

$$x = y = 0$$
(9)

and another at the point

$$x = \frac{aB-Ab}{Ac-aC} = p ,$$
(10)

$$y = \frac{aB-Ab}{bC-Bc} = q .$$
(11)

These, then, remain equilibrium points, though their stability requires new investigation. This will be given in due course. Here it remains to be noted that the equations (5), (6) remain unchanged in form upon introduction of new variables

$$x' = x - p ,$$ (12)

$$y' = y - q .$$ (13)

Or, what amounts to the same thing, equations (5), (6), as they stand, may be applied either to deviations from the equilibrium at 0, 0, or to deviations from the equilibrium at p,q . The values of the constants A, B, C, a, b, c will, of course, be different in the two cases.

Particular solution. A rather obvious expedient in seeking a solution of the system (5), (6) is to make the trial substitution

$$x = P_1 e^{\lambda t} + P_{11} e^{2\lambda t} + P_{111} e^{3\lambda t} \dots ,$$ (14)

$$y = Q_1 e^{\lambda t} + Q_{11} e^{2\lambda t} + Q_{111} e^{3\lambda t} \dots .$$ (15)

This substitution is found to satisfy the system (5), (6), provided that

$$\begin{vmatrix} (A-\lambda) & Be^{-\lambda u} \\ ae^{-\lambda v} & (b-\lambda) \end{vmatrix} = 0$$ (16)

and that, further,

$$\frac{Q_1}{P_1} = \frac{\lambda - A}{Be^{-\lambda u}} = \frac{ae^{-\lambda v}}{\lambda - b} , \tag{17}$$

$$P_{11} = \frac{P_1 Q_1 \{(2\lambda - b)Ce^{-2\lambda u} + Bce^{-2\lambda(u+v)}\}}{3\lambda^2 - (A+b)\lambda - aB(e^{-2\lambda(u+v)} - 1)} , \tag{18}$$

$$Q_{11} = \frac{P_1 Q_1 \{(2\lambda - A)ce^{-2\lambda v} + bCe^{-2\lambda(u+v)}\}}{3\lambda^2 - (A+b)\lambda - aB(e^{-2\lambda(u+v)} - 1)} , \tag{19}$$

with similar equations relating to P_{111}, Q_{111}, etc.

Equation (9), in developed form, is

$$\lambda^2 - (A+b)\lambda + Ab = aBe^{-(u+v)\lambda} = aBe^{-k\lambda} , \tag{20}$$

a transcendental equation having, in general, an infinite number of roots λ_1, λ_2, ... The trial substitution (14), (15) has therefore furnished an infinite number of particular solutions, of which the second, for example, is

$$x = P_2 e^{\lambda_2 t} + P_{22} e^{2\lambda_2 t} + \dots , \tag{21}$$

$$y = Q_2 e^{\lambda_2 t} + Q_{22} e^{2\lambda_2 t} + \dots . \tag{22}$$

General solution. These particular solutions can be united into a general solution, which, however, will contain also product terms as follows:

$$x = P_1 e^{\lambda_1 t} + P_2 e^{\lambda_2 t} + \dots + P_{11} e^{2\lambda_1 t} + P_{12} e^{(\lambda_1 + \lambda_2)t} + \dots$$
$$+ P_{22} e^{2\lambda_2 t} + \dots \tag{23}$$

$$y = Q_1 e^{\lambda_1 t} + \dots . \tag{24}$$

Exponential coefficients λ . It should be particulalry observed that the exponential coefficients λ are fully determined by the linear terms in (5), (6). This enables us to obtain from those terms alone certain significant information regarding characteristic properties of the solution (23), (24).

Real roots λ . From (20) it is clear that the real roots λ are represented by the intersections of a conic with an exponential curve. In general, therefore, there might be either three, two distinct, two coincident, one, or no such real roots. In any numerical example the real roots are easily located either graphically, or, in certain cases, by solving the transcendental equation (20) by successive approximations. This will be illustrated in dealing with a numerical case below.

Complex roots λ . The right hand member of (20) may be expanded into a power series with real coefficients. That equation then appears as a transcendental equation with real coefficients. The complex roots of such an equation occur in conjugate pairs. Let such a pair be

$$\lambda_1 = m + in , \tag{25}$$

$$\lambda_2 = m - in . \tag{26}$$

Substituting these values in (20) we find

$$(m - A + in)(m - b + in) = aBe^{-k(m+in)} , \tag{27}$$

$$(m - A - in)(m - b - in) = aBe^{-k(m-in)} . \tag{28}$$

Multiplication of (27) by (28) gives

$$\{(m - A)^2 + n^2\} \ \{(m - b)^2 + n^2\} = a^2B^2e^{-2km} . \tag{29}$$

Hence

$$|(m - A) (m - b)| < |aB|e^{-km} , \tag{30}$$

a relation which enables us to draw certain conclusions regarding the magnitude of the real parts of the complex roots, as will be seen later, in dealing with a numerical case.

Relation between real and imaginary parts of roots λ . Solving (29) for n we find

$$n^2 = -\frac{(m-A)^2 + (m-b)^2}{2}$$
$$+ \sqrt{\frac{\{(m-A)^2 - (m-b)^2\}^2}{4} + a^2B^2e^{-2km}} . \qquad (31)$$

For large numerical values of m , when m is negative, n approaches the value

$$n = \sqrt{|aB|}\ e^{-km/2} \qquad (32)$$

If it is desired, *vice versa,* to express m explicitly in terms of n , this is most easily effected by writing (27) in trigonometric form

$$(m - A + in)(m - b + in) = aBe^{-km} (\cos kn - i \sin kn) . \qquad (33)$$

Forming, on each side of equation (33), the ratio of the imaginary to the real part, we obtain, after a simple transformation,

$$m = \frac{A+b}{2} - \frac{n}{\tan n} - \sqrt{\frac{n^2}{\sin^2 n} + \left(\frac{A-b}{2}\right)^2} . \qquad (34)$$

For purposes of computation equation (31) was found more convenient than (34).

To determine each complex root a second, independent relation between m and n is required. Such a relation follows immediately from the imaginary part of (33), which gives

$$\frac{2m - (A+b)}{e^{km}} = -\frac{aB \sin kn}{n} . \qquad (35)$$

The equations (31) or (34) and (35) together determine a series of values of m and n . The character of the roots so defined will be considered more in detail in dealing with a numerical example.

Multiplicative constants P, Q

Arbitrary conditions. Not all the constants P_1, P_2, ..., Q_1, Q_2, ... are independent. Of 2n constants with single subscripts P_1, P_2, ..., P_n ; Q_1, Q_2, ..., Q_n, one half, that is to say n , are arbitrary. The remainder are fixed by relations of the form (17) as soon as the values of n of these constants are given. At the same time all the coefficients with multiple subscripts $m \le n$ are fixed by relations of the form (18), (19) or of corresponding form for the higher terms.

There still remain, nevertheless, an infinite number of arbitrary constants to be determined by some condition not considered thus far.

Such conditions are furnished by the fact that a system of equations of the form (5), (6) leaves wholly undetermined the form of the functions

$$x = f_1(t) \tag{36}$$

$$y = f_2(t) , \tag{37}$$

between the limits

$$t = -v \quad \text{and} \quad t = 0 , \tag{38}$$

$$t = -u \quad \text{and} \quad t = 0 , \tag{39}$$

respectively. Any arbitrary form can therefore be imposed upon these functions in those time intervals. The course of events is then fully determined for all subsequent time. Just as, in the case of a system obeying laws expressed by ordinary differential equations, the course of events depends on the initial state *at an instant* of time, so here the course of events depends on the previous history of the system *during a finite interval* of time. This is a state of affairs characteristic of what might be termed, in a general sense, hysteresis effects.

No attempt will be made here to determine in a perfectly general way the relation between the constants P, Q and the arbitrary functions $f_1(t)$, $f_2(t)$ in the prescribed intervals u, v. We shall restrict ourselves to a consideration of the region near equilibrium, where terms of second and higher degree are negligible. In that case we can apply directly an expansion given by Picard[1] and based on a method due to Cauchy, as follows:

Let λ_1, λ_2, ... be the roots of the equation for λ

$$\varphi(\lambda) + \psi(\lambda) = 0 \tag{40}$$

and let φ, ψ satisfy the conditions

1. $$\lim_{\lambda=\infty} \frac{\psi(-\lambda)}{\varphi(-\lambda) + \psi(-\lambda)} = c,$$

2. $$\lim_{\lambda=\infty} \frac{\psi(\lambda)}{\varphi(\lambda) + \psi(\lambda)} e^{\lambda(t-t_2)} = 0,$$

3. $$\lim_{\lambda=\infty} \frac{\varphi(\lambda)}{\varphi(\lambda) + \psi(\lambda)} = C,$$

4. $$\lim_{\lambda=\infty} \frac{\varphi(-\lambda)}{\varphi(-\lambda) + \psi(-\lambda)} e^{\lambda(t_1-t)} = 0. \tag{41}$$

Then any arbitrary function $f(t)$, between the prescribed limits t_1, t_2, can be expanded into an exponential series

$$f(t) = P_1 e^{\lambda_1 t} + P_2 e^{\lambda_2 t} + \dots, \tag{42}$$

and the coefficients P are given by

$$P_i = - \frac{\psi(\lambda_i)}{\varphi'(\lambda_i) + \psi'(\lambda_i)} \int_{t_1}^{t_2} e^{-\lambda_i \mu} f(\mu) d\mu. \tag{43}$$

In the present case, according to (20), (40), we have

$$\varphi(\lambda) = \lambda^2 - (A + b)\lambda + Ab , \qquad (44)$$

$$\psi(\lambda) = - aBe^{-k\lambda} , \qquad (45)$$

and the conditions (1) to (4) are satisfied. Hence

$$P_i = - \frac{aBe^{-k\lambda_i} \int_{-v}^{u} e^{-\lambda_i t} f(t)dt}{2\lambda_i - (A + b) + aBke^{-k\lambda_i}} \qquad (46)$$

where $f(t)$ is given from $t = - v$ to $t = + u$.

This expansion can be applied to the problem of determining x and y for $t > 0$ when $x = f_1(t)$ is given from $t = - v$ to $t = 0$ and $y = f_2(t)$ is given from $t = - u$ to $t = 0$. For, equation (5), omitting the term of second degree, can be integrated when y is known as a function of t . This gives

$$\left. \begin{array}{l} x = x_0 e^{at} + e^{at} \int_0^t bf_2(t-u)e^{-at}dt \\[1em] = f_1(t) \end{array} \right\} \qquad (47)$$

We are now able to form the integral appearing in (46). For from $t = - v$ to $t = 0$ $f_1(t)$ is given; and from $t = 0$ to $t = + u$ it is found by (47). The integral in (46) is thus completely determined, for the function $f_1(t)$, that is to say, for the variable x . In precisely analogous manner the terms of a series for y can be determined. From the manner in which these series have been obtained it is evident that they represent a solution of the system of equations (5), (6), and that they, furthermore, satisfy the condition of giving prescribed values to x and y within the prescribed limits of t .

B. Numerical Part

As a numerical example of the application of the method set forth above, one case has been picked out among those discussed in detail in the first article of this series In that article, however, the incubation lag was left out of account. It is of interest to observe in what respects and to what extent the taking into account of this lag affects the results.

The case picked out for consideration here is that in which, in the notation of the first paper,

$$a = 64 , \qquad A = 19.42 \qquad\qquad (48)$$

that is, there are present 19.42 mosquitoes per head of the human population. All other conditions, and all characteristic constants, have been assumed to have the same values as before. The period of incubation u of the malaria parasite has been assumed to be .5 month in man, and .6 month in the mosquito, so that (u + v) = 1.1; these figures are based on Prevention of Malaria, p. 653.

Characteristic equation

Real roots. For purposes of computing the real roots, the characteristic equation for λ

$$\lambda^2 - (A + b)\lambda + Ab - aBe^{-(u+v)\lambda} = 0 \qquad\qquad (20)$$

may be treated as follows:

Let

$$\omega^2 - (A + b)\omega - Ab - aB = 0 \qquad\qquad (49)$$

The quantity ω thus defined is evidently the value which would be found for λ if the lag were disregarded, or if it were zero. This quantity ω has therefore already been computed in the third paper of this series, where it was denoted by λ .

We now write, however,

$$\lambda = \omega + \varepsilon . \tag{50}$$

Substituting this in (20) we find, in view of (49),

$$\varepsilon^2 - (A + b - 2\omega)\varepsilon + aB(1 - e^{-(u+v)\lambda}) = 0 , \tag{51}$$

$$\varepsilon = \frac{\varepsilon^2 + aB\{1 - e^{-(u+v)\lambda}\}}{A + b - 2\omega} . \tag{52}$$

If ε is small, this gives, for a first approximation,

$$\varepsilon = \frac{aB\{1 - e^{-(u+v)\omega}\}}{A + b - 2\omega} . \tag{53}$$

This furnishes a value ε_1 and $\lambda = \omega + \varepsilon_1$, which, substituted in (52), give, in second approximation,

$$\varepsilon = \varepsilon_2 , \tag{54}$$

$$\lambda = \omega + \varepsilon_2 \tag{55}$$

and so on.

Stable equilibrium. Computations were carried out in detail for the singular point

$$\left. \begin{array}{l} x_1 = p = 0.294 , \\[2mm] x_2 = q = 0.350 , \end{array} \right\} \tag{56}$$

which is of principal interest, since it was found in Part I, disregarding the lag, that this point corresponds to a stable equilibrium.

Numerically equation (20) here appears as

$$\lambda^2 + 3.6837\lambda + 1.0986 = 0.7614e^{-1.1\lambda} \tag{57}$$

and the successive approximations give

	ε	ε^2	λ
1	0.0	0.0	$-0.09391 = \omega$
2	.0237	.000562	$-.0702$
3	.0173	.00300	$-.0766$
4	.0191	.000363	$-.0753$
5	.0186	.0003456	$-.07519$
6	.01872	.0003503	$-.07523$
7	.01868	.0003489	$-.07522$

Graphic construction for real root. The real root of (20) appears as the intersection of the conic (parabola)

$$y = \lambda^2 + 3.6837\,\lambda + 1.0986 \qquad (58)$$

with the exponential curve

$$y = 0.7614 e^{-1.1\lambda} . \qquad (59)$$

The graph of these curves is shown in Figure 1. It is interesting in this graph, to observe the effect of the incubation lag upon the real roots of the characteristic equation for λ . If the lag is disregarded, or is zero, two real roots are found for λ , corresponding to the two intersections of the straight line

$$y = 0.7614 \qquad (60)$$

with the conic (58). The effect of the lag is to sharply tilt upward the left portion of the curve (60) so that one of the real roots is lost altogether (or, to be more precise, is removed to infinity), while the other is not very greatly altered, being changed from -0.09391 to -0.07522 . It should be particularly

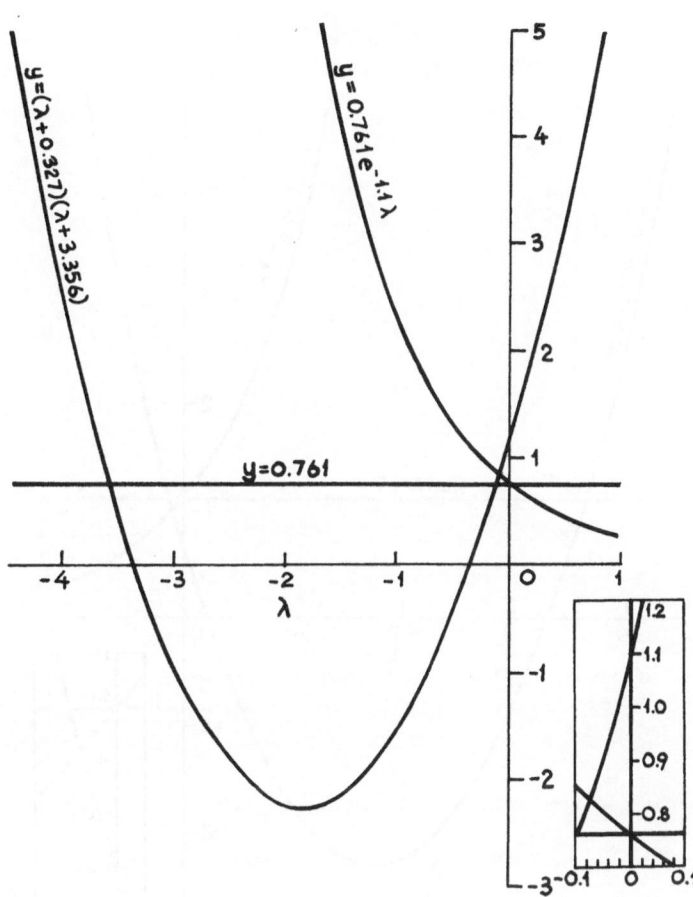

Fig. 1. Graphic construction for real root of characteristic equation
referred to second singular point.

362

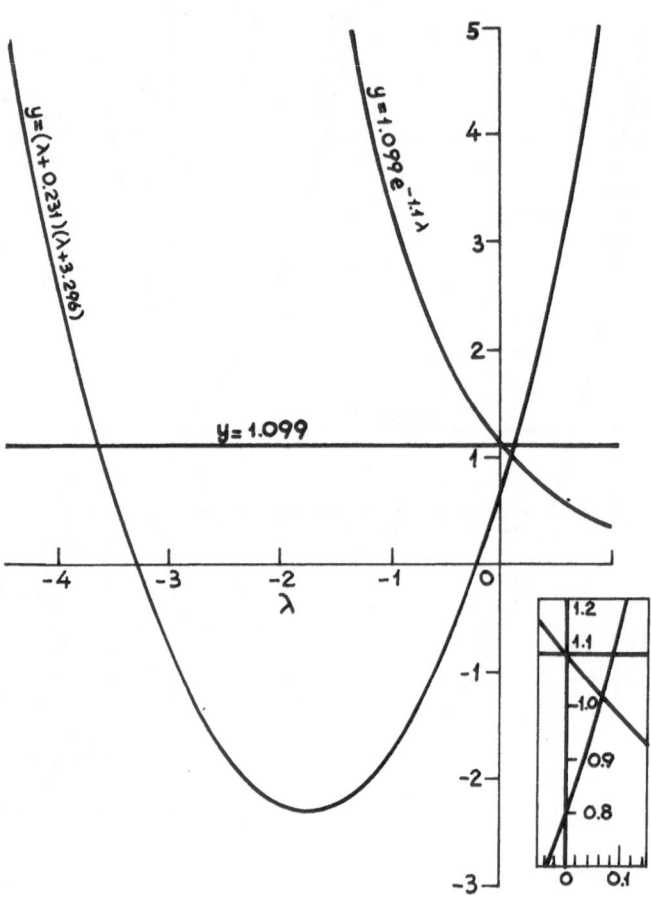

Fig. 2. Graphic construction for real root of characteristic equation
referred to origin.

noted that the sign of this root remains negative, as before. This is important, since the sign of the real parts of the roots λ determine the stability or instability of the equilibrium.

Complex roots

General properties. Applying to the numerical case under consideration the inequality (30), we find, for the real part m of any root λ ,

$$|(m + 0.3273)(m + 3.3564)| < 0.7614e^{-1.1m} , \qquad (61)$$

whereas, for the real root λ_r ,

$$(\lambda_r + 0.3273)(\lambda_r + 3.3564) = 0.7614e^{-1.1\lambda_r} . \qquad (62)$$

A glance at Figure 1 makes it immediately apparent that condition (61) can be satisfied only if m is negative and numerically greater than λ_r . We have, therefore, these significant conclusions:

1. There is one, and only one, finite real root λ_r , which is negative.
2. All the complex roots have their real parts m negative, and numerically greater than λ_r .

These conclusions have a twofold significance, as follows:

Since the real parts of all the roots are negative, the equilibrium is stable. In this case, then, the lag introduced by the period of incubation not only leaves the position of equilibrium unchanged, as compared with the equilibrium computed without regard to the lag, but it also leaves the stability of the equilibrium unaltered.

Since the real root is numerically less than the real part of any of the complex roots, the final stages of the process represented by the system of equations (5), (6) will be controlled essentially by the real root; the oscillatory terms in the solution (14), (15), that is to say, the terms in complex λ's , being damped out more rapidly than the aperiodic term with the real λ .

Complex roots. Numerical values. The numerical values of the complex roots of (20) may be determined either by a process of successive trials, or by graphic methods, or by a combination of these two methods.

For purposes of computation it is expedient to write

$$L = -\lambda - 1.8418 , \qquad (63)$$

$$m' = -m - 1.8418 , \qquad (64)$$

where m denotes, as before, the real part of λ, and m' denotes the real part of L. The characteristic equation (56) then appears in the simplified form

$$L^2 - 2.2939 = 0.7614e^{1.1(L+1.8418)} . \qquad (65)$$

Putting

$$L_1 = m' + in , \qquad L_2 = m' - in \qquad (66)$$

we have

$$m'^2 - n^2 - 2.2939 + 2im'n = 0.7614e^{1.1(m'+in+1.8418)} , \qquad (67)$$

$$m'^2 - n^2 - 2.2939 - 2im'n = 0.7614e^{1.1(m'-in-1.8418)} . \qquad (68)$$

Multiplying (67) by (68) and simplifying gives

$$n^2 = -(m' + 2.2939) + \sqrt{(0.7614)^2 e^{2.2(m'+1.8418)} + 9.175m'} . \qquad (69)$$

On the other hand, selecting the imaginary terms in (67) we have

$$2m'e^{-1.1m'} = \frac{5.775 \sin 1.1n}{n} . \qquad (70)$$

The graph of (69) is shown in Figure 3. On Figure 4 appear two graphs; the first of these represents the left-hand member of (70) plotted as a function of m'. The second curve of Figure 4 is obtained by computing the right-hand member of (70) for successive values of n, reading from the curve of Figure 3 the corresponding values of m', and then plotting the values of the right-hand member of (70) against the corresponding values of m' thus read from Figure 3.

The intersections of the descending limbs of the second curve of Figure 4 with the first curve give the values of m', the real parts of the complex roots of (65). The corresponding imaginary parts can then be read from the curve Figure 3.

In this way were found the following complex roots:

h	λ	m'	1.1n in Degrees	(2h-1)π	Difference	
1	$-2.274 + 2.629i$	0.432	165° 42'	180		
2	$-4.142 + 8.078i$	2.300	509° 20'	540	30°	40'
3	$-5.085 + 13.866i$	3.243	873° 56'	900	26°	04'
4	$-5.700 + 19.637i$	3.858	1237° 38'	1260	22°	22'

It will be observed that the real part of these complex roots is not only greater, numerically, than the real root $\lambda_r = -0.0752$, but even the least of the real parts of the complex roots is 30 times greater (numerically) than λ_r. The result must be that in the course of events the contribution of the complex roots very rapidly subsides to an insignificant amount. Furthermore, although subsequent events are, strictly, dependent upon the previous history of the system, yet, owing to this rapid subsidence of all terms except those containing only the real root, a condition must very soon be reached, after a first brief "transient" period, when the state of the system depends *practically* only on two fundamental constants P_1, Q_1, and is, to this extent, independent of the previous history

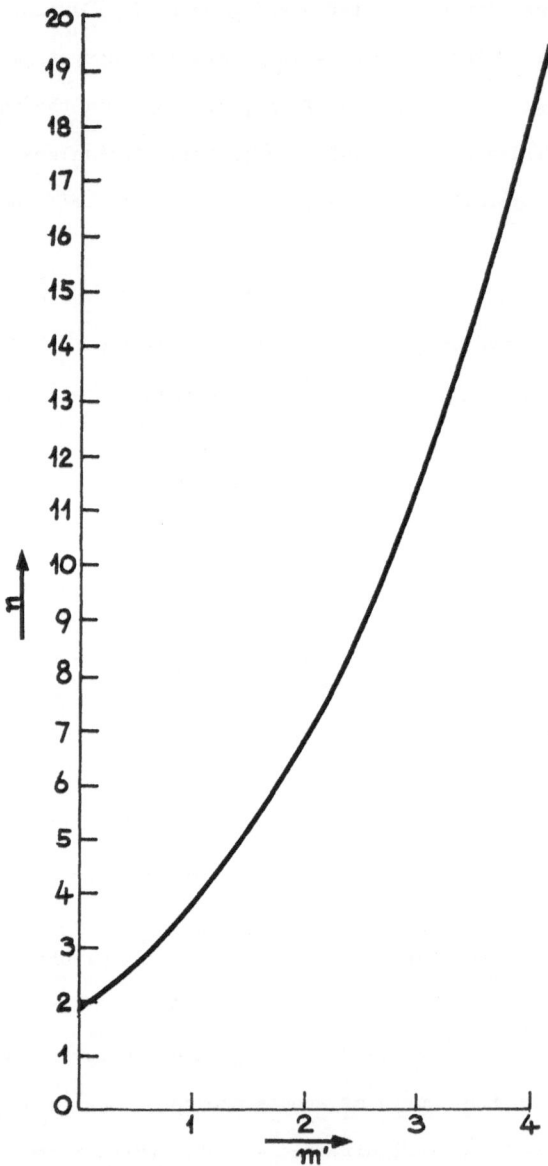

Fig. 3.   Graphic construction for complex root of characteristic
          equation referred to second singular point.   Graph of
          equation (69).

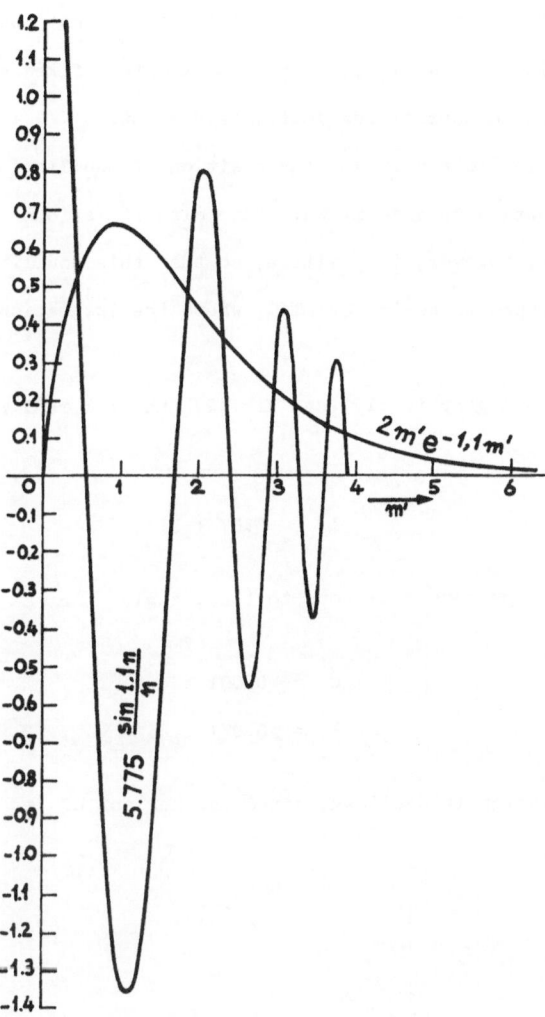

Fig. 4. Graphic construction for complex roots of characteristic equation
referred to second singular point. Solution of equation (70).

of the system. We have here a state of affairs closely analogous to that subjected to mathematical analysis some years ago by the present authors,[2] namely that which must occur in any population starting from an arbitrarily imposed age distribution. Proof was offered, on that occasion, that such a population settles down, under ordinary conditions, to an approximately fixed age distribution independent of any abnormalities initially present.

Unstable equilibrium. For the position of equilibrium at the origin only the real roots were determined. Here also it is found that there is only one real root, which, however, is positive, so that this equilibrium is unstable, just as it was found to be in Part III, where the incubation lag was left out of account.

It was found graphically (see Fig. 2) that the real root is located near $\lambda = 0.07$. Writing

$$\lambda = 0.07 + \varepsilon$$

it was found by a process analogous to (52), neglecting ε^2, that

$$\varepsilon = 0.001 ,$$
$$\lambda = 0.071 .$$

Further approximation is needless, since the neglect of ε^2 is justified by the value found.

Footnotes:

[1] Picard, E., Traité d'Analyse, Vol. 2, 1893, pp. 167-175.

[2] Phil. Mag., April, 1911, p. 435; Proc. Natl. Acad. Sci. 1922, p. 339.

SYMBIOSIS, PARASITISM

AND

EVOLUTION

V. A. KOSTITZIN

To my wife

TABLE OF CONTENTS

MATHEMATICAL NOTES

I. GENERALITIES

1. My purpose here is to elucidate the mathematical theory of symbiosis and parasitism with some remarks on the theory of evolution. The equations which I consider are little more than an initial and very gross approximation. The difficulties one encounters in trying to construct a rational or mathematical biology are enormous, and the prejudices almost discouraging. The first and the largest of these difficulties is the almost complete absence of quantitative data. With the exception of a few fortunate cases of practical importance, one finds no data, even imprecise and scattered in time, on animal or plant associations. Consequently, all attempts at theory are severely limited by the impossibility of experimental verification.

There are also large mathematical difficulties. Even the simplest biological problem can give rise to equations which are difficult to manage. One is often tempted to use simplifications in order to arrive at approximate solutions. However, such solutions may have little correspondence with biological reality, and one derives conclusions from them which are often biologically absurd.

Such difficulties, although real and serious, are not insurmountable. In fact, such difficulties must surely mark the birth of any rational science. We find them again here with all the prejudices, all the objections, and all the groping. History repeats itself.

2. The credit for having first translated a biological problem into equations belongs to A. J. Lotka (8). Following a series of notes by W. R. Thompson (15) on the action of human parasites, Lotka carried out his mathematical analysis by constructing a system of first order differential equations and then displaying the nature of their solutions. All of the terms to be found in later publications, including the principle of encounters, already appear in these original equations. Thanks to the great Italian mathematician V. Volterra (16), the brilliant developments of such theories are well known. The importance of Lotka's work, then, exceeds by far the limits of the special problems he attempted to solve.

As an initial orientation, it will be useful to follow Lotka'a reasoning (9).
Assume with W. R. Thompson that there are two interacting species: the host--an
insect pest--and a parasite introduced to destroy it. The parasite lays a certain
number of eggs on the host. Of these, a number h, on the average, will first
develop into larvae and then molt into free-living adults. Each deposition of eggs
by a parasite leads ultimately to the loss of one individual of the host species.
Making use of such hypotheses Lotka treated the vital balance of the two species
in the form of equations.

Let N_1 be the number of parasite-free hosts, and N_2 the number of free-living,
adult parasites looking for a host on which to lay eggs. The number N_1 increases
by $n_1 N_1$ births per unit time. It decreases by $m_1 N_1$ deaths which are not caused by
the parasites, and by $\alpha N_1 N_2$ deaths which are caused by the parasites. All told, the
total change of N_1 per unit time is represented by the following equation

$$\frac{dN_1}{dt} = n_1 N_1 - m_1 N_1 - \alpha N_1 N_2 = N_1 (n_1 - m_1 - \alpha N_2). \tag{1}$$

On the other hand, the birth rate of the parasite is $h\alpha N_1 N_2$ and the number of
their deaths is $m_2 N_2$. The total rate of change of the number of parasites per unit
time is then represented by the equation

$$\frac{dN_2}{dt} = h\alpha N_1 N_2 - m_2 N_2 = N_2 (h\alpha N_1 - m_2). \tag{2}$$

This system of first order differential equations allows us to compute N_1 and N_2
as a function of time, t, once their initial values are specified.

3. A few remarks are necessary before discussing these equations. The term
$\alpha N_1 N_2$ corresponds to the principle of encounters or adverse interaction. A term
of this sort may be introduced whenever coexisting species exhibit some form of
adverse interaction. In addition to adverse interactions among species, one must
also consider adverse interactions among individuals within the same species. One
could then complete the equations (1) and (2) by the addition of quadratic terms,
so that they become

$$\frac{dN_1}{dt} = (n_1 - m_1)N_1 - \alpha N_1 N_2 - \beta_1 N_1^2$$

$$\frac{dN_2}{dt} = h\alpha N_1 N_2 - m_2 N_2 - \beta_2 N_2^2. \qquad\qquad (3)$$

Such quadratic terms are all the more necessary since, if the parasites were not present, equation (1) would become

$$\frac{dN_1}{dt} = (n_1 - m_1)N_1$$

in which case the host species would increase without bounds if $n_1 > m_1$ and disappear if $n_1 < m_1$. The introduction of the new quadratic term is, therefore, quite appropriate. One could easily criticize both the structure of these terms, and more generally, the principle of encounters. However, two facts are evident: (1) the structure chosen by Lotka is the simplest and most manageable, and (2) the absence of the new quadratic terms will always lead to imprecise conclusions.

We could as well find fault with the birth and death rates, n and m, respectively. We know that the death rate is never constant, and that it depends on age. In man, for example, mortality is very high among infants and young children. Later it declines, reaches a minimum, and then rises again in later adulthood. In certain cases the hypothesis of constant mortality does not effect the results. In other cases, to the contrary, it produces a flagrant contradiction of reality and common sense. This could be improved by introducing age as a complementary, independent variable, at the cost of an extraordinary increase in complexity. This procedure was employed to study a case of parasitism (5), based on the numerical data of C. Pérez (11). Alternatively, one could subdivide the species into two or more age classes, and assume that the death rate is constant within each class. The latter procedure could be utilized to study symbiosis.

There is a further, serious objection against the supposed constancy of the coefficients, hereafter called vital coefficients. Such vital coefficients are affected by either periodic or non-periodic variations. The origins of such variations can be found in phenomena connected with climate, physiology, astronomy, etc., and they are by no means negligible. In certain cases the variation can be eliminated by substituting average values for the vital coefficients. In other cases, to the contrary, the variations themselves are of greater interest, since changes in a species may arise primarily from their effects. In addition, one must not forget that parasitism and symbiosis always deeply affect the vital coefficients of such interacting species, and, therefore, the stability of their associations.

Regarding periodical phenomena, one can point to a profound difference between the equations representing biological phenomena, and those representing the motion of a system of material points. In problems of mechanics an approximate concordance of the periods generally implies that a system is unstable. In biological problems, on the other hand, an approximate concordance of the vital periods in an association is considered to be an indication of stability. One sees that mechanical metaphores are not always applicable in the dynamics of biological systems.

4. Let us now study the solution of equations (1) and (2), whose integral is

$$N_1{}^{m_2}N_2{}^{n_1-m_1} = N_1{}^{m_2}(0)N_2{}^{n_1-m_1}(0)e^{h\alpha N_1 + \alpha N_2 - h\alpha N_1(0) - \alpha N_2(0)} , \qquad (4)$$

$N_1(0)$ and $N_2(0)$ representing the initial values of N_1 and N_2.

Suppose that the birth rate of the host, n_1, is higher than its death rate, m_1. It is easy to see that the curves C_1, C_2, ... (Fig. 1) representing equation (4) are closed. The numbers N_1 and N_2 move, then, through periodic variations. In this case the activity of the insect parasite is not enough to completely destroy the host. It just limits the indefinite increase which would take place if the host species were alone.

It is interesting to compare this result with the results of equations (3). Two cases may arise:

I. Assume that $\dfrac{n_1 - m_1}{\beta_1} > \dfrac{m_2}{h\alpha}$. There are three equilibrium points:

1) $N_1 = 0$ $\qquad\qquad\qquad\qquad\qquad N_2 = 0$

2) $N_1 = \dfrac{n_1 - m_1}{\beta_1}$ $\qquad\qquad\qquad\quad N_2 = 0$

3) $N_1 = \dfrac{\beta_2(n_1 - m_1) + \alpha m_2}{\beta_1\beta_2 + \alpha^2 h}$, $\qquad N_2 = \dfrac{h\alpha(n_1 - m_1) - m_2\beta_1}{\beta_1\beta_2 + \alpha^2 h}$

among which only one, the last, is stable. If there were no parasites the final state would be $N_1 = \frac{n_1 - m_1}{\beta_1}$. The introduction of the parasites decreases the number of hosts to

$$N_1 = \frac{\beta_2(n_1 - m_1) + \alpha m_2}{\beta_1 \beta_2 + \alpha^2 h} < \frac{n_1 - m_1}{\beta_1}.$$

II. Alternatively, let $\frac{n_1 - m_1}{\beta_1} < \frac{m_2}{h\alpha}$. In this case there are only two equilibrium points:

1) $N_1 = 0$, $\qquad\qquad\qquad N_2 = 0$;

2) $N_1 = \frac{n_1 - m_1}{\beta_1}$, $\qquad\qquad N_2 = 0$.

When $n_1 - m_1 > 0$ the former is unstable and the latter is stable. In this case not only do the parasites fail to decrease the number of their insect hosts, but the parasites eventually disappear completely.

In this way the introduction of parasites is not enough to completely annihilate the hosts. It is necessary to compare this mathematical deduction with reality. In effect, it is not rare to witness the total disappearance, through the action of parasites, of one or another group of animals or plants. This fact, however, can be reconciled with equations (1), (2), (3), and their consequences. In these equations we have necessarily implied the continuity of all variables. In reality N_1 and N_2 are, by their very nature, integer numbers, and they also change only by integer values.

The total destruction of a species is for all practical purposes achieved as soon as the density of its distribution decreases below a limit which can be easily determined in each special case. In a species with separate sexes, for instance, this limit is attained when only one individual (a male or an unfertilized female) is left in an area of such a size that an encounter with a conspecific of the opposite sex is impossible. Figure 1 shows that this possibility can arise for any curve C which passes sufficiently close to the N_2 axis.

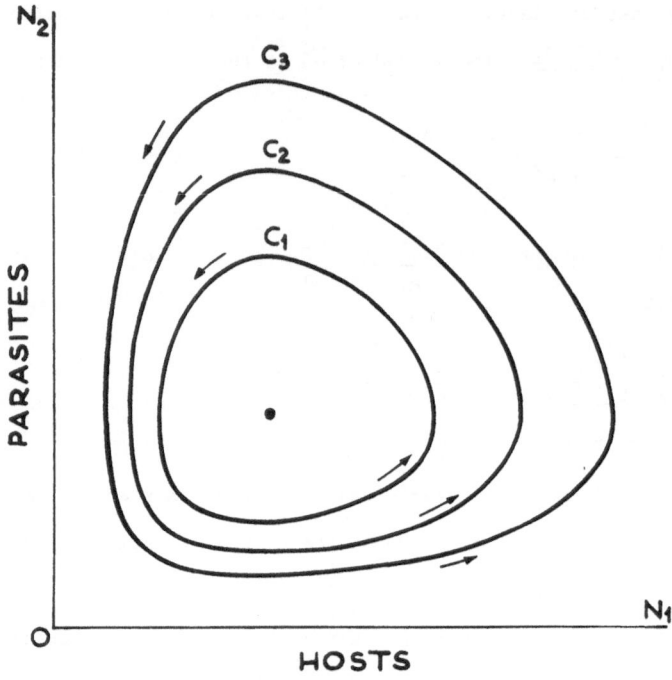

Fig. 1.

II. SYMBIOSIS: CONSTANT VITAL COEFFICIENTS

1. It is impossible to represent the whole variety of symbiotic relationships
(2) by a single type of equation. Certainly there is an enormous difference
between the case of lichens, for instance, and that of Pagurus and Adamsia. One
can, nevertheless, point to a whole series of transitional forms between such
extreme cases. Such problems are all the more important since a number of
biologists (e.g., Famintzine, Merejkovski, Portier) consider all living matter to
be a case of symbiosis among cellular elements. On the other hand, there are
certain sociological theories which look for analogies with human society in animal
and plant associations. Certain biologists with sociological tendencies even go so
far as to state that there can be no evolution without mutualism. Be it as it may,
the question has its importance, and a mathematical study is strongly indicated to
add substance to the analogous mathematical studies in sociology and econometrics.

2. In certain cases it is feasible to assume that the vital coefficients in an association of species do not depend on age. Let x_1, x_2 be the number of free-living individuals in two associated species, x the number of symbiotic couples, n_1 and n_2 the birth rates of the two species in the free state, and ν_1, ν_2 the birth rates in the symbiotic state. Also let m_1 and m_2 be the mortality rates in the free state, while μ_1 and μ_2 are the corresponding rates in the symbiotic state. One can assume that some increase, νx, in the number of symbiotic couples is due to new symbiotic events, that is, associations in statu nascendi. Similarly, there is a complementary mortality of symbiotic couples μ in addition to the specific mortalities μ_1 and μ_2. It might well be, however, that the death of one of the individual partners need not imply the death of the other. The increase in x_1 from such events will then be

$$(n_1 - m_1)x_1 + (\nu_1 - \nu + \beta_1)x.$$

The increase in x_2 will be

$$(n_2 - m_2)x_2 + (\nu_2 - \nu + \beta_2)x;$$

and for x,

$$[\nu - (\mu + \mu_1 + \mu_2)]x.$$

In addition to these linear terms one must also introduce second order terms which take into account: (1) that the encounters between free-living individuals of the two species give rise to a certain number $\alpha x_1 x_2$ of symbiotic couples; and (2) that the competition between the species and among individuals also gives rise to a certain number of second order terms.

By incorporating all such terms, the following differential equations are obtained:

$$\left.\begin{array}{l} x_1' = (n_1 - m_1)x_1 + (\nu_1 - \nu + \beta_1)x - d_1 x_1^2 - e_1 x_1 x - (c_1 + \alpha)x_1 x_2 \\[2mm] x_2' = (n_2 - m_2)x_2 + (\nu_2 - \nu + \beta_2)x - d_2 x_2^2 - e_2 x_2 x - (c_2 + \alpha)x_1 x_2 \\[2mm] x' = (\nu - \mu_1 - \mu_2 - \mu)x - \delta x^2 - \varepsilon_1 x_1 x - \varepsilon_2 x_2 x + \alpha x_1 x_2. \end{array}\right\} \qquad (5)$$

These equations are difficult to manage, and the study of their solutions will have to be qualitative (see Mathematical Note III).

One must first derive the equilibrium points by solving the system of algebraic equations:

$$
\left.
\begin{array}{l}
(n_1 - m_1)x_1 + (\nu_1 - \nu + \beta_1)x - d_1 x_1^2 - e_1 x_1 x - (c_1 + \alpha)x_1 x_2 = 0 \\[2mm]
(n_2 - m_2)x_2 + (\nu_2 - \nu + \beta_2)x - d_2 x_2^2 - e_2 x_2 x - (c_2 + \alpha)x_1 x_2 = 0 \\[2mm]
(\nu - \mu_1 - \mu_2 - \mu)x - \delta x^2 - \epsilon_1 x_1 x - \epsilon_2 x_2 x + \alpha x_1 x_2 = 0
\end{array}
\right\} \qquad (6)
$$

Such equations generally admit a set of eight solutions, which is to say that there are eight equilibrium states. Among them only those in the non-negative region of the (x_1, x_2, x) space are relevant. Among these equilibrium states, three are self-evident:

1) $x_1 = 0, \qquad x_2 = 0, \qquad x = 0.$

In this case all groups go extinct. For this to occur three conditions are necessary:

$$
n_1 < m_1, \qquad n_2 < m_2, \qquad \nu < \mu_1 + \mu_2 + \mu,
$$

namely, mortality must be higher than natality in all the groups.

2) $x_1 = \dfrac{n_1 - m_1}{d_1}, \qquad x_2 = 0, \qquad x = 0.$

In this case one species goes completely extinct when the following conditions are satisfied:

$$
n_1 - m_1 > 0, \qquad \nu - \mu - \mu_1 - \mu_2 < \frac{\epsilon_1(n_1 - m_1)}{d_1},
$$

$$
n_2 - m_2 < \frac{(c_2 + \alpha)(n_1 - m_1)}{d_1} \qquad \nu_2 - \nu + \beta_2 > 0
$$

that is, the state is stable when the birth rate of the first species is higher than its death rate, and the birth rates of the other groups are sufficiently small. Notice that the assumption $\nu_2 - \nu + \beta_2 < 0$ is absurd, since ν is always smaller than ν_2.

3) $x_1 = 0, \qquad x_2 = \dfrac{n_2 - m_2}{d_2}, \qquad x = 0.$

This case is wholly analogous to the preceding one.

3. Special case 1. A very important case is the one in which the associated species do not survive in the free state. In this case x_1 and x_2 are zero while $\underline{x}$ may be greater than zero. Equations (6) show that this can only occur if

$$
\nu_1 + \beta_1 = \nu, \qquad \nu_2 + \beta_2 = \nu. \qquad (7)
$$

Now, it is necessarily true that $\nu_1 \geq \nu$, $\nu_2 \geq \nu$, which means that $\beta_1 = 0$ and $\beta_2 = 0$. In this case, then, the death of one of the members of a symbiotic couple also implies the death of the other. The equations (7) become

$$\nu = \nu_1 = \nu_2. \tag{8}$$

In this way there is complete harmony between the birth rates of the two associated groups, which would undoubtedly be the outcome of a long evolution, but which is mathematically represented by sudden or continuous changes of the vital coefficients. Now, in a relatively high number of cases, the equality (8) is not satisfied despite the fact that x_1 and x_2 are virtually zero. In this case the equations (5) can no longer be applied, and one must instead use the equations (10).

Suppose that the conditions (7) and (8) are satisfied, and refer back to equations (6). One has

$$x_1[(n_1 - m_1) - d_1 x_1 - e_1 x - (c_1 + \alpha)x_2] = 0$$

$$x_2[(n_2 - m_2) - d_2 x_2 - e_2 x - (c_2 + \alpha)x_1] = 0$$

$$nx - \delta x^2 - \varepsilon_1 x_1 x - \varepsilon_2 x_2 x + \alpha x_1 x_2 = 0 \quad ,$$

by setting $n = \nu - \mu - \mu_1 - \mu_2$. Let us determine the eight equilibrium points. First of all, those which have been studied already are:

1) $x_1 = 0$, $\qquad x_2 = 0$, $\qquad x = 0$;

2) $x_1 = \dfrac{n_1 - m_1}{d_1}$, $\qquad x_2 = 0$, $\qquad x = 0$;

3) $x_1 = 0$, $\qquad x_2 = \dfrac{n_2 - m_2}{d}$ $\qquad x = 0$,

and we will not deal with them any longer. The next equilibrium point is for us the most interesting:

4) $x_1 = 0$, $\qquad x_2 = 0$, $\qquad x = \dfrac{n}{\delta}$.

One must necessarily have $n > 0$, $\nu > \mu + \mu_1 + \mu_2$, and the stability of this equilibrium is assured if

$$n > 0, \qquad\qquad n_1 - m_1 < \frac{e_1 n}{\delta} , \qquad n_2 - m_2 < \frac{e_2 n}{\delta} . \tag{9}$$

In other words, the symbiotic state must be more favorable than the free state for both species. One can easily see that the stability of this equilibrium point, $x_1 = 0$, $x_2 = 0$, $x = \frac{n}{\delta}$, is incompatible with the stability of the point $x_1 = 0$,

$x_2 = 0$, $x = 0$.

Let us continue the study of the equilibrium points:

5) $x_1 = 0$, $\qquad n_2 - m_2 = d_2 x_2 + e_2 x$, $\qquad n = \delta x + \varepsilon_2 x_2$.

This implies $n > 0$, $n_2 - m_2 > 0$ and

$$x = \frac{n d_2 - \varepsilon_2(n_2 - m_2)}{\delta d_2 - \varepsilon_2 e_2} > 0, \qquad x_2 = \frac{\delta(n_2 - m_2) - n e_2}{\delta d_2 - \varepsilon_2 e_2}.$$

It will be stable when

$$(c_1 + \alpha)x_2 + e_1 x < n_1 - m_1$$

$$\delta d_2 - e_2 \varepsilon_2 > 0.$$

from which

$$\delta(n_2 - m_2) > n e_2.$$

6) In the same way, the equilibrium point

$$x_1 d_1 + e_1 x = n_1 - m_1, \qquad x_2 = 0, \qquad n = \delta x + \varepsilon_1 x_1$$

is stable when the following inequalities are satisfied:

$$(c_2 + \alpha)x_1 + e_2 x < n_2 - m_2, \qquad \delta d_1 - e_1 \varepsilon_1 > 0, \qquad \delta(n_1 - m_1) > n c_1.$$

7, 8) These two equilibrium points satisfy the equations

$$n_1 - m_1 = d_1 x_1 + e_1 x + (c_1 + \alpha)x_2 > 0$$

$$n_2 - m_2 = d_2 x_2 + e_2 x + (c_2 + \alpha)x_1 > 0$$

$$nx = \delta x^2 - \varepsilon_1 x_1 x - \varepsilon_2 x_2 x + \alpha x_1 x_2 = 0.$$

One of the two is stable; the other is unstable.

It is necessary to clarify the conditions for which different equilibrium states are stable at the same time. The stability of the equilibrium $x_1 = 0$, $x_2 = 0$, $x = 0$ is clearly incompatible with that of all the others. Both states (2) and (3) are stable if the following inequalities

$$n_1 - m_1 > 0, \qquad n_2 - m_2 > 0, \qquad n < \varepsilon_1 \frac{n_1 - m_1}{d_1}, \qquad n < \varepsilon_2 \frac{n_2 - m_2}{d_2}$$

$$n_2 - m_2 < (c_2 + \alpha)\frac{n_1 - m_1}{d_1}, \qquad n_1 - m_1 < (c_1 + \alpha) \cdot \frac{n_2 - m_2}{d_2},$$

$$d_1 d_2 < (c_1 + \alpha)(c_2 + \alpha)$$

are satisfied. The equilibrium (4) will be also stable if

$$\delta d_1 < e_1 \varepsilon_1, \qquad \delta d_2 < e_2 \varepsilon_2.$$

On the other hand, the stability of equilibria (5) and (6) is not compatible with that of (4). Even without pursuing this study further, one sees that the numerical growth of the two associated species can take several directions. Which direction is taken will depend both on the vital coefficients and on the initial numbers. The values of the equilibria, however, will depend only on the vital coefficients.

4. <u>Special case 2</u>. Let us examine a very simple case, with the following assumptions:

1) There is no competition between the species, so that
$$c_1 = 0, \qquad c_2 = 0;$$

2) The birth rate, ν, of symbiotic couples is zero;

3) The birth and death rates of free-living individuals are the same as those of symbiotically paired individuals;
$$n_1 = \nu_1, \qquad n_2 = \nu_2, \qquad m_1 = \mu_1, \qquad m_2 = \mu_2;$$

4) The complementary mortality, μ, of symbiotic couples is zero;

5) The coefficients of competition of the symbiotic couples are zero:
$$\delta = 0, \qquad \varepsilon_1 = 0, \qquad \varepsilon_2 = 0;$$

6) The coefficients of intraspecific competition are the same in the two species:
$$d_1 = e_1, \qquad d_2 = e_2.$$

Under such conditions the equations representing the problem take the very simple form
$$x_1' = (x_1 + x)[(n_1 - m_1) - d_1 x_1] + (m_1 + m_2)x - \alpha x_1 x_2$$
$$x_2' = (x_2 + x)[(n_2 - m_2) - d_2 x_2] + (m_1 + m_2)x - \alpha x_1 x_2$$
$$x' = -(m_1 + m_2)x + \alpha x_1 x_2.$$

There are four equilibrium points

1) $x_1 = 0, \qquad x_2 = 0, \qquad x = 0,$

2) $x_1 = a_1 = \dfrac{n_1 - m_1}{d_1}, \qquad x_2 = 0, \qquad x = 0,$

3) $x_1 = 0$, $\qquad x_2 = a_2 = \dfrac{n_2 - m_2}{d_2}$, $\qquad x = 0$,

4) $x_1 = a_1$, $\qquad x_2 = a_2$, $\qquad x = a = \dfrac{\alpha a_1 a_2}{m_1 + m_2}$.

The first three points are unstable, but the fourth is stable.

Suppose now that no symbiosis took place until a certain time t_o. The numerical change in the two species is then described by the differential equations

$$x_1' = x_1(n_1 - m_1) - d_1 x_1^2$$
$$x_2' = x_2(n_2 - m_2) - d_2 x_2^2.$$

These equations are easily solved, and one finds that, for sufficiently large t, x_1 tends toward the limit $a_1 = \dfrac{n_1 - m_1}{d_1}$, and x_2 towards the limit $a_2 = \dfrac{n_2 - m_2}{d_2}$. Now, suppose that such a limit state already exists at time t_o, and that at this moment symbiosis became possible. There is an initial decline in the numbers of free individuals, but, eventually, x_1 and x_2 return to their initial values.

To simplify matters we assumed that symbiosis appeared suddenly. However, it would be an easy matter to consider the case in which α was a function of time, and equal to zero for $t < t_o$.

III. SYMBIOSIS: VITAL COEFFICIENTS VARY WITH AGE.

1. In the preceding chapter we assumed that the vital coefficients were constant. We will examine now the important case in which these coefficients change with age. This change can be continuous, as in mammals, or discontinuous, as in animals with discrete metamorphosis. In both cases each species can be subdivided into two or more groups, each having different vital coefficients. In this way equations are obtained which are more complex than (5), but also more flexible.

For the sake of simplicity assume that each species is made up of two groups:

1) A group of younger individuals, y, which do not reproduce;

2) A more mature group, z, which increases per unit time by the addition of ky individuals coming from the first group.

Suppose, furthermore, that the terms which account for competition are zero for the first group. Under these conditions it is easy to derive the following system of differential equations:

$$y_1' = n_1 z_1 + (\nu_1 - \nu)x - (m_1 + k_1)y_1 - \alpha y_1 y_2 - \gamma_1 y_1 z_2 \qquad (10)$$

$$y_2' = n_2 z_2 + (\nu_2 - \nu)x - (m_2 + k_2)y_2 - \alpha y_1 y_2 - \gamma_2 y_2 z_1$$

$$z_1' = k_1 y_1 + \beta_1 x - \tau_1 z_1 - \gamma_2 y_2 z_1 - d_1 z_1^2 - e_1 z_1 x - (\beta + c_1)z_1 z_2$$

$$z_2' = k_2 y_2 + \beta_2 x - \tau_2 z_2 - \gamma_1 y_1 z_2 - d_2 z_2^2 - e_2 z_2 x - (\beta + c_2)z_1 z_2$$

$$x' = nx - dx^2 - \varepsilon_1 z_1 x - \varepsilon_2 z_2 x + \alpha y_1 y_2 + \gamma_1 y_1 z_2 + \gamma_2 y_2 z_1 + \beta z_1 z_2.$$

In the last equation, the four final terms take into account the increase in the number of symbiotic couples through encounters of y_1 with y_2, y_1 with z_2, y_2 with z_1, and z_1 with z_2.

To begin with notice the existence of the following equilibrium points:

1) $y_1 = 0$, $\qquad y_2 = 0$, $\qquad z_1 = 0$, $\qquad z_2 = 0$, $\qquad x = 0$.

This state is stable when the inequalities:

$$n < 0 ; \qquad k_1 n_1 < (m_1 + k_1)\tau_1 ; \qquad k_2 n_2 < (m_2 + k_2)\tau_2$$

are satisfied, that is when the death rates are sufficiently large.

2) $x = 0$, $\qquad y_1 = 0$, $\qquad z_1 = 0$, $\qquad y_2 = \dfrac{n_2 z_2}{m_2 + k_2}$,

$$z_2 = \frac{k_2 n_2 - \tau_2 (m_2 + k_2)}{(m_2 + k_2)d_2}.$$

First of all one must have

$$k_2 n_2 > \tau_2 (m_2 + k_2)$$

that is, the equilibrium only exists if the death rate of the surviving species is sufficiently small.

3) $x = 0$, $\quad y_1 = \dfrac{n_1 z_1}{m_1 + k_1}$, $\quad y_2 = 0$, $\quad z_1 = \dfrac{k_1 n_1 - \tau_1 (m_1 + k_1)}{(m_1 + k_1)d_1}$, $\quad z_2 = 0$.

This case is analogous to the previous one.

Rather than continuing to investigate the high number of equilibrium states, it might be convenient to study a few, simple special cases.

2. Special case 1. In this case the number of free-living individuals in each species tends toward 0 while the number of symbiotic couples, x, tends toward $\frac{n}{\delta} > 0$. The equations (10) show that the two fundamental conditions, already established in analogous cases in the previous chapter, still hold. One has then,

$$\beta_1 = 0, \qquad \beta_2 = 0; \qquad \nu_1 = \nu_2 = \nu,$$

which, in this case, mean that:

1) The death of one member of a couple necessarily entails the death of
   the other;

2) An evolutionary process, which undoubtedly was quite long, resulted
   in equal birth rates for each of the two species.

The conditions for stability are

$$n > 0, \qquad (m_1 + k_1)\left(\tau_1 + \frac{e_1 n}{\delta}\right) > n_1 k_1, \qquad (m_2 + k_2)\left(\tau_2 + \frac{e_2 n}{\delta}\right) > n_2 k_2.$$

which may have a variety of biological interpretations.

3. **Special case 2.** Assume that z_1 and z_2 are practically zero, and that the
symbiotic birth rate, ν, is zero. This case arises when the mortalities, m_1 and m_2,
are so large that young individuals which are not protected by symbiosis fail to
attain maturity. Then the equations (10) take the following form:

$$y_1' = \nu_1 x - m_1 y_1 - \alpha y_1 y_2$$

$$y_2' = \nu_2 x - m_2 y_2 - \alpha y_1 y_2$$

$$x' = -\tau x - \delta x^2 + \alpha y_1 y_2$$

with the equilibrium point

$$m_1 y_1 = x(\nu_1 - \tau - \delta x) > 0$$

$$m_2 y_2 = x(\nu_2 - \tau - \delta x) > 0.$$

Set $u = \tau + \delta x$, $\sigma = \dfrac{\delta m_1 m_2}{\alpha}$. It is easy to see that u must satisfy the 3rd
degree equation

$$F(u) = (u - \tau)(u - \nu_1)(u - \nu_2) - \sigma u = 0. \qquad (11)$$

To be explicit, assume that $\nu_1 < \nu_2$. One will then have $\tau < \nu_1 < \nu_2$. When σ is
sufficiently small, equation (11) will have three real roots u_1, u_2, u_3, satisfying
the following inequality

$$\tau < u_1 < u_2 < \nu_1 < \nu_2 < u_3.$$

Of these roots, u_1 corresponds to an unstable equilibrium, u_2 to a stable one,
with u_3 being outside the positive region of the (y_1, y_2, x) space. We have shown
the possibility of a stable equilibrium, even when $\nu = 0$. This case is particularly

interesting when dealing with metamorphosing animals.

4. Special case 3. One can treat in the same way the asymmetrical case in
which z_2 is practically zero while z_1 is not. This can occur, for instance, when
one of the two associated species can only exist free in the larval stage. Assume,
on the other hand, that: 1) the symbiotic birth rate, ν, is zero, and 2) the
competition coefficients satisfy the simple conditions

$$d_1 = e_1 = \sigma\delta, \qquad \delta = \varepsilon_1;$$

3) encounters leading to the formation of symbiotic couples only take place among
young individuals; and 4) the mortality of free-living adults of the first species
is $\sigma\tau$.

The equations (10) take the form

$$y_1' = n_1 z_1 + n_1 x - (m_1 + k)y_1 - \alpha y_1 y_2$$

$$y_2' = n_2 x - m_2 y_2 - \alpha y_1 y_2$$

$$z_1' = k y_1 - \sigma\tau z_1 - \sigma\delta z_1^2 - \sigma\delta z_1 x$$

$$x' = -\tau x - \delta x^2 - \delta z_1 x + \alpha y_1 y_2$$

Set

$$u = \tau + \delta x + \delta z_1, \qquad M = \frac{[(m_1 + k)\sigma - k]m_2 \delta}{\alpha\sigma}$$

Assuming that the symbiotic state is more favorable than the free state, one will
have $\sigma > 1$ and, therefore, $M > 0$. One can compute the equilibrium point by solving
the equation

$$F(u) = (u - t)(u - n_1)(u - n_2) - Mu = 0$$

which is wholly analogous to equation (11) but for one detail: the hypotheses
$\nu_2 > \nu_1$ and $n_2 > n_1$ have somewhat different meanings. In any event one finds, as
in the preceding example, three values of u: $u_1 < u_2 < u_3$ among which only one,
u_2, corresponds to a stable equilibrium.

5. These examples show that the method outlined above can take into account
several forms of symbiosis, beginning with very close and stable associations, as
in the lichens, and leading to cases in which symbiosis is hardly present at all.

The difficulty in studying the stability of the equilibrium states is not insurmountable. Side by side with cases leading to equilibria, one can also imagine cyclic cases, in which the variables oscillate in a periodic or quasi-periodic fashion.

One can also pose and solve more complex problems in which the vital coefficients vary in either sudden or continuous fashion. Such evolutionary processes can lead to new stable equilibria, as well as to periodic or quasi-periodic processes.

IV. PARASITISM

1. One may consider parasitism to be a special case of symbiosis, since it is enough to modify the hypothesis for the vital coefficients in a way that only one of the associated species is favored.

Nevertheless, the problem of parasitism presents a number of interesting peculiarities which deserve an independent study (14). Such is the case studied by A. J. Lotka (8) in which the "symbiosis" is lethal for one of the components. Side by side with this many other forms are possible (2) such as the presence of one or more parasites which, sparing the host, severely alter its vital coefficients. It is from this point of view that the case of multiple parasitism becomes particularly interesting.

2. Multiple parasitism. Multiple parasitism can arise in a variety of ways. One can concieve of it either as a multiple infestation or as the multiplication of a single parasite after entering the body of its host. The two hypotheses lead to different mathematical considerations.

Suppose that a host species, x, is composed of x_0 healthy individuals, of x_1 individuals with one parasite, of x_2 individuals with two parasites, etc. Then,

$x = \sum_{k=0}^{\infty} x_k$ and $z = \sum_{k=1}^{\infty} k x_k$ will respectively represent the total numbers of host and

of parasite. Any biological hypotheses concerning the host and the parasite can be expressed in mathematical terms.

Suppose, for instance, that the adult parasites cannot survive in the free state, but that their larvae do live free for some time, and that they enter their hosts through chance encounters. Suppose, on the other hand, that such larvae are only capable of free life for short periods of time. Under such conditions a number $\Sigma z c_k x_k$ of the hosts is infested per unit time.

The coefficients of contamination, c_k, will generally depend on k, namely on the number of parasites in a host. They are constant if parasitism does not weaken the host. They are zero for $k > 0$ if the presence of a parasite immunizes the host against further infestations. They grow with k if the host, weakened by successive infestations, becomes less and less resistant.

The birth rates n_k of hosts infected by k parasites can be constant if the presence of the parasites does not affect reproduction. On the other hand, it is zero for $k > 0$ if there is parasitic castration, etc. One can make analogous hypotheses on the mortalities, m_k, and on the competition coefficients, δ_{ks}, of hosts affected by k parasites, relative to those affected by s parasites. On the other hand, one can take into account the birth and death rates of parasites within a host, which change the number of parasites, k, it carries. Let $\alpha_k x_k$ be the number of hosts passing from group k to group (k − 1) because of the death of one parasite, and $\beta_k x_k$ the number of hosts passing from group k to group (k + 1) by parasitic reproduction. Taking all into account, one obtains the following differential equations:

$$x_0' = \sum_{s=0}^{\infty} n_s x_s - \tau_0 x_0 + \alpha_1 x_1 - c_0 z x_0 - x_0 \sum_{s=0}^{\infty} \delta_{0s} x_s$$

$$x_1' = -x_1(\tau_1 + \alpha_1 + \beta_1 + c_1 z) - x_1 \sum_{s=0}^{\infty} \delta_{1s} x_s + \alpha_2 x_2 + c_0 z x_0$$

$$x_2' = -x_2(\tau_2 + \alpha_2 + \beta_2 + c_2 z) - x_2 \sum_{s=0}^{\infty} \delta_{2s} x_s + c_1 z x_1 + x_1 \beta_1 \qquad (12)$$

$$\cdots\cdots\cdots\cdots\cdots$$

$$x_k' = -x_k(\tau_k + \alpha_k + \beta_k + c_k z) - x_k \sum_{s=0}^{\infty} \delta_{ks} x_s + \alpha_{k+1} x_{k+1} + x_{k-1}(\beta_{k-1} + c_{k-1} z).$$

It is a system with an infinite number of equations for an infinite number of unknown functions. This system is not linear, and its solution presents almost insurmountable difficulties.

3. <u>Special case 1</u>. Suppose that:

    1) there is no multiple contamination:

$$c_1 = c_2 = \ldots = 0$$

    2) the parasites do not multiply within the hosts:

$$\beta_1 = \beta_2 = \ldots = 0$$

In this case there will only be two groups of hosts, healthy and contaminated ones. The differential equations (12) become

$$x_0' = (n_0 - \tau_0)x_0 + (n_1 + \alpha_1)x_1 - \delta_{00}x_0^2 - (c_0 + \delta_{01})x_0 x_1 \\ x_1' = x_1[-(\tau_1 + \alpha_1) + (c_0 - \delta_{10})x_0 - \delta_{11}x_1].$$ (13)

Suppose, at first, that $c_0 < \delta_{10}$, namely the contaminated hosts are clearly handicapped in competition with healthy hosts. In this case x_1' is always negative, and x_1 tends to zero. The first of equations (13) shows that x_0 tends to $\frac{n_0 - \tau_0}{\delta_{00}}$ when $n_0 - \tau_0 > 0$, and to 0 in the opposite case.

We will not deal with the case $c_0 > \delta_{10}$. The study of the intersections between the straight line

$$-(\tau_1 + \alpha_1) + (c_0 - \delta_{10})x_0 - \delta_{11}x_1 = 0$$

and the hyperbola

$$(n_0 - \tau_0)x_0 + (n_1 + \alpha_1)x_1 - \delta_{00}x_0^2 - (c_0 + \delta_{01})x_0 x_1 = 0$$

is not difficult. Mathematical note III gives all the information which is useful to this end.

4. <u>Special case 2</u>. Suppose that:

    1) The natality, n_k, does not depend on k:

$$n_k = n \quad (k = 0, 1, 2, \ldots);$$

    2) The mortality rates, τ_k, are a linear function of k:

$$\tau_k = \tau + \theta k \quad (k = 0, 1, 2, \ldots);$$

    3) The competition coefficients, δ_k, are all the same:

$$\delta_{ks} = \delta;$$

    4) The coefficient of contamination, c_k, does not depend on k:

$$c_k = c;$$

5) The coefficient of mortality of the parasites is zero:

$$\alpha_k = 0.$$

This hypothesis does not mean that parasites are immortal, but only that they do not die before their hosts;

6) The internal reproduction of parasites follows a simple, although arbitrary, law:

$$\beta_k = k\tau_k - \lambda k = k(\theta k + \tau - \lambda);$$

that is, we suppose that the complementary death rate of the hosts, θ, comes from internal reproduction by the parasites.

Under such conditions the equations (12) take the form

$$\left.\begin{aligned}
x_0' &= -x_0 P_0 + nx \\
x_k' &= -P_k x_k + Q_k x_{k-1} (k = 1, 2, 3, \ldots),
\end{aligned}\right\} \tag{14}$$

setting

$$P_k = \tau_k + \beta_k + cz + \delta x, \qquad Q_k = \beta_{k-1} + cz.$$

It is easy to see that x and $z = \Sigma k x_k$ satisfy a system of two equations

$$\left.\begin{aligned}
x' &= x(n - \tau) - \theta z - \delta x^2 \\
z' &= z[(c - \delta)x - \lambda].
\end{aligned}\right\} \tag{15}$$

Substituting the values of x and z determined by these equations into equations (14), one successively finds x_0, x_1, x_2, $\ldots$ which take the following form

$$\left.\begin{aligned}
x_0 &= x_0(o)e^{-\int_0^t P_0(s)ds} + n\int_0^t x(u)e^{-\int_u^t P_0(s)ds}du \\
x_k &= x_k(o)e^{-\int_0^t P_k(s)ds} + \int_0^t Q_k(u)x_{k-1}(u)e^{-\int_u^t P_k(s)ds}du \;.
\end{aligned}\right\} \tag{16}$$

To begin with, let us study the equations (15). At equlibrium the point $(\underline{x}, \underline{z})$ is at the intersection of the parabola

$$\theta z = x[(n - \tau) - \delta x]$$

and the straight line $x = \frac{\lambda}{c-\delta}$. Let us denote the coordinates of this point by N and S, and set, for simplicity, $M = \frac{n-\tau}{\delta}$. We easily find

$$N = \frac{\lambda}{c-\delta}, \qquad \theta S = \delta N(M - N).$$

The equilibrium point (N, S) is stable if

$$c - \delta > 0, \qquad \lambda > 0, \qquad \frac{M}{2} < N < M.$$

One of the two other equilibrium points,

$$x = 0, \qquad z = 0$$

is only stable for $n - \tau < 0$; the other,

$$x = \frac{n - \tau}{\delta} > 0, \qquad z = 0$$

is only stable for $N > M$. Therefore, it is not possible for more than one of these states to be stable.

In the case $M = 2N$ the system

$$x' = -\delta(x - N)^2 - \theta(z - S)$$

$$z' = (c - \delta)z(x - N)$$

has the integral

$$(x - N)^2 = \theta\left(\frac{S}{S} - \frac{2}{c + \delta} z - Hz^{-\frac{2\delta}{c - \delta}}\right),$$

in which the constant H is determined by the initial values of x and z. The final states of the two associated species are completely different according to whether H is positive or negative. For positive H the point (x, z) follows a closed curve around the point (N, S), so that the solution is periodic. For negative H the process ends up in the complete destruction of the hosts.

It is interesting to note that the periodicity of of x, z does not entail that of x_0, x_1, x_2, ... In this case the coefficients P_k, Q_k in the equations (14) are periodic. Consider the first of them

$$x_0' = -x_0 P_0 + nx$$

and denote by p its period. One can show (cf., Mathematical note II) that x_0 is periodic for any initial value, $x_0(0)$, if P_0 and x satisfy the following integral conditions

$$\int_0^p P_0(s)ds = 0, \qquad \int_0^p x(u)e^{-\int_u^p P_0(s)ds}du = 0.$$

It is clear that such conditions are not satisfied, and, therefore, $x_0(t)$ can only be periodic when its initial value is

$$x_0(0) = \frac{n\int_0^p x(u)e^{-\int_u^p P_0(s)ds}du}{1 - e^{-\int_0^p P_0(s)ds}}.$$

Analogous conditions are found for the variables x_1, x_2, ...:

$$x_k(0) = \frac{\int_0^P Q_k(u)x_{k-1}(u)e^{-\int_u^P P_k(s)ds}du}{1 - e^{-\int_0^P P_k(s)ds}} \; .$$

Only when the x_0, x_1, x_2, ... take such initial values are the functions $x_0(t)$, $x_1(t)$, $x_2(t)$, ... periodical.

After this digression, let us return to the equilibrium case. The equations (16) show that if x and z tend toward an equilibrium point, the same happens to x_0, x_1, x_2, ... One has, in fact,

$$x_0(\infty) = \frac{nx(\infty)}{P_0(\infty)} \; , \qquad x_0(\infty) = \frac{Q_k(\infty)x_{k-1}(\infty)}{P_k(\infty)}$$

or

$$\frac{Q_k(\infty)}{P_k(\infty)} < 1.$$

The series $\sum_{k=0}^{\infty} x_k(\infty)$ converges as does the series $\sum_{k=1}^{\infty} kx_k(\infty)$. Therefore, a limit equilibrium point exists.

This example shows that the final state of a parasitized species can depend not only on the vital coefficients, but also on the initial values of the variables. One sees, on the other hand, that the stable equilibrium corresponds to the vital coefficients satisfying rather restrictive inequalities, and that, if such inequalities are not satisfied, the associated species tend quite rapidly to extinction.

5. Special case 3: Parasitic castration. Let us further simplify the conditions, and suppose that the death rate of a host does not depend on the number of parasites it carries:

$$\tau_k = \tau \qquad (k = 0, 1, 2, 3, \ldots)$$

All the other conditions of the preceding section remain unchanged except those concerning the birth rates of the hosts and the parasites. We assume that $\beta_k = \beta k$. Regarding n_k we will compare two hypotheses:

1) that of parasitic castration:

$$n_0 = n, \qquad n_k = 0, \qquad (k = 1, 2, 3, \ldots)$$

2) that of the previous section:

$$n_k = n \qquad (k = 0, 1, 2, 3, \ldots)$$

In the case of parasitic castration the equations (12) become

$$\left.\begin{array}{l} x_0' = x_0(n - \tau - cz - \delta x) \\[2ex] x' = nx_0 - \tau x - \delta x^2 \\[2ex] z' = z[(\beta - \tau) + (c - \delta)x] \end{array}\right\} \qquad (17)$$

$$x_k' = -P_k x_k + Q_k x_{k-1} \qquad (18)$$

in which

$$P_k = \tau + \beta k + cz + \delta x, \qquad Q_k = \beta(k - 1) + cz.$$

The equations (17) show that an equilibrium point exists if $\frac{\tau - \beta}{c - \delta} > 0$. Denote by N_0, N, S the coordinates of this equilibrium point. A first necessary condition for its stability is

$$c - \delta > 0, \qquad \tau - \beta > 0.$$

One easily finds the second condition:

$$\delta N(\delta N + \tau)(2\delta N + \tau) > c(c - \delta)nSN_0 .$$

Unless both conditions are satisfied, this equilibrium point is unstable.

Suppose, on the contrary, that there is no parasitic castration. The equations (17) are replaced by

$$x_0' = nx - x_0(\tau + cz + \delta x)$$

$$x' = x(n - \tau - \delta x)$$

$$z' = z[(c - \delta)x - (\tau - \beta)]$$

while equations (18) remain unchanged. One easily finds

$$x = \frac{x(0)(n - \tau)e^{(n-\tau)t}}{n - \tau - \delta x(0) + \delta x(0)e^{(n-\tau)t}}$$

so that, in any case, $\lim\limits_{t \to \infty} x = \frac{n - \tau}{\delta}$. One also finds

$$z = z(0)e^{-(\tau - \beta)t}\left[\frac{n - \tau - \delta x(0) + \delta x(0)e^{(n - \tau)t}}{n - \tau}\right]^{\frac{c - \delta}{\delta}} .$$

One sees that z tends toward zero when

$$(n - \tau)(c - \delta) < \delta(\tau - \beta)$$

and toward infinity in the opposite case. The limit of z is finite only in the exceptional, and most improbable case that

$$(n - \tau)(c - \delta) = \delta(\tau - \beta).$$

When $(n - \tau)(c - \delta) < \delta(\tau - \beta)$, x_0 tends toward $\frac{n - \tau}{\delta}$, and x_1, x_2, ... tend toward zero. When $(n - \tau)(c - \delta) > \delta(\tau - \beta)$, x_0 tends toward zero.

There is, then, a considerable difference between the case of parasitic castration, and that of uniform natality. In the former case a "peaceful" coexistence is possible through some sort of equilibrium between the hosts and their parasites. In the latter case there are only two alternatives: Either the parasites exterminate their hosts, or they disappear. It would be easy to study the intermediate case of incomplete parasitic castration.

6. <u>Special case 4</u>. Let us consider now a case which is slightly more complex than the previous ones. Suppose that:

   1) the birth rate does not depend on k:

   $$n_k = n \qquad (k = 0, 1, 2, \ldots)$$

   2) the death rate is a linear function of k:

   $$\tau_k = \tau + \theta k \qquad (k = 0, 1, 2, \ldots)$$

   3) the presence of parasites weakens the hosts, making them more susceptible to further contamination:

   $$c_k = c + \varepsilon k \qquad (k = 0, 1, 2, \ldots)$$

   4) the competition coefficients remain constant:

   $$\delta_{ks} = \delta$$

   5) $\beta_k = k\tau_k + \alpha_k - \lambda k \qquad (k = 0, 1, 2, \ldots)$; this hypothesis is arbitrary, but it can be interpreted to mean that there is a correlation between the complementary mortality, θ, and the relative multiplication coefficients of the parasites $\beta_k - \alpha_k$.

Regarding the mortality of the parasites, α_k, we will leave it completely undetermined.

In such conditions one has the following differential equations:

$$x' = (n - \tau)x - \theta z - \delta x^2$$

$$z' = z[(c - \delta)x + \varepsilon z - \lambda] \tag{19}$$

$$x_0' = \alpha_1 x_1 + nx - x_0 P_0$$

$$x_k' = -x_k P_k + x_{k-1} Q_k + \alpha_{k+1} x_{k+1} \tag{20}$$

in which, as usual,

$$P_k = \tau_k + \alpha_k + \beta_k + c_k z + \delta x, \qquad Q_k = \beta_{k-1} + c_{k-1} z .$$

The study of equations (20) is slightly more complex than equations (15). To begin with, there are the two equilibrium points:

$$x = 0, \qquad z = 0 \qquad \text{and} \qquad x = \frac{n - \tau}{\delta}, \qquad z = 0.$$

The point $x = 0$, $z = 0$ is only stable for $n - \tau < 0$, $\lambda > 0$. The point $x = \frac{n - \tau}{\delta}$, $z = 0$ is stable when

$$n - \tau > 0, \qquad \frac{(c - \delta)(n - \tau)}{\delta} < \lambda.$$

The other two equilibrium points are the intersections of the parabola $\theta z = (n - \tau)x - \delta x^2$ with the straight line $(c - \delta)x + \varepsilon z - \lambda = 0$. Denoting these points by (N_1, S_1), (N_2, S_2), let us briefly review the different cases which might arise:

1) $\lambda < 0$, $c - \delta > 0$; the points (N_1, S_1) and (N_2, S_2) are outside the positive region; the hosts are completely exterminated by the parasites.

2) $\lambda > 0$, $c - \delta < 0$; the fate of the species depends on the initial conditions; two alternatives are possible: the complete disappearance of the host, or the complete disappearance of the parasites, with x tending toward $\frac{n - \tau}{\delta}$.

3) $\lambda < 0$, $c - \delta < 0$; $\frac{\lambda}{c - \delta} < \frac{n - \tau}{\delta}$; the points (N_1, S_1) and (N_2, S_2) are outside the positive region, and the hosts are completely exterminated by the parasites.

4) $\lambda > 0$, $c - \delta > 0$; $\frac{\lambda}{c - \delta} > \frac{n - \tau}{\delta}$; suppose, for easy reference, that $N_1 < N_2$. The points (N_1, S_1) will be stable if a few more conditions, which we do not specify, are satisfied. In this case there are two alternatives: the equilibrium point (N_1, S_1) is stable, or the

equilibrium point $\left(\dfrac{n-\tau}{\delta}\ ,\ 0\right)$ is stable. If the conditions mentioned above are not satisfied, one has, again, two alternatives: either the complete disappearance of the hosts, or a stable equilibrium at the point $\left(\dfrac{n-\tau}{\delta},\ 0\right)$.

5) $\lambda > 0$, $c - \delta > 0$; $\dfrac{\lambda}{c-\delta} < \dfrac{n-\tau}{\delta}$; this case differs from the previous one since the point $\left(\dfrac{n-\tau}{\delta}\ ,\ 0\right)$ is no longer stable, and therefore, the parasites cannot disappear.

6) $\lambda < 0$, $c - \delta < 0$; $\dfrac{\lambda}{c-\delta} > \dfrac{n-\tau}{\delta}$; this gives rise to two alternatives: either a stable equilibrium at $\left(\dfrac{n-\tau}{\delta},\ 0\right)$, or the complete disappearance of the hosts, exterminated by ever increasing numbers of parasites.

One sees the large diversity of possibilities which can arise for both the hosts and for the parasites. Their final state depends on the relationships among the vital coefficients, and upon the initial numbers.

Assume now that x and z have attained the stable state (N_1, S_1), and calculate the corresponding values of x_0, x_1, x_2, ... Setting $A_k = \dfrac{P_k}{\alpha_{k+1}}$, $B_k = \dfrac{Q_k}{\alpha_{k+1}}$, the

equations (20) give

$$x_1 = A_0 x_0 - \frac{nx}{\alpha_1}, \qquad x_{k+1} = A_k x_k - B_k x_{k-1} \qquad (k = 1, 2, 3, \ldots)$$

It is easy to see that their solutions can be represented by continuous fractions

$$\frac{x_1}{x_0} = \cfrac{B_1}{A_1 - \cfrac{B_2}{A_2 - \cdots}}\ , \qquad \frac{x_{k+1}}{x_k} = \cfrac{B_{k+1}}{A_{k+1} - \cfrac{B_{k+2}}{A_{k+2} - \cdots}}$$

One finds, on the other hand,

$$x_0 = \cfrac{\dfrac{nx}{\alpha_1}}{A_0 - \cfrac{B_1}{A_1 - \cfrac{B_2}{A_2 - \cdots}}}$$

from which one can successively compute x_0, x_1, x_2, ...

We are now interested in the limit value of $\dfrac{x_{k+1}}{x_k}$. From our hypotheses

$$\alpha_k = a_0 + a_1 k + a_2 k^2, \qquad \beta_k = b_0 + b_1 k + b_2 k^2;$$

the coefficients a_0, a_1, a_2, b_0, b_1, b_2 are connected by the relations

$$a_0 = b_0, \qquad a_1 = b_1 + \lambda - \tau, \qquad a_2 = b_2 - \theta.$$

In such conditions

$$\lim_{k \to \infty} A_k = \frac{a_2 + b_2}{a_2} = a > 1, \qquad \lim_{k \to \infty} B_k = \frac{b_2}{a_2} = b < 1.$$

Therefore, for a sufficiently large k,

$$x_{k+1} = a x_k - b x_{k-1}$$

$$\frac{x_{k+1}}{x_k} = \frac{a}{2} - \left| \frac{a}{2} - 1 \right|.$$

For $a > 2$ the ratio $\dfrac{x_{k+1}}{x_k}$ tends toward 1, and for $a < 2$ it tends toward $b < 1$. In the case of the hermit crabs, _Pagurus_, parasitized by _Chlorogaster_ (11), $a = \dfrac{5}{3}$, $b = \dfrac{2}{3}$.

V. MULTIPLE PARASITISM: PAGURUS AND CHLOROGASTER

1. I shall now consider the only case of parasitism for which precise figures are available: the hermit crabs, _Pagurus_, parasitized by _Chlorogaster_. The _Chlorogaster_ is a rhizocephalate barnacle, generally characterized by the simultaneous presence of several external visceral sacs on each infested hermit crab. Each sac has its own system of roots, which on first inspection appear to be independent of one another. C. Pérez (11) examined over 2,000 hermit crabs in order to determine the distribution of the number of visceral sacs on infested individuals.

In order to explain this form of multiple parasitism, Pérez (11) suggests two possible origins: 1) Multiple contamination of the same host by as many larvae as there are observed visceral sacs; 2) A single contamination followed by a certain process which leads to a multiplicity of visceral sacs. This second alternative can be realized in a number of ways (10):

1) One might imagine that an original infestation is due to a single
   larva, but, early in its internal development, the parasitic mass
   breaks up into a number of fragments. Each fragment then develops
   autonomously, side by side, and in synchrony with the others. This
   is the hypothesis of Smith (Pérez, 10).

2) One might also imagine that the parasitic mass comes from the penetration of the host by a single larva, which itself remains undivided. It gives rise, however, to a continuous system of roots from which successive groups of external sacs bud in a discontinuous manner. Pérez (12) arrived at this hypothesis of budding as a result of his most recent anatomical studies of infested hermit crabs.

2. Let us consider the hypothesis of multiple contamination. Very simple considerations show that it must be discarded. Consider, in fact, the figures in Table 1. The distribution of the number of visceral sacs has a sharp maximum, while the probability of multiple contamination decreases with the number of contaminations. Assume that the probability of a single contamination is $P_1 < 1$. If the contaminated crabs are neither weakened nor immunized by the first parasite, the probability of double contamination will be $P_1^2 < P_1$, of triple $P_1^3 < P_1^2 < P_1$, etc. If, on the other hand, the crabs are weakened, the probability of double contaminations will be $P_1 P_2 < P_1$, in which $P_2 > P_1$. The same reasoning gives a probability of triple contamination $P_1 P_2 P_3 < P_1 P_2 < P_1$, etc. If one assumes that the numbers of crabs with 1, 2, 3, ... sacs are proportional to such probabilities, one would obtain a decreasing sequence of numbers, in clear contradiction with Table 1. One arrives at the same conclusion through equations (12), by setting at equilibrium $\alpha_k = 0$, $\beta_k = 0$. One has, in fact,

$$\frac{x_k}{x_{k-1}} = \frac{c_{k-1} z}{\tau_k + c_k z + \sum_{s=0}^{\infty} \delta_{ks} x_s} .$$

Assuming that $c_k > c_{k-1}$ (corresponding to the hypothesis of weakening) one sees that $x_k < x_{k-1}$. We have seen in the previous chapter that this inequality also holds for less restrictive hypotheses on α_k and β_k. The hypothesis of multiple contamination is insufficient, then, to explain Pérez's figures. We shall suppose for simplicity that multiple contamination never takes place. In fact, the proportion of infested crabs is less than ten percent. The probability of double contamination already will be quite small and that of a triple contamination, exceedingly small. One can, therefore, without risking appreciable error, assume in the following that each infestation is due to a single larva.

Table 1

P N (p)	1	2	3	4	5	6	7	8	9	10	11
N (p) males obs.	4	7	9	24	29	25	11	10	6	3	2
N (p) males calc.	2.3	7.8	15.7	24.5	29.0	24.2	15.5	8.8	5.1	3.0	1.9
N (p) fem. obs.	2	10	17	30	29	17	7	5	--	2	--
N (p) fem. calc.	3.0	10.0	20.0	29.3	29.3	19.5	10.8	5.7	3.2	1.9	--
N (p) total obs.	6	17	26	54	58	42	18	15	6	5	--
N (p) total calc.	5.3	17.8	35.7	53.8	58.3	43.7	26.3	14.5	8.3	4.9	--

3. According to the hypothesis of Smith, the number of visceral sacs in a crab should be random, and unrelated to the age of the host. Without knowing why the parasitic mass undergoes fragmentation, it is impossible to predict a priori the distribution of the number of fragments in contaminated crabs. Does this distribution have one maximum, or two, or is it decreasing? Nothing is known.

On the contrary, the hypothesis of budding does suggest a certain relationship between the number of sacs and the mean age of the host. In this case very simple hypotheses suffice to give a rational explanation to the figures of Pérez.

Let us denote by u the age of a group of sacs in a hermit crab, or what is much the same, the age of the parasite. This age is clearly related somehow to the age of the host, and it is necessarily always less. Then, if the number of sacs increases with the age of the parasite, those crabs carrying a higher number of sacs will also, on the average, be older.

Let N_c be the total number of contaminated hosts, and $N_c \phi(u) du$ the number of parasites in the age interval $(u, u + du)$. This number decreases with increased age owing to both the mortality of the crabs, and the intrinsic mortality of their parasites. Denote by $x(u)$ the combined rate of mortality; then clearly

$$x(u) = - \frac{\phi'(u)}{\phi(u)} ; \qquad \phi(u) = \phi(o) e^{-\int_o^u x(s) ds}$$

Let us introduce the distribution function for the parasites

$$N_c \phi(u) \psi(u, p) du \, dp$$

in which p denotes the number of visceral sacs. This function gives the number of parasites which have $(p, p + dp)$ sacs, and an age between $(u, u + du)$. This function is wholly analogous to functions employed in demography, such as the distribution function for a population with two characteristics. Assume that this function $\psi(u, p)$ has the form

$$\psi(u, p) = \frac{1}{\beta(u)} V\left(\frac{p}{\beta(u)}\right)$$

and, finally, denote by $N(p)dp$ the number of hosts carrying $(p, p + dp)$ sacs. One has then

$$N_c = \int_0^\infty N(p)dp, \qquad 1 = \int_0^\infty V(p)dp.$$

One easily derives the equation

$$N(p) = N_c \int_0^\infty \frac{\phi(u)}{\beta(u)} V\left(\frac{p}{\beta(u)}\right) du$$

from which, knowing all other functions, the distribution function is determined. One has, in fact,

$$V(p) = \frac{1}{2\pi p} \int_{-\infty}^{+\infty} p^{-iz} F(z)dz$$

setting

$$F(z) = \frac{\int_0^\infty N(p)p^{iz}dp}{\int_0^\infty \phi(u)[\beta(u)]^{iz}du} \ .$$

The functions $\beta(u)$ and $x(u)$ are not known. All that we can say about the combined mortality $x(u)$ is that it increases with the age of the parasites, even more so since their presence considerably weakens the host. In the absence of precise information one can assume, for example, that $\int_0^u x(s)ds = Bu^\eta$, that is, $x(u) = B\eta u^{\eta-1}$. The function $\beta(u)$ poses a more subtle problem. For a given u, $\beta(u)$ is proportional to the mean number of sacs found on crabs which have been parasitized for a time u. If the budding hypothesis is true, $\beta(u)$ must be an increasing function of u. Assume, for instance, that $\beta(u) = Au^\varepsilon$. On the other hand the distribution of parasitic sacs is very well represented by the function

$$N(p) = \frac{N(z) \cdot \frac{n}{m}\left(\frac{p}{z}\right)^m}{\left(\frac{p}{z}\right)^n + \frac{n - m}{m}}$$

in which z is the number of visceral sacs which corresponds to the maximum of $N(p)$. As shown in Table 1 this approximation is a very satisfactory one. In it one has chosen as parameters:

Males: n = 7   m = 1.75   z = 5.0   N(z) = 29

Females: n = 7   m = 1.75   z = 4.5   N(z) = 31

Assuming $n = \dfrac{\eta}{\epsilon}$, $m = \dfrac{1 - \epsilon}{\epsilon}$, it is very easy to compute the functions $V(p)$ and $\psi(u, p)$. One finds

$$\psi(u, p) = \frac{[\beta(u)]^{n-m-1} \; n^{\frac{m+1}{n}} \; e^{-\frac{n-m}{m}[\beta(u)]^{n} \, p^{-n}}}{(n - m)^{\frac{m+1}{n} - 1} \; p^{n-m} \; \Pi\left(-\dfrac{m + 1}{n}\right)} \quad .$$

This is a common distribution function in mathematical statistics. Qualitatively, it represents well the probable distribution of hermit crabs with respect to u and p, assuming that the number of sacs increases, up to a point, with age. It fits very well the distribution observed by Pérez. The hypotheses of single contamination and of subsequent budding are in good agreement with the facts.

4. Fluctuations. Unfortunately the data by Pérez are not sufficient to permit a study of the fluctuations around an equilibrium, or even to ascertain what the equilibrium might be. Nevertheless, the relative abundance of infested hermit crabs changes from year to year, and it is the impression of Pérez that such fluctuations are real.

VI. EVOLUTION

1. The considerations developed in the preceding chapters show that the mathematical apparatus is flexible enough to handle the most different situations. It is always possible to improve the treatment either by introducing new terms, or by considering changes in the vital coefficients. The example of the hermit crabs illustrates a possibility which previously had not been utilized: that of considering frequency distributions in place of simple variables, such as the number of individuals of a species or a group.

In addition to time one can also introduce space and consider biological associations in their different habitats, recreating mathematically, so to speak, their natural milieu (13). One could point to a whole series of biogeographical problems which can be handled in this way (6, 7).

The method which we have followed is that developed by A. J. Lotka (9) and V. Volterra (16). The reader who is familiar with Volterra's works will have noticed that our interpretations of the vital coefficients is slightly different from his. We do not impose on these coefficients the restrictions introduced by Volterra. We have seen that a given term can have the most varied biological interpretations.

In the case of a single species, and the most simple biological equation

$$x' = nx - mx - dx^2$$

the term dx^2 can have two different interpretations: 1) a decrease in natality, or 2) an increase in mortality arising from the increase in population size. In the problem of symbiosis the term $(c_1 + \alpha)x_1x_2$ had a double meaning: 1) formation of symbiotic couples, and 2) competition between species.

In such conditions one would actually misrepresent Volterra's thinking, (16-17), if one were to state, for example, that there is a fundamental distinction between associations with even and odd numbers of species. One only finds this distinction when the vital coefficients satisfy very special conditions (17), and it does not exist otherwise. Since there is an extreme wealth of types of associations (3), one can be sure that there will always be associations which satisfy such special conditions, and other associations which do not. Actually, the exact number of species in an association is of little relative interest. We have seen, in fact, that the vital coefficients must be considered to be functions of age, and this in turn can lead us to subdivide each species into an arbitrary number of groups.

2. The foremost important problem which can be treated mathematically is that of evolution. More often than not this problem has been introduced in the form of changes in the vital coefficients. It has, nevertheless, an aspect which is unrelated to such changes. We have seen, in fact, that there can be a number of different stable equilibria for the same values of the vital coefficients. Depending on initial conditions, an association will move toward one or another of these equilibria. In many cases the outcome is a matter of chance.

One can easily imagine (13) a habitat composed of different patches, and separated by gaps of different sorts. In one of these patches a certain parasitic species may tend toward a stable coexistence with its host. In another patch the association may tend toward a different stable state, in which the parasite is excluded, etc. Natural selection will then tend to stablize these differences.

On the other hand, side by side with stable equilibria, there will also be unstable equilibria, and processes which lead to the complete disappearance of the host. One can also assert that the probability of such a process is generally higher than the probability that the association will tend toward a stable state, especially when the association involves several species. One will always have, side by side, groups which tend to disappear, and others which tend to persist in a stable state. The former constitute the milieu; the latter guarantee evolution.

3. We have seen on the other hand that stable regimes are possible when the vital coefficients satisfy certain inequalities. Assume, for example, that an association x_1, x_2, ..., x_n satisfies the differential equations

$$x_1' = \phi_1(x_1, x_2, \ldots, x_n \mid \alpha_1, \alpha_2, \ldots, \alpha_k)$$
$$\cdots \cdots \cdots \cdots \cdots \cdots \cdots \cdots \cdots$$
$$x_n' = \phi_n(x_1, x_2, \ldots, x_n \mid \alpha_1, \alpha_2, \ldots, \alpha_k)$$

in which α_1, α_2, ..., α_k are vital coefficients. Denote by a_1, a_2, ..., a_n the values of x_1, x_2, ..., x_n in one of the stable states. These values are functions of the vital coefficients:

$$a_1 = \psi_1(\alpha_1, \alpha_2, \ldots, \alpha_k), \ldots, a_n = \psi_n(\alpha_1, \alpha_2, \ldots, \alpha_k).$$

The characteristic equation for this state

$$\begin{vmatrix} \dfrac{\partial \phi_1}{\partial a_1} - \omega, & \dfrac{\partial \phi_1}{\partial a_2}, & \cdots, & \dfrac{\partial \phi_1}{\partial a_n} \\[2ex] \dfrac{\partial \phi_2}{\partial a_1}, & \dfrac{\partial \phi_2}{\partial a_2} - \omega, & \cdots, & \dfrac{\partial \phi_2}{\partial a_n} \\[2ex] \cdots & \cdots & \cdots & \cdots \\[2ex] \dfrac{\partial \phi_n}{\partial a_1}, & \dfrac{\partial \phi_n}{\partial a_2}, & \cdots, & \dfrac{\partial \phi_n}{\partial a_n} - \omega \end{vmatrix} = 0$$

will generally have n different roots ω_1, ω_2, ..., ω_n which are also functions of the vital coefficients

$$\omega_1 = f_1(\alpha_1, \ldots, \alpha_k), \ldots, \omega_n = f_n(\alpha_1, \ldots, \alpha_k).$$

The equilibrium is stable when ω_1, ω_2, ..., ω_n are negative, or complex with negative real parts; that is to say, when the vital coefficients are located in certain domains of the parameter space $(\alpha_1, \alpha_2, ..., \alpha_k)$. Such domains might just as well be disconnected. All such hypotheses are very plausible, and we have given a number of examples in which they were realized. On the other hand, the vital coefficients are just averages: each group of organisms is composed of individuals which are very different from the point of view of competition, stability, reproductive capacity, etc.

If a group is one way or another subdivided into several subgroups with different vital coefficients, the point in the parameter space $(\alpha_1, \alpha_2, ..., \alpha_k)$ which corresponds to each subgroup may be in one of the regions of stability as well as in one of the regions of instability. Now, similar situations are, in fact, realized when the habitat of an association is fragmented into small regions which are more or less isolated from one another. As a result there will be certain regions containing stable associations which are separated by empty or sparcely populated spaces.

In a habitat with a single occupant that moves around a lot, geographical isolation becomes difficult, and, naturally, evolution will be less rapid. Even in such cases, however, physiological isolation, or the effects of symbiosis or parasitism, can play a large role. We have seen for instance that parasitic castration can give rise to a stable equilibrium where such an equilibrium is otherwise impossible. In other situations the reverse might as well hold true.

We will not insist further on additional mathematical aspects of the problem of evolution (1, 4). We have tried to show that any hypothesis whatsoever, no matter what the biological problem, can be expressed in mathematical form. For the time being, and lacking sufficient numerical data, one must be content with the qualitative study of the equations. One can foresee that the availability of such data will greatly modify the rather a priori equations we have constructed up to now. The example of the hermit crabs is significant in this respect. Never mind! Mathematical analysis is capable of proposing other equations, other solutions, other methods.

MATHEMATICAL NOTES

I. Systems of first order differential equations

Let

$$x_1' = \phi_1(x_1, x_2, \ldots, x_n), \ldots, x_n' = \phi_n(x_1, x_2, \ldots, x_n) \qquad (I)$$

be a system of first order differential equations. If the system of equations

$$\phi_1(x_1, x_2, \ldots, x_n) = 0, \ldots, \phi_n(x_1, x_2, \ldots, x_n) = 0$$

admits of a solution

$$x_1 = a_1, \qquad x_2 = a_2, \qquad \ldots, \qquad x_n = a_n$$

the point $(a_1, a_2, \ldots, a_n)$ is an equilibrium point.

In fact, if one takes such values as initial values, the system (I) is solved by them, and x_1, x_2, $\ldots$, x_n remain constant. This equilibrium state may be stable or unstable. It is stable when small perturbations of the initial values lead to functions x_1, x_2, $\ldots$, x_n which oscillate around the equilibrium a_1, a_2, $\ldots$, a_n without going outside a region of sufficiently small radius around the point $(a_1, a_2, \ldots, a_n)$. It is unstable when the smallest perturbation of the initial values lead to an increasing departure of the point $(x_1, x_2, \ldots, x_n)$ from the region $(a_1, a_2, \ldots, a_n)$. The study of the stability of an equilibrium can be reduced to that of the small oscillations around it. Set

$$x_1 = a_1 + \xi_1, \ldots, x_2 = a_2 + \xi_2, \ldots, x_n = a_n + \xi_n.$$

Substitute these values into the equation (I), and suppose that the functions ϕ_1, ϕ_2, $\ldots$, ϕ_n can be developed in a Taylor series around the point $A(a_1, a_2, \ldots, a_n)$. As a first approximation, one has

$$\xi_1' = \frac{\partial \phi_1}{\partial a_1} \xi_1 + \frac{\partial \phi_1}{\partial a_2} \xi_2 + \ldots + \frac{\partial \phi_1}{\partial a_n} \xi_n$$
$$\cdots\cdots\cdots\cdots\cdots\cdots\cdots\cdots\cdots$$
$$\xi_n' = \frac{\partial \phi_n}{\partial a_1} \xi_1 + \frac{\partial \phi_n}{\partial a_2} \xi_2 + \ldots + \frac{\partial \phi_n}{\partial a_n} \xi_n .$$

These are linear functions in the variables ξ_1, ξ_2, $\ldots$, ξ_n, which are solved by exponential functions. Suppose, in fact, that

$$\xi_1 = A_{11} e^{\omega_1 t} + A_{12} e^{\omega_2 t} + \ldots +, A_{1n} e^{\omega_n t}$$
$$\cdots\cdots\cdots\cdots\cdots\cdots\cdots\cdots$$
$$\xi_n = A_{n1} e^{\omega_1 t} + A_{n2} e^{\omega_2 t} + \ldots + A_{nn} e^{\omega_n t} ,$$

in which ω_1, ω_2, ..., ω_n satisfy the algebraic equation of degree $\underline{n}$

$$\begin{vmatrix} \dfrac{\partial \phi_1}{\partial a_1} - \omega, & \dfrac{\partial \phi_1}{\partial a_2}, & \cdots, & \dfrac{\partial \phi_1}{\partial a_n} \\[2ex] \dfrac{\partial \phi_2}{\partial a_1}, & \dfrac{\partial \phi_2}{\partial a_2} - \omega, & \cdots, & \dfrac{\partial \phi_2}{\partial a_n} \\[1ex] \cdots & \cdots & \cdots & \cdots \\[1ex] \dfrac{\partial \phi_n}{\partial a_1}, & \dfrac{\partial \phi_n}{\partial a_2}, & \cdots, & \dfrac{\partial \phi_n}{\partial a_n} - \omega \end{vmatrix} = 0.$$

This equation is called the characteristic equation. The exponents ω_1, ω_2, ..., ω_n can be positive or negative, real or imaginary. If they are all negative, or with negative real parts, the equilibrium is stable since the fluctuations ξ_1, ξ_2, ..., ξ_n tend toward zero. If even one of these exponents is positive, or has a positive real part, it is sufficient to make the equilibrium unstable.

II. The periodicity of the solution of linear first order differential equations with periodic coefficients

Let

$$x' + p(t)x = q(t)$$

be an equation of this sort. One has

$$p(\pi + t) = p(t), \qquad q(\pi + t) = q(t),$$

where π denotes the period. The function $x(t)$ has the same period, no matter what the initial value, $x(0)$, if the conditions

$$\int_0^\pi p(t)dt = 0, \qquad \int_0^\pi q(t)e^{-\int_t^\pi p(u)du} dt = 0$$

are satisfied. If this is not the case, $x(t)$ can still be periodic when $x(0)$ has the following, special value

$$x(0) = \frac{\int_0^\pi q(t)e^{-\int_t^\pi p(u)du} dt}{1 - e^{-\int_0^\pi p(u)du}} .$$

III. The qualitative study of differential equations

The occasionally insurmountable difficulties associated with analytical solutions to differential equations limits us to their qualitative study. We shall

give some examples for systems of differential equations with two variables. Let

$$x' = \phi(x, y), \qquad y' = \psi(x, y)$$

be such a system, and a, b the coordinates of an equilibrium point. The character-

istic equation is

$$\begin{vmatrix} \dfrac{\partial \phi}{\partial a} - \omega & \dfrac{\partial \phi}{\partial b} \\[3mm] \dfrac{\partial \psi}{\partial a} & \dfrac{\partial \psi}{\partial b} - \omega \end{vmatrix} = 0$$

or

$$\omega^2 - \omega C + D = 0$$

in which

$$C = \frac{\partial \phi}{\partial a} + \frac{\partial \psi}{\partial b} , \qquad D = \frac{\partial \phi}{\partial a} \frac{\partial \psi}{\partial b} - \frac{\partial \phi}{\partial b} \frac{\partial \psi}{\partial a} .$$

In the case

$$C < 0, \qquad D < 0, \qquad E = C^2 - 4D > 0$$

one has a <u>node</u>, that is to say, both solutions of the characteristic equation are

negative, and the point (x, y) tends toward (a, b) following an arc of finite length.

In the case

$$C < 0, \qquad D > 0, \qquad E < 0$$

one has a <u>focus</u>, that is to say both solutions of the characteristic equation are

complex with negative real parts, and the point (x, y) tends toward (a, b), tracing

a spiral-like curve.

In the case

$$C = 0, \qquad D > 0, \qquad E < 0$$

one has a <u>cycle</u>, that is to say both solutions of the characteristic equation are

imaginary, and the point (x, y) traces a closed curve around the point (a, b).

Denote by l the angular coefficient of the tangent to the curve $\phi = 0$ at the

point (a, b), by m the analgous coefficient for the curve $\psi = 0$, by S the tangent

of the angle between such two tangents, and by F the expression

$$F = \frac{\partial \phi}{\partial a} \frac{\partial \psi}{\partial a} + \frac{\partial \phi}{\partial b} \frac{\partial \psi}{\partial b} .$$

One has

$$1 = - \frac{\partial \phi}{\partial a} : \frac{\partial \phi}{\partial b} \ , \qquad m = - \frac{\partial \psi}{\partial a} : \frac{\partial \psi}{\partial b} \ , \qquad G = \frac{m - 1}{1 + ml} = \frac{D}{F} \ .$$

In Table 2 one can find all the necessary information regarding the nature of the point (a, b). Such a table makes it possible to determine rapidly the character of such an equilibrium.

Table 2

Case	$\frac{\delta\phi}{\delta a}$	$\frac{\delta\phi}{\delta b}$	$\frac{\delta\psi}{\delta a}$	$\frac{\delta\psi}{\delta b}$	C	D	E	F	G	Characteristics
1	+	+	+	+	+			+		Unstable equilibrium
2	−	+	+	+		−				Unstable
3	+	−	+	+	+	+				Unstable
4	+	+	−	+	+	+				Unstable
5	+	+	+	−		−				Unstable
6_1	−	−	+	+	−	+	−	−	−	Stable focus
6_2	−	−	+	+	−	+	+	−	−	Stable node
6_3	−	−	+	+	0	+	−	−		Periodic cycle
6_4	−	−	+	+	+	+		−		Unstable
6_5	−	−	+	+		−		−		Unstable
7_1	−	+	−	+	−	+	−	+	+	Stable focus
7_2	−	+	−	+	−	+	+	+	+	Stable node
7_3	−	+	−	+	0	+	−	+	+	Periodic cycle
7_4	−	+	−	+	+	+		+	+	Unstable
7_5	−	+	−	+		−		−	−	Unstable
8_1	−	+	+	−	−	−		−	+	Unstable
8_2	−	+	+	−	−	+	+	−	−	Stable node
9	+	−	−	+	+			−		Unstable
10_1	+	−	+	−	−	+	−	+	+	Stable focus
10_2	+	−	+	−	−	+	+	+	+	Stable node
10_3	+	−	+	−	0	+	−	+	+	Periodic cycle
10_4	+	−	+	−	+	+		+	+	Unstable
10_5	+	−	+	−		−		+	−	Unstable
11_1	+	+	−	−		−		−	+	Unstable
11_2	+	+	−	−	−	+	−	−	−	Stable focus
11_3	+	+	−	−	−	+	+	−	−	Stable node
11_4	+	+	−	−	0	+	−	−	−	Periodic cycle
11_5	+	+	−	−	+	+		−	−	Unstable
12	+	−	−	−		−				Unstable

Case	$\frac{\delta\phi}{\delta b}$	$\frac{\delta\phi}{\delta b}$	$\frac{\delta\psi}{\delta b}$	$\frac{\delta\psi}{\delta b}$	C	D	E	F	G	Characteristics
13_1	−	+	−	−	−	+	−	−	−	Stable focus
13_2	−	+	−	−	−	+	+	−	−	Stable node
14_1	−	−	+	−	−	+	−	+	+	Stable focus
14_2	−	−	+	−	−	+	+	+	+	Stable node
15	−	−	−	+		−				Unstable
16_1	−	−	−	−	−			+	−	Unstable
16_3	−	−	−	−	−	+	+	+	+	Stable node

BIBLIOGRAPHY

1. BAILEY (V. A.), Quarterly J. of Math., Oxf. Ser. 2, pp. 68-77, 1931.

2. CAULLERY (M.), Le parasitisme et la symbiose, Paris, Doin, 1925.

3. CAULLERY (M.), Le problème de l'évolution, Paris, Payot, 1931.

4. HALDANE (J. B. S.), The causes of evolution, London, 1932.

5. KOSTITZIN (Mme J. et V. A.), C. R. 193, pp. 86-88, 1931.

6. KOSTITZIN (V. A.), C. R. 195, pp. 1219-1222, 1932.

7. KOSTITZIN (V. A.), C. R. 196, pp. 214-215, 1933.

8. LOTKA (A. J.), Journ. of Wash. Ac. of Sc., 13, pp. 152-158, 1923.

9. LOTKA (A. J.), Eléments of physical biology, Baltimore, 1925.

10. PÉREZ (Ch.), C. R. 187, pp. 771-773, 1928.

11. PÉREZ (Ch.), C. R. 192, pp. 1274-1276, 1931.

12. PÉREZ (Ch.), C. R. 193, pp. 195-197, 1931.

13. PRENANT (M.), Géographie des animaux, Paris, Colin, 1933.

14. RABAUD (E.), Parasitisme et évolution (Revue Philos., 105, pp. 18-81, 1928).

15. THOMPSON (W. R.), C. R. 174, pp. 1201-1204, 1433-1435, 1647-1649, 1922.

16. VOLTERRA (V.), Lecons sur la théorie mathématique de la lutte pour la vie, Paris, 1931.

17. VOLTERRA (V.), Giorn. 1st. Ital. Attuari, 2, pp. 295-355, 1931.

PART IV

GENOTYPIC SELECTION AND EVOLUTION

INTRODUCTION

According to Darwin and Wallace, the distribution and abundance of organisms can be explained primarily in terms of competition, predation, and other biotic interactions. The same factors should also explain the majority of evolutionary changes within species. Non-biotic factors would influence the outcome of biotic interactions but, usually, their direct effects would be marginal (cf., Darwin, 1859, Chapters III and IV).

Volterra and Lotka explicitly applied Darwin's theory only to explain the distribution and abundance of organisms. Kostitzin, in simultaneously dealing with ecology and evolution, was much closer to Darwin's original approach. Thus, when treating competitive interactions Kostitzin usually referred to different "groups." He repeatedly stressed that the same general treatments apply, depending on the particular situation, to different species or to discontinuous phenotypic or genetic variants within the same species.

In chapter 15 of "Biologie mathématique" and in five short papers, which we also present, Kostitzin dealt explicitly with the selective effects of competition and other biotic interactions on diploid, interbreeding organisms. It is unfortunate that his biological research virtually ended at the outbreak of World War II, while his research on natural selection was still at its initial stages. Much at the same time Volterra had also begun to work on evolutionary problems (see D'Ancona, 1954, page 193), but he died shortly afterwards. It is not clear whether Kostitzin's loss of interest in biological applications was due to his brief internment in a concentration camp, or to Volterra's death, or to his renewed interest in astro- and geophysics (cf. Scudo and Ziegler, 1976).

Kostitzin was also very interested in the obscure relationships between phenotypic and genotypic variation. As an example he discussed the two different "size races" of _Didinium nasutum_, a predatory ciliate. When the prey become scarce, the large "race" rapidly disappears, and only the small "race" is able to survive. The larger form is again produced as soon as the prey become abundant.

Kostitzin's interest in mainly phenotypic variations seems to be connected to his luke-warm attitude towards "transformism." This view, originating primarily

with A. R. Wallace, maintains that evolution results only from the selective accumulation of small, randomly produced variations in the "germ plasm." After analyzing
the reformulation of this view in Mendelian terms, now called "neo-Darwinism" or
the "synthetic theory," the best that Kostitzin could say about it was: "It is
based on a number of axioms and postulates, and it does not present any internal
contradiction" (1938e).

It was most appropriate that Volterra should sponsor for publication in the
Comptes Rendus Kostitzin's first paper (1936b) on the effects of intraspecific
competition in the case of a pair of alleles. Kostitzin dealt with this problem
in greater depth in chapter 15 of "Biologie mathématique" (1937). In this work
Kostitzin provides a rough analysis of the conditions for stability of the resulting
equilibria in the special cases in which the genotypes differ only for their rates
of mortality or their competitive abilities, or in which the crosses have different
fertilities. Kostitzin also considers a simplified case in which the same genetic
system is affected by the appearance of a new species of prey.

In the short paper which follows, Kostitzin sets up general differential
equations for the selective effects of intraspecific interactions for any number,
n, of dimorphic loci in populations of interbreeding diploids. The coefficients
of Mendelian heredity which appear in such equations are discussed in the two
following papers, the latter dealing also with sex-linked loci. The fourth paper
deals with these differential equations under the assumption of panmixia. It
discusses in some detail the singular points of lower order, corresponding to
s < n dimorphic loci. Kostitzin also gives an heuristic analysis of their stability,
with particular reference to 0-th order singular points, corresponding to
monomorphism at all n loci. The tone of the last paper in this section "Sélection
naturelle et transformation des espèces du point de vue analytique, statistique et
biologique" (1938e) closely reminds us of "Sur la ségrégation physiologique et la
variation des espèces" (1940b), suggesting that the latter paper may have been
written at an earlier time. The many misprints in both papers may be connected with
Kostitzin's detention or, alternatively, with the waning of his interest in biology.

This last paper begins by pointing out the consequences of the stability conditions worked out in 1938c. The size of the domain in the hyperspace of the vital coefficients for which a singular point is stable, tends to be progressively smaller as the order of the singular points increases. Since the vital coefficients tend to vary with time, the maintenance of a complex polymorphism by one "factor" would be a priori most unlikely. He then points out that, more often than not, new mutants would be selected against just because of their initial low frequency.

This reasoning applies to genetic variants affecting the relative performance of individuals with respect to each other. Polymorphism for such a trait can be maintained "from within" a population, by some form of heterozygote advantage, under very stringent conditions. On the other hand, spatial variations in the habitat of a species would tend to select different monomorphic states in different areas of the range. Local polymorphism for such a trait could then more easily result from migration.

Other considerations apply to genetic variants affecting the performance of individuals with respect to different "resources" or to different interacting species. Such variants can be far more easily maintained at a stable polymorphism, even with no heterozygote advantage. This directly follows by interpreting the general theory of competition and predation in terms of haploid genotypes.

Kostitzin was well aware, especially through Haldane (cf., "Symbiose, parasitisme et évolution"), of the emerging theories on evolutionary change in terms of gene frequencies and relative fitnesses. It is significant that he chose to deal with the same problem in terms of numbers and explicit ecological advantages, and that he did not relate his approach to frequency models. This is especially interesting since the two approaches seem to lead to strikingly different conclusions which were also worrying Lotka (cf. Preface, Plate III), and which have not yet been reconciled.

MATHEMATICAL BIOLOGY: Chapter 15

Evolution

V. A. Kostitzin

1. **Life and Entropy**

Before attacking the study of evolution we must reach an agreement on the
exact meaning of this word. All of the existing definitions vary between two
extremes. Some apply the word to any process in nature whatsoever, contrasting
evolution with lack of change. Other definitions look for "the formula of
progress," denoting by evolution anything which is progressive and ascendant, as
opposed to leveling. The former definitions try to preserve an objective attitude;
the latter view the universe through rose-colored, or dark glasses, according to
their inspiration.

It seems to me that one can give to this word a quantitative meaning by
defining evolution as those processes which lead from a more probable state to a
different, less probable one, or which are directed against an increase in entropy.
One may object that, by applying this definition to partial processes, or to certain
ascendant phases of cyclical processes, one risks being lost in vagueness. Well,
this must be avoided. It is very difficult to establish the distinction, as
difficult as in the paradox of "the heap of corn." Yet the heap exists, and it is
very different from a grain. Evolution exists, and it is very different from any
other special process, even an ascendant one.

Evolution: A local phenomenon. Entropy: A universal phenomenon. One must
now establish a very important distinction--that between local and universal
tendencies. Universal tendencies would undoubtedly be directed towards a more
probable state, if this state had not already been attained. Local tendencies can
be ascendant or descendant, and undoubtedly the former are far less probable than
the latter. Still, their probability is not zero. In other words, the
realization of an improbable state is not impossible, and the lack of realization
of such a state is very improbable. Evolution must then be a very rare phenomenon,
highly localized in space and in time.

Very rare combinations which are realized in a more or less regular manner may
form a kind of universe of their own, superimposed on chaos. If the appearance of
matter is due to this game of combinations, then each particle is a local phenomenon.
The ensemble of such particles, however, can have a universal character. It can

last for a time long enough to allow the realization of far less probable, but not impossible events, such as living matter. It follows that life is a link in an ascending chain of very unlikely events, the realization of which is in strict agreement with the laws of probability.

This ascending movement has not come to an end. Its latest stage is the one leading from the appearance of life to the appearance of human consciousness.

This forward march is in fact accompanied by a large wastage. The inevitable counterparts of harmony are all the imperfections of Nature: all the failed beings, all the germs that perished before maturation, before fertilization, before hatching. Each living being is, in some way, a big winner in the lottery of life. Considered from this point of view, the vital process becomes a stunning confirmation of the law of entropy, rather than an exception to it.

2.   The Struggle for Existence and Natural Selection

In the process of evolution which has transformed species, what is the exact role of the struggle for existence? In the preceding chapters we have examined several cases in which a slight disadvantage led to the complete disappearance of one or more of the groups comprising the population of a biotope. We have seen that such disadvantages could be of very diverse nature, but they could always be computed as some increase in mortality. In this way the mechanism for eliminating the least fit seems completely clear when applied to organisms which belong to an already existing species. But, how does it act on related groups belonging to the same species, and giving fertile hybrids?

An example taken from Gause has shown us a population composed of two races of Didinium nasutum feeding on Paramecium. Everything went smoothly in this microcosm as long as there was sufficient food. But the diminution of paramecia led to a rapid disappearance of the giant race of predators, while the dwarf race was able to survive for some time. What happens to possible survivors of the dwarf race when food reappears? The population of Didinium is restored through a mechanism which eludes us, but which may be related to the behavior of chromosomes at conjugation. Some sort of giant mutation may reappear, and multiply. One might suppose that size is determined by Mendelian traits, and that the larger race reappears by segregation among small hybrids which originated by conjugations between the two races.

In what follows we are not concerned with the mechanism by which mutations appear. We shall consider a population which, to begin with, is composed of two pure groups and a hybrid group. Each of these groups will have its own vital coefficients.

In the case of an isolated, homogeneous species, the logistic equation

$$p' = np - mp - \nu p^2 - \mu p^2 \tag{1}$$

describes the numerical increase of this species. The coefficients of mortality, m and μ, do not require any special treatment. Each group has its own set of such coefficients. However, the coefficients of fertility, n and ν, may vary among the crosses according to which of the three groups form the mating pair. In the simplest case of a species with separate sexes there are nine types of crossings, and each type may have its own characteristic fertility. Let p_1 and p_2 be the effective numbers of the two pure groups, and p_3 that of the hybrid group. If there is no special rule for the crossing among individuals of such groups, the number of births from the crossings between the groups p_i and p_k would be, in the absence of limiting factors, $n_{ik}p_ip_k/p$ where

$$p = p_1 + p_2 + p_3 . \tag{2}$$

Suppose that the action of limiting factors on the females decreases the natality, n_{ik}, by

$$\sigma_k = \nu_{1k}p_1 + \nu_{2k}p_2 + \nu_{3k}p_3$$

and that the analogous action on the males has an effect

$$\tau_i = \rho_{i1}p_1 + \rho_{i2}p_2 + \rho_{i3}p_3 .$$

Under these conditions, the effective coefficient of fertility takes the form

$$N_{ik} = n_{ik} - \sigma_k - \tau_i . \tag{3}$$

On the other hand, the groups are distributed in the progeny of such crosses according to Mendel's laws. One finally gets the differential equations of the problem:

$$p'_1 = \frac{N_{11}P_1^{\,2}}{p} + \frac{(N_{13} + N_{31})P_1P_3}{2p} + \frac{N_{33}P_3^{\,2}}{4p} - m_1P_1 -$$

$$-P_1(\mu_{11}P_1 + \mu_{12}P_2 + \mu_{13}P_3)$$

$$p'_2 = \frac{N_{22}P_2^{\,2}}{p} + \frac{(N_{23} + N_{32})P_2P_3}{2p} + \frac{N_{33}P_3^{\,2}}{4p} - m_2P_2 -$$

$$-P_2(\mu_{21}P_1 + \mu_{22}P_2 + \mu_{23}P_3) \qquad\qquad (4)$$

$$p'_3 = \frac{(N_{12} + N_{21})P_1P_2}{p} + \frac{(N_{13} + N_{31})P_1P_3}{2p} + \frac{(N_{23} + N_{32})P_2P_3}{2p} +$$

$$+\frac{N_{33}P_3^{\,2}}{2p} - m_3P_3 - P_3(\mu_{31}P_1 + \mu_{32}P_2 + \mu_{33}P_3) \ .$$

The problem of selection appears in this form when dealing with two groups which give fertile hybrids. If other species inhabit the same biotope, one must complete system (4) by adding new terms and new equations.

Consideration of equations (4) shows that:

(1) When conditions are unfavorable to the hybrids, the more numerous pure group will survive; it follows that a new mutant, which must necessarily be rare, will have a very small chance of surviving.

(2) In conditions that are highly favorable to the hybrids the three groups can co-exist, and tend towards a stable equilibrium.

(3) In the case in which the hybrids are intermediate between the two pure groups, only the pure group with the more favorable vital coefficients survives.

We shall verify these conclusions in a few special and relatively simple cases.

Special case 1. Suppose that all the coefficients ν_{ik} and ρ_{ik} are zero, and the three groups only differ in their mortalities. Equations (4) simplify to

$$p'_1 = \frac{n\left(P_1 + \frac{P_3}{2}\right)^2}{p} - m_1P_1 - hp_1p$$

$$p'_2 = \frac{n\left(P_2 + \frac{P_3}{2}\right)^2}{p} - m_2P_2 - hp_2p \qquad\qquad (5)$$

$$p'_3 = \frac{2n\left(P_1 + \frac{P_3}{2}\right)\left(P_2 + \frac{P_3}{2}\right)}{p} - m_3P_3 - hp_3p \ .$$

In the case of a low birth rate, smaller than the mortalities:

$$n < m_1, \qquad n < m_2, \qquad n < m_3 \tag{6}$$

all three groups tend towards zero:

$$\lim_{t \to \infty} p_1 = \lim_{t \to \infty} p_2 = \lim_{t \to \infty} p_3 = 0 . \tag{7}$$

If the death rate of the first pure group is low:

$$n > m_1, \qquad m_3 > m_1 \tag{8}$$

the hybrids and the second pure group disappear, and one has

$$\lim_{t \to \infty} p_1 = \frac{n - m_1}{h} , \qquad \lim_{t \to \infty} p_2 = 0, \qquad \lim_{t \to \infty} p_3 = 0 . \tag{9}$$

Conversely, if the second pure group has a low mortality:

$$n > m_2, \qquad m_3 > m_2 \tag{10}$$

the hybrids and the first pure group disappear:

$$\lim_{t \to \infty} p_1 = 0, \qquad \lim_{t \to \infty} p_2 = \frac{n - m_2}{h} , \qquad \lim_{t \to \infty} p_3 = 0 . \tag{11}$$

This means that of the two pure groups, the one which has a positive coefficient of increase, and a mortality lower than the hybrids, will survive. What happens, then, when the inequalities (8) and (10) are simultaneously satisfied? As previously stated, the behavior of this case depends on the initial values, $p_1(0)$, $p_2(0)$, and $p_3(0)$, and the least numerous pure group disappears together with the hybrids. In conditions which highly favor the hybrids:

$$m_3 << m_1, \qquad m_3 << m_2 \tag{12}$$

a co-existence of the three groups is possible. Finally, when m_3 lies between m_1 and m_2, the pure race with a positive coefficient of increase and the smaller mortality will be the only one to survive.

Special case 2. Consider another case in which the coefficients ν_{ik} and ρ_{ik} are zero, the mortalities and limiting factors are again the same for all the groups, but fertility varies according to the crossing. Suppose, to fix our ideas, that

$$n_{11} = n_1, \qquad n_{22} = n_2,$$

and that

$$n_{ik} = \frac{n_{ii} + n_{kk}}{2} . \tag{13}$$

That is, the fertility of a hybrid cross is the average of the fertility in the pure groups. In this way one has

$$n_{33} = n_{12} = n_{21} = \frac{n_1 + n_2}{2}$$

$$n_{13} = n_{31} = \frac{3n_1 + n_2}{4} \qquad\qquad (14)$$

$$n_{23} = n_{32} = \frac{3n_2 + n_1}{4} \;.$$

Equations (4) then become

$$p'_1 = \frac{n_1 p_1^2}{p} + \frac{(3n_1 + n_2) p_1 p_3}{4p} + \frac{(n_1 + n_2) p_3^2}{8p} - m p_1 - h p_1 p$$

$$p'_2 = \frac{n_2 p_2^2}{p} + \frac{(3n_2 + n_1) p_2 p_3}{4p} + \frac{(n_1 + n_2) p_3^2}{8p} - m p_2 - h p_2 p \qquad (15)$$

$$p'_3 = \frac{(3n_1 + n_2) p_1 p_3}{4p} + \frac{(3n_2 + n_1) p_2 p_3}{4p} + \frac{(n_1 + n_2) p_1 p_2}{p} +$$

$$\frac{(n_1 + n_2) p_3^2}{4p} - m p_3 - h p_3 p \;.$$

In this case the three equilibrium states are incompatible with one another. In the first case, in which the fertilities are low:

$$n_1 < m, \qquad n_2 < m \qquad\qquad (16)$$

all three groups disappear. In the second case, the first pure group has a positive rate of increase and a higher fertility than the second pure group:

$$n_1 > m, \qquad n_1 > n_2 \qquad\qquad (17)$$

at the limit only the first pure group survives:

$$\lim_{t \to \infty} p_1 = \frac{n_1 - m}{h} \;, \qquad \lim_{t \to \infty} p_2 = 0, \qquad \lim_{t \to \infty} p_3 = 0 \;. \qquad (18)$$

The third case is the converse of the second:

$$n_2 > m, \qquad n_2 > n_1$$

$$\lim_{t \to \infty} p_1 = 0, \qquad \lim_{t \to \infty} p_2 = \frac{n_2 - m}{h} \;, \qquad \lim_{t \to \infty} p_3 = 0 \;. \qquad (19)$$

Therefore, in this special case, only the most fertile pure group survives, provided its coefficient of increase is positive.

Special case 3. Suppose that fertility and mortality are the same for the three groups and for all the crosses, and that the limiting coefficients satisfy the following relations:

$$\rho_{ik} = \nu_{ik} = 0, \qquad \mu_{1k} = \mu_1, \qquad \mu_{2k} = \mu_2, \qquad \mu_{3k} = \mu_3 .$$

Equations (4) then become

$$
\left.
\begin{aligned}
p'_1 &= \frac{n\left(p_1 + \dfrac{p_3}{2}\right)^2}{p} - mp_1 - \mu_1 p_1 p \\[2ex]
p'_2 &= \frac{n\left(p_2 + \dfrac{p_3}{2}\right)^2}{p} - mp_2 - \mu_2 p_2 p \\[2ex]
p'_3 &= \frac{2n\left(p_1 + \dfrac{p_3}{2}\right)\left(p_2 + \dfrac{p_3}{2}\right)}{p} - mp_3 - \mu_3 p_3 p .
\end{aligned}
\right\} \tag{20}
$$

Their behavior depends only on the limiting coefficients of the hybrid group, μ_3. Suppose that $n - m > 0$, and that μ_3 lies between μ_1 and μ_2. In this case only the more resistant pure group will survive.

Suppose, next, that

$$\mu_3 > \mu_2, \qquad \mu_3 > \mu_1, \qquad n - m > 0 .$$

In this case the least numerous pure group will rapidly disappear along with the hybrid group.

Finally, when the hybrids are highly favored:

$$\mu_3 \ll \mu_1, \qquad \mu_3 \ll \mu_2, \qquad n - m > 0$$

the three groups can co-exist. Thus the final result of this discussion is in agreement with our earlier conclusions.

3.   Indirect Selection

Suppose that a species, p, composed of two pure groups, p_1 and p_2, and their hybrids, p_3, lives in a biotope under conditions which favor the hybrids. An equilibrium is established and it persists until the appearance of a new species, x, which upsets the relationships in this microcosm.

The resulting problem is too complex, both from the mathematical and from the biological point of view, to be approached here in its general form. We shall, therefore, confine ourselves to the study of a special case. Suppose, to fix our ideas, that the species p feeds on the new species x, and that the system is described by the following equations:

$$p'_1 = \frac{n\left(p_1 + \frac{p_3}{2}\right)^2}{p} - mp_1 - hp_1p + \ell_1 p_1 x$$

$$p'_2 = \frac{n\left(p_2 + \frac{p_3}{2}\right)^2}{p} - mp_2 - hp_2p + \ell_2 p_2 x$$

$$p'_3 = \frac{2n\left(p_1 + \frac{p_3}{2}\right)\left(p_2 + \frac{p_3}{2}\right)}{p} - Mp_3 - hp_3p + \ell_3 p_3 x$$

$$x' = Ex - Hx^2 - \lambda x (\ell_1 p_1 + \ell_2 p_2 + \ell_3 p_3)$$

$$\left. \right\} \quad (21)$$

in which

$$M \ll m, \quad \text{and} \quad \varepsilon = n - m > 0 \qquad (22)$$

Suppose at first that the fertility, n, is sufficiently large, and that the coefficients of utilization of the food satisfy the following inequalities:

$$\ell_1 \gg \ell_3, \qquad hE > \varepsilon \lambda \ell_1 .$$

These conditions favor the first pure group, and they eventually lead to the disappearance of the second pure group and the hybrids. Only the first pure group will survive, eventually reaching the equilibrium with the food-species:

$$\lim p_1 = \frac{E\ell_1 + \varepsilon H}{\lambda \ell_1^2 + Hh}$$

$$\lim p_2 = 0, \qquad \lim p_3 = 0$$

$$\lim x = \frac{hE - \varepsilon \lambda \ell_1}{\lambda \ell_1^2 + Hh} .$$

$$\left. \right\} \quad (23)$$

When the second pure group is favored:

$$\ell_2 \gg \ell_3, \qquad hE > \varepsilon \lambda \ell_2$$

one has

$$\lim p_1 = 0$$

$$\lim p_2 = \frac{E\ell_2 + \varepsilon H}{\lambda \ell_2^2 + Hh}$$

$$\lim p_3 = 0$$

$$\lim x = \frac{hE - \varepsilon \lambda \ell_2}{\lambda \ell_2^2 + Hh} .$$

$$\left. \right\} \quad (24)$$

When the growth rate is insufficient:

$$\varepsilon < -\frac{\ell_k E}{H}, \qquad (k = 1, 2, 3) \tag{25}$$

the predatory species disappears:

$$\lim p_1 = \lim p_2 = \lim p_3 = 0, \qquad \lim x = \frac{E}{H}. \tag{26}$$

Finally, when ℓ_3 is sufficiently large, and ε is positive, the co-existence of the three predatory groups with their prey is not impossible. Thus the ultimate fate of the system depends primarily on the hybrid group, and in particular on its coefficient of food utilization, ℓ_3. If ℓ_3 is between ℓ_1 and ℓ_2 only the pure group which better utilizes its food will survive. If the hybrids utilize their food less well than the two pure groups, the result depends on the initial composition of the predatory species. If the hybrids utilize their food better than either pure group, then co-existence of the three groups with the prey species is possible. Needless to say, that supplementary conditions expressed by inequalities are necessary in every case for such conclusions to be valid.

All of these phenomena are still possible when ε is negative, that is to say when species x is almost the only source of food for species p, provided, however, that the inequalities (25) are not satisfied.

In this way the reality of the selection mechanism can no longer be disputed. In the case of two very similar species, which nevertheless cannot hybridize, the probability that the less favored species will disappear is far greater than the probability that the two species will co-exist. In the case of two interbreeding races the fate of these races depends on the vital coefficients of the hybrids. If the hybrids are less favored than the two pure races it is very probable that the less numerous race will disappear. This conclusion determines the fate of all new mutations which produce feeble hybrids. In the opposite case the co-existence of the three groups becomes quite probable. In the intermediate case only the more favored pure race survives. This is what happens when all the characters of the hybrids are either dominated by one of the two pure races, or they are intermediate between the characters of the two pure races.

The terms "more favored" and "less favored" are very vague. We have tried to replace them by inequalities connecting the vital coefficients, and to study separately the action of the three major groups of factors. The conclusions we have obtained are in full agreement with the observed reality.

Two points remain obscure: the causes of the appearance of mutations, and the number of different varieties which can be present in a species. The two problems are not mutually independent. Everything would seem to depend on the initial and final distributions of genes in the chromosomes. Certain physical and chemical factors may act in this process. One wonders whether, in critical circumstances, a sudden change in external factors might cause abnormal distributions, some of which may be advantageous and stable.

It would seem, on the other hand, that there is a considerable number of stable genetic distributions, and that the phenotype may vary in an almost continuous manner. We shall not enter into the details of such chromosomal mechanisms. What is important to us is the possibility of almost unlimited variation in a species through the interplay of two independent mechanisms: the appearance of mutations, and selection.

General Differential Equations for the Problem of Natural Selection

V. A. Kostitzin

In this note I establish the general differential equations for the problems of natural selection. These equations hold for a very large array of hypotheses on the probabilities of the crossings.

1. **Probabilities of the crossings ω_{hk}.** Let p be a population composed of n groups p_1, p_2, ..., p_n. Denote by ω_{hk} the probability that the fertile state of a female p_h is due to a male p_k. Such probabilities are functions of p_1, p_2, ..., p_n, connected by the obvious relationships:

$$\omega_{h1} + \omega_{h2} + \dots + \omega_{hn} = 1 \ . \tag{1}$$

They possess a few very simple properties:

(1) Suppose that the group p_k increases beyond any limit, while the other groups remain limited. In this case the probability ω_{hk}, which clearly supposes that this type of crossing is effectively possible, tends toward certainty, and one then has

$$\lim_{p_k \to \infty} \omega_{hk} = 1 \ , \qquad \lim_{p_k \to \infty} \omega_{hm} = 0 \qquad (m \neq k) \ . \tag{2}$$

(2) Suppose, on the other hand, that the group p_h increases beyond any limit. In this case the probability ω_{hh} that a female p_h will be fertilized by a male p_h tends toward certainty, and one has

$$\lim_{p_h \to \infty} \omega_{hh} = 1 \ , \qquad \lim_{p_h \to \infty} \omega_{hk} = 0 \qquad (h \neq k) \ . \tag{3}$$

This certainty is nevertheless a potential one, and a certain number of p_h females will nevertheless be fertilized by p_k males. This number is necessarily proportional to p_k, which gives

$$\lim_{p_h \to \infty} p_h \omega_{hk} = a_{hk} p_k \tag{4}$$

in which the a_{hk} are constants.

(3) The disappearance of a group p_k implies the disappearance of all the corresponding probabilities, so that

$$\lim_{p_k \to 0} \omega_{hk} = 0 \ . \tag{5}$$

(4)   In the absence of crossings between the groups p_h and p_k one has

$$\omega_{hk} = \omega_{kh} = 0 \ ,$$

and all the probabilities ω_{hm} are independent of p_k, as the probabilities ω_{km} are all independent of p_h.

(5)   In the absence of crossings between a group, p_h, and the rest of the population, one clearly has

$$\omega_{hh} = 1 \ , \qquad \omega_{hk} = 0 \qquad (h \neq k) \ . \tag{6}$$

(6)   In such conditions one can suppose that the probability ω_{hk} is a decreasing function of any one p_m, $m \neq k$, and it is an increasing function of p_k:

$$\frac{\partial \omega_{hk}}{\partial p_k} > 0 \ , \qquad \frac{\partial \omega_{hk}}{\partial p_m} < 0 \qquad (m \neq k) \ . \tag{7}$$

2.   **Coefficients of heredity.**   Under the assumptions above one can express by $n_{hk} p_h \omega_{hk}$ the rate of increase of the population p, due to the crossings between p_h females and p_k males.   Here n_{hk} denotes a number proportional to the fertility of these crossings.   It appears to be well established that, on the average, the increase is distributed among the groups of the population p following a numerical law which is relatively independent of the external conditions, and of the values of p_1, p_2, ..., p_n.   The contribution to the group p_m is expressed by the product $\lambda^m_{hk} p_h n_{hk} \omega_{hk}$.   The constant numbers λ^m_{hk}, which we call the coefficients of heredity, satisfy the following relations:

$$\lambda^1_{hk} + \lambda^2_{hk} + \ldots + \lambda^n_{hk} = 1 \ . \tag{8}$$

In certain simple cases, as in that of Mendel's laws, the coefficients of heredity can be easily calculated by probability considerations, based on the repartitioning of chromosomes during the reduction divisions in the germ cells.

3.   **Differential equations.**   Suppose now that, aside from considerations of heredity, each group p_h has its own coefficient of mortality, m_h, and is in competition with other groups.   Giving to such factors their standard mathematical form, and adding the contributions of the crossings and of asexual forms of reproduction, one gets the following differential equations:

$$p_i' = \Sigma\Sigma n_{hk} p_h \omega_{hk} \lambda^i_{hk} - m_i p_i - p_i \Sigma \mu_{ik} p_k \qquad (i = 1,2,\ldots,n). \tag{9}$$

These equations are very flexible, and they permit the study of even extreme and limit cases. Thus, in the absence of crossings, one gets from (6) Volterra's differential equations:

$$p_i' = n_i p_i - m_i p_i - p_i \Sigma \mu_{ik} p_k .$$
(10)

In the case of <u>normal panmixis</u>, in which all the groups participate on an equal basis, the following probabilities of the crossings,

$$\omega_{hk} = \frac{p_k}{p_1 + p_2 + \ldots + p_n} = \frac{p_k}{p} ,$$
(11)

give the selection equations,

$$p_i' = \frac{1}{p} \Sigma \Sigma n_{hk} p_h p_k \lambda^i_{hk} - m_i p_i - p_i \Sigma \mu_{ik} p_k ,$$
(12)

which I have studied in my recent publications (1).

(1) <u>Comptes rendus</u>, <u>203</u>:156-157 (1936); <u>Biologie mathématique</u>, Hermann, Paris (1937).

On the Mendelian Coefficients of Heredity

V. A. Kostitzin

The general differential equations for the problem of natural selection (1)

contain two systems of coefficients which have a statistical origin: the

probabilities of the crossings, ω_{hk}, and the coefficients of heredity, λ^i_{hk}. The

latter express the repartitioning of the products of a crossing (p_h female x p_k

male) among the pure groups or the hybrids which compose a population p_1, p_2, ...,

p_n:

$$\text{Products of the crossing } (p_h \times p_k) = \sum_{i=1}^{n} \lambda^i_{hk} p_i \; . \tag{1}$$

We are going to study such coefficients in the Mendelian case, under the

assumption that the characters involved are not linked to the sex chromosomes. This

hypothesis is quantitatively expressed by the symmetry of the coefficients of

heredity,

$$\lambda^i_{hk} = \lambda^i_{kh} \; , \qquad (p_h \times p_k) = (p_k \times p_h) \; , \tag{2}$$

and it allows us to disregard sexual differences within the groups.

1. **Cellular structure**. Let then p_1, p_2, ..., p_n be a population initially

composed of two or more pure groups within the same species. After a few generations

the population will contain all the hybrids which can originate from the crosses

$(p_h \times p_k)$.

Following the chromosomal theory of heredity we assume that a germ cell before

maturation contains two copies of each chromosome on which the characters are

carried. Consider a character α which can occur in the form A as well as in the

form a, and which is localized on an autosome (a pair of alleles in the usual

terminology). Since in the cells this character is carried by two chromosomes,

three combinations are possible: The homozygote of type $\varepsilon_1 = $ (AA), the homozygote

of type $\varepsilon_2 = $ (aa), and the heterozygote of type $\varepsilon_3 = $ (Aa).

One can then characterize the chromosomal structure of a germ cell before the

reduction division by the symbol:

$$E = (\varepsilon^1_{i_1}, \quad \varepsilon^2_{i_2}, \quad \varepsilon^3_{i_3}, \quad \ldots, \quad \varepsilon^m_{i_m}) \; . \tag{3}$$

The upper indices enumerate the characters, and the lower indices denote the genotypes of these characters (for example, i_1 denotes the genotype of character 1, i_2 that of character 2, etc., where $i = 1$, 2 or 3).

2. <u>The maturation divisions and the following combinations</u>. In the maturation divisions the cell loses at random one-half of its chromosomes, retaining only one of each kind. The crosses, by combining reduced cells, reconstitute diploid cells. One can express the operation and its results by the following symbolic relations:

$$\varepsilon_1 \times \varepsilon_1 = (\varepsilon_1); \qquad \varepsilon_2 \times \varepsilon_2 = (\varepsilon_2);$$
$$\varepsilon_3 \times \varepsilon_3 = \frac{1}{4}(\varepsilon_1) + \frac{1}{4}(\varepsilon_2) + \frac{1}{2}(\varepsilon_3); \qquad \varepsilon_1 \times \varepsilon_2 = \varepsilon_3; \qquad (4)$$
$$\varepsilon_1 \times \varepsilon_3 = \frac{1}{2}(\varepsilon_1) + \frac{1}{2}(\varepsilon_3); \qquad \varepsilon_2 \times \varepsilon_3 = \frac{1}{2}(\varepsilon_2) + \frac{1}{2}(\varepsilon_3).$$

These relations express Mendel's laws, and give the probable repartitioning of the products of all the crosses in the simplest case of a character with two alleles. In the case of more than one character, $m > 1$, the algorithm must be independently applied to each character through symbolic multiplication, and further grouping of the terms. To give an idea of this process, consider the case of a difference for only two independent characters. There will be nine genotypes, the four homozygotes

$$P_1 = (\varepsilon_1^1, \varepsilon_1^2), \quad P_2 = (\varepsilon_2^1, \varepsilon_2^2), \quad P_3 = (\varepsilon_1^1, \varepsilon_2^2), \quad P_4 = (\varepsilon_2^1, \varepsilon_1^2), \qquad (5)$$

the four simple heterozygotes

$$P_5 = (\varepsilon_1^1, \varepsilon_3^2), \quad P_6 = (\varepsilon_3^1, \varepsilon_1^2), \quad P_7 = (\varepsilon_3^1, \varepsilon_2^2), \quad P_8 = (\varepsilon_2^1, \varepsilon_3^2), \qquad (6)$$

and a double heterozygote

$$P_9 = (\varepsilon_3^1, \varepsilon_3^2). \qquad (7)$$

The heterogeneity is measured by the number of symbols ε_3 appearing in the formula for a given genotype. To compute the result of a crossing, it is enough to apply the very simple symbolic operations in formulas (1) and (4). One has, for instance,

$$P_1 P_3 = (\varepsilon_1^1, \varepsilon_1^2) \times (\varepsilon_1^1, \varepsilon_2^2) = (\varepsilon_1^1 \times \varepsilon_1^1, \varepsilon_1^2 \times \varepsilon_2^2) = (\varepsilon_1^1, \varepsilon_3^2) = P_5,$$

which, then, gives

$$\lambda_{13}^5 = 1, \qquad \lambda_{13}^i = 0 \qquad (i \neq 5).$$

In the slightly more complex case

$$P_5 P_9 = (\varepsilon_1^1, \varepsilon_3^2) \times (\varepsilon_3^1, \varepsilon_3^2) = (\varepsilon_1^1 \times \varepsilon_3^1, \varepsilon_3^2 \times \varepsilon_3^2) =$$
$$(\frac{1}{2}\varepsilon_1^1 + \frac{1}{2}\varepsilon_3^1, \frac{1}{4}\varepsilon_1^2 + \frac{1}{4}\varepsilon_2^2 + \frac{1}{2}\varepsilon_3^2) =$$

$$\frac{1}{8}(\varepsilon_1{}^1,\varepsilon_1{}^2) + \frac{1}{8}(\varepsilon_1{}^1,\varepsilon_2{}^2) + \frac{1}{4}(\varepsilon_1{}^1,\varepsilon_3{}^2) + \frac{1}{8}(\varepsilon_3{}^1,\varepsilon_1{}^2) + \frac{1}{8}(\varepsilon_3{}^1,\varepsilon_2{}^2)$$

$$+ \frac{1}{4}(\varepsilon_3{}^1,\varepsilon_3{}^2) = \frac{1}{8}\ P_1 + \frac{1}{8}\ P_3 + \frac{1}{4}\ P_5 + \frac{1}{8}\ P_6 + \frac{1}{8}\ P_7 + \frac{1}{4}\ P_9\ .$$

one finds

$$\lambda_{59}{}^1 = \lambda_{59}{}^3 = \lambda_{59}{}^6 = \lambda_{59}{}^7 = \frac{1}{8}\ ;$$

$$\lambda_{59}{}^5 = \lambda_{59}{}^9 = \frac{1}{4}\ ;\qquad \lambda_{59}{}^2 = \lambda_{59}{}^4 = \lambda_{59}{}^8 = 0.$$

(1)   Comptes rendu, 206:570-572 (1938).

The Differential Equations for the Problem of Natural Selection

in the Case of Mutations on Sexual Chromosomes

V. A. Kostitzin

In a very high number of cases the cellular structure of the two sexes is not the same with respect to the sex chromosomes. In this respect two different sex determinations are possible: either the male is heterogametic and the female is homogametic:

$$\text{Male} = (XY), \qquad \text{Female} = (XX) \tag{1}$$

or the converse:

$$\text{Male} = (XX), \qquad \text{Female} = (XY) \tag{2}$$

We refer here to the former case (1), since the population behavior of case (2) is essentially the same as for case (1). Consider a sex-linked mutant of the normal X chromosome, denoted by x. In this case five cellular configurations are possible:

$$
\left.
\begin{aligned}
\eta_1 &= (XX) - \text{homozygote non-mutant female} \\
\eta_2 &= (XY) - \text{non-mutant male} \\
\eta_3 &= (xx) - \text{homozygote mutant female} \\
\eta_4 &= (xY) - \text{mutant male} \\
\eta_5 &= (xX) - \text{heterozygote female.}
\end{aligned}
\right\} \tag{3}
$$

The results of the crosses among these five different genotypes can be described by the following symbolic formulas:

$$
\left.
\begin{aligned}
\eta_1\eta_2 &= \tfrac{1}{2}(\eta_1) + \tfrac{1}{2}(\eta_2), & \eta_1\eta_4 &= \tfrac{1}{2}(\eta_2) + \tfrac{1}{2}(\eta_5), \\
\eta_3\eta_2 &= \tfrac{1}{2}(\eta_4) + \tfrac{1}{2}(\eta_5), & \eta_3\eta_4 &= \tfrac{1}{2}(\eta_3) + \tfrac{1}{2}(\eta_4), \\
\eta_5\eta_2 &= \tfrac{1}{4}(\eta_1) + \tfrac{1}{4}(\eta_2) + \tfrac{1}{4}(\eta_4) + \tfrac{1}{4}(\eta_5), \\
\eta_5\eta_4 &= \tfrac{1}{4}(\eta_2) + \tfrac{1}{4}(\eta_3) + \tfrac{1}{4}(\eta_4) + \tfrac{1}{4}(\eta_5),
\end{aligned}
\right\} \tag{4}
$$

which are analogous to formulas (4) in my note on the Mendelian coefficients of heredity for the autosomal case (1). If, besides this sex-linked character, m autosomal characters are also polymorphic for a pair of alleles, one can represent the possible cellular configurations by the symbolic form:

$$H = (\eta_j, \; \varepsilon_{i_1}^{1}, \; \varepsilon_{i_2}^{2}, \; \ldots, \; \varepsilon_{i_m}^{m}) \qquad (j = 1,2,3,4,5; \; i_1, i_2, \ldots, i_m = 1,2,3) \tag{5}$$

analogous to formula (3) of the same note. The results of the crossings are easily calculated through symbolic multiplications utilizing the formulas (1) and (4) of my cited note. The number n of groups of type (5) is $5 \cdot 3^m$, and each sex-linked group must appear separately. In this way, within each homozygous group one distinguishes between two sexual groups. As in the case of autosomal variations there are also in this case subsystems whose structure is necessarily more complex. As an example consider the system of order 2. One will have the following fifteen genotypes:

$$\left.\begin{array}{lll}
P_1 = (\eta_1, \varepsilon_1) \;, & P_6 = (\eta_1, \varepsilon_2) \;, & P_{11} = (\eta_1, \varepsilon_3) \;, \\
P_2 = (\eta_2, \varepsilon_1) \;, & P_7 = (\eta_2, \varepsilon_2) \;, & P_{12} = (\eta_2, \varepsilon_3) \;, \\
P_3 = (\eta_3, \varepsilon_1) \;, & P_8 = (\eta_3, \varepsilon_2) \;, & P_{13} = (\eta_3, \varepsilon_3) \;, \\
P_4 = (\eta_4, \varepsilon_1) \;, & P_9 = (\eta_4, \varepsilon_2) \;, & P_{14} = (\eta_4, \varepsilon_3) \;, \\
P_5 = (\eta_5, \varepsilon_1) \;, & P_{10} = (\eta_5, \varepsilon_2) \;, & P_{15} = (\eta_5, \varepsilon_3) \;,
\end{array}\right\} \qquad (6)$$

among which there are four types of homozygous females, P_1, P_3, P_6, P_8, four types of homozygous males, P_2, P_4, P_7, P_9, four types of females heterozygous at one locus, P_5, P_{10}, P_{11}, P_{13}, two types of males heterozygous at one locus, P_{12}, P_{14}, and a single double heterozygous female, P_{15}. Regarding the singular points, there are: (1) the extinction of all the types; (2) the points (P_1, P_2), (P_3, P_4), (P_6, P_7), (P_8, P_9), representing the survival of what one might call different "pure races," with the extinction of all the other types; (3) points of survival of the subsystems $(P_1, P_2, P_3, P_4, P_5)$, $(P_6, P_7, P_8, P_9, P_{10})$ homogeneous with respect to the autosomal characters; (4) points of survival of the subsystems $(P_1, P_2, P_6, P_7, P_{11}, P_{12})$, $(P_3, P_4, P_8, P_9, P_{13}, P_{14})$ homogeneous with respect to the sex-linked trait; (5) equilibrium points with all fifteen types.

The differential equations for natural selection (2) still hold, but now the double sum is only extended to the crosses which involve one male and one female genotype. Also, the probabilities of the crossings, ω_{hk}, now depend only on the P_k males. The internal stability of the "pure races" is guaranteed if their respective natalities are sufficiently large. Such "pure races" are stable with respect to selection on the "missing" alleles if the possible heterozygotes have sufficiently high mortality. Concerning the other singular points one can only repeat what was said in my previous note (3), (Editors' note: cf. the following paper by Kostitzin).

(1)   Comptes rendus, 206:883–885 (1938).

(2)   Comptes rendus, 206:570–572 (1938).

(3)   Comptes rendus, 206:976–978 (1938).

On the Singular Points of the Differential Equations

in the Problem of Natural Selection

V. A. Kostitzin

1. Let $(p_1, p_2, \ldots, p_n)$ be an ensemble of n genotypes of the same species which differ among themselves for m independent characters $(\alpha_1, \alpha_2, \ldots, \alpha_m)$, each of which is represented by a pair of alleles. Assuming that all of the possible combinations of these characters are effectively realized, one will have, as in my previous note (1), a genetic system composed of $n = 3^m$ different genotypes. The dynamics of this system can be represented by the system of differential equations (2):

$$p_i' = \sum_{h=1}^{n} \sum_{k=1}^{n} n_{hk} \lambda_{hk}^{i} \, p_h \omega_{hk} - m_i p_i - p_i \sum_{k=1}^{n} \mu_{hk} p_k \qquad (i = 1,2,\ldots,n). \qquad (1)$$

As also specified in my previous note, there can be, corresponding to m pairs of alleles, 2^m different genotypes homozygous for each character, $m \cdot 2^{m-1}$ different genotypes heterozygous for only one character, and so on. In general there will be n_s different genotypes:

$$n_s = \frac{m! \; 2^{m-s}}{s! \, (m-s)!} \, , \qquad (2)$$

each genotype heterozygous for s characters, and homozygous for all the others.

2. <u>Subsystems of order s</u>. Each simple heterozygote results from the crossing of two homozygotes, and forms with them a subsystem of the first order which is homozygous for all the characters but one. In the same way one can form subsystems of order s composed of 3^s genotypes, and homozygous for all but s of the characters. The number of these different subsystems is given by the formula (2). My previous note gives an example of a system of order 2. Each subsystem constitutes a complete ensemble in the sense that crossings among the genotypes of this ensemble do not introduce any new genotypes. It then follows that in the cross of two genotypes, p_h and p_k, belonging to the same system, the Mendelian coefficients λ_{hk}^{i} will be zero if the index i does not belong to one of the genotypes in the subsystem.

3. <u>Singular points</u>. This classification is of great importance for the study of the singular points of system (1). In fact, <u>all</u> <u>the</u> <u>singular</u> <u>points</u> <u>of</u> <u>all</u> <u>the</u>

subsystems of any order less than m are also singular points of the complete system (1). On the other hand, for any of the singular points, the genotypes which comprise the equilibrium at these points must constitute a complete ensemble. It follows that the singular points of system (1) must necessarily belong to one of the following categories: (1) The single point of extinction of the whole population; (2) The 2^m points corresponding to the survival of a single homozygous genotype and the extinction of all the others; (3) The points corresponding to the survival of the genetic subsystems which represent an ensemble of some of the genotypes and the extinction of all the others; (4) The equilibrium points for the ensemble of all n genotypes.

4. Probabilities of the crossings. The preceding results do not depend on the form of the probabilities of the crossings, ω_{hk}. However, the study of the stability of the singular points is impossible without more concrete hypotheses on these functions. One can suggest a very simple form for such functions which satisfies all the conditions imposed on them. To this end, set:

$$\omega_{hk} = \frac{c_{hk}p_k}{\pi_h} \quad , \qquad \pi_h = \sum_{s=1}^{n} c_{hs}p_s \, . \tag{3}$$

The coefficients c_{hk} express the attraction between p_h females and p_k males. In the case of normal panmixis one again finds the values considered earlier. In the complete absence of crossings between individuals of different genotypes one obtains Volterra's biological equations. Thus, this particular form of the ω_{hk} is still sufficiently flexible.

5. Stability of the singular points. The extinction of all the genotypes is incompatible with all the other singular points since it is only realized when all the mortalities are sufficiently high:

$$m_i > n_{ii} \qquad (i = 1, 2, \ldots, n). \tag{4}$$

Let us briefly examine the stability of one of the singular points corresponding to the extinction of all the genotypes except for one homozygous genotype. Let p_i be this homozygote which attains a final population size:

$$\lim p_i = \frac{n_{ii} - m_i}{\mu_{ii}} > 0 \ . \tag{5}$$

The characteristic equation for this equilibrium has $\dot{n}$ roots. First of all, one finds for $s = i$:

$$\rho_i = - (n_{ii} - m_i) \ . \tag{6}$$

Then, one has for all the indices s corresponding to the homozygous genotypes:

$$\rho_s = - m_s - \mu_{si} p_i \ , \tag{7}$$

and, for all the indices corresponding to heterozygous genotypes:

$$\rho_s = \lambda_{si}^s \left(n_{si} + \frac{n_{is} c_{is}}{c_{ii}} \right) - m_s - \mu_{si} p_i \ . \tag{8}$$

The equilibrium is stable if all the latent roots ρ_s are negative. This condition is satisfied if the intrinsic birth rate n_{ii} of the surviving homozygous genotype, and all the mortalities of the heterozygous genotypes, are sufficiently large. All of the equilibrium points of this type are compatible among themselves, and, if the stability conditions are satisfied for yet another homozygous genotype, the final composition of the population will depend on the initial conditions. In all the cases so far considered the singular points of this type are incompatible with the stability of all the others. Concerning this last case, the stability conditions for a subsystem of order s become more and more complex and difficult to meet as s increases. This fact is of great biological importance.

(1) Comptes rendus, 206:883-885 (1938).

(2) Comptes rendus, 206:570-572 (1938).

Natural Selection and Transformation of the Species from the Mathematical, Statistical and Biological Points of View

V. A. Kostitzin

In this note I will compare three aspects, mathematical, statistical and biological, of the problem of the transformation of species. The transformation is assumed to result from competition among different genetic types or "races" of the same species.

1. <u>Relative probabilities of the singular points</u>. The stable singular points of the differential equations of this problem correspond to the final equilibrium states in the struggle for life among the groups which compose a single population. Such stability conditions determine a domain in the hyperspace of the vital coefficients. The size of this domain becomes progressively smaller as the order of the singular point increases (1). This remark gives us a rough idea of the relative probabilities of the singular points. These probabilities are the same as the <u>a priori</u> probabilities of different adaptive solutions in the struggle for life. One gets an even better picture of the problem by taking into consideration the statistical distribution of the vital coefficients.

2. <u>The instability of the vital coefficients, and its effect on the stability of the singular points</u>. The vital coefficients vary considerably from one time to another, and this variation alters the singular points and their domains of stability. In fact, the ensemble of vital coefficients for a population in a given biotope will remain for only insignificantly short periods of time inside the domains of stability for singular points of order greater than 1. Therefore, one might consider as nonexistent any equilibrium for a genetic system with more than one segregating character. The equilibrium points of order greater than 1 are not highly probable, but, considering the ensemble of all the biotopes occupied by a given species, they are not impossible from a statistical point of view. Thus, the local instability of vital conditions opposes the formation of structurally complex genetic equilibria, and the mechanism of natural selection actually favors the

simplification of a population's genetic structure.

3. <u>Conditions for the maintenance of homozygosity</u>. Consider, then, the more probable hypothesis: that of a monomorphic population. Suppose that in a homogeneous population a different mutation appears for each of m autosomal characters. In this way 3^m genotypes become possible, of which 2^m are homozygous, and ($3^m - 2^m$) are heterozygous to various degrees. Let p_i be one of the homozygous types. It is intrinsically stable if the difference ($n_{ii} - m_i$) is positive. The conditions for its stability with respect to mutations (1):

$$\rho_s = \lambda_{si} {}^s \left(n_{si} + \frac{n_{is}c_{is}}{c_{ii}} \right) - m_s - \mu_{si}p_i < 0 \tag{1}$$

must be satisfied for all the s corresponding to the heterozygotes. Such conditions might be simultaneously satisfied by several homozygotes, and in this case the outcome depends on the initial conditions.

Now, <u>in the majority of the cases, mutants are very rare, and, owing to the numerical advantage of the original genotype, the latter has an even higher probability of survival if it also satisfies the stability conditions (1)</u>. Such results also hold in the case of mutations affecting sex-linked characters (2).

4. <u>The appearance of mutations</u>. All our reasoning is based on the chromosomal theory of heredity, and on the combinatorics which result from it. Such relatively simple rules are only an external expression of a much deeper process which is still entirely unknown. The complexity of the cellular edifice, and of the phenomena which derive from it, force us to predict that irregularities during the reduction divisions are inevitable. Must one attribute to such irregularities the appearance of mutations? If this is the case mutations due to <u>undirected chance</u> must almost always be disadvantageous. Evolution will be historically explained by the accumulation of favorable variations, and the elimination of all the others. The cell has become a well structured ensemble with respect to the development and defense of the multicellular organism. It is difficult to imagine that pure chance might frequently give mutations which are better organized, or better adapted to the ensemble of the preexisting conditions of life.

5. **Variation of species.** One sees that generally the stability of existing species is very well protected. But rare possibilities eventually materialize, and it may occur that one of the conditions of homozygosity (1) will no longer be satisfied by the original genotype, and, at the same time, that at least one of the new homozygous types will satisfy all of the stability conditions. In this case one will have a local variation of the species. In this way the mathematical model outlined above accounts very well, from the biological point of view, for natural selection and the variation of species. It permits us to reconcile the stability of species with their transformation which, after all, is inevitable (3). From the point of view of logic, then, such a model constitutes a rational theory of Darwinian evolution. It is based on a number of axioms and postulates, and it does not present any internal contradiction.

---

(1) Comptes rendus, 206:976-978 (1938).

(2) Comptes rendus, 206:1273-1275 (1938).

(3) One might see in the Encyclopedie Francaise, vol 5, pp. 5.82.3-5.82.8 (1938) a slightly modernized exposition of the major objections against transformism.

PART V

LIFE AND THE EARTH

INTRODUCTION

In this section we present a short monograph by Kostitzin "Évolution de
l'atmosphère, circulation organique, époques glaciaires" (1935). To our knowledge
this is the first comprehensive treatment of both the long range effects of life on
the development of the atmosphere and of the earth's crust, and of the effects of
short term geological phenomena on life. It follows previous treatments by
Kostitzin of geophysical and abiotic aspects of ecology (1932, 1933). The monograph
begins with a general review of the available information and some of the theories
concerning the circulation of oxygen, carbon, and nitrogen, and discusses the long
term changes in their abundance in the atmosphere and in the soil. Chapter II
discusses such theories in the light of a simple model incorporating a system of
linear and quadratic differential equations. Kostitzin pointed out the stabilizing
effect of living matter with respect to sudden variations in the composition of
the atmosphere, which arise from intense outbreaks of volcanism or from glaciations.
To our knowledge this treatment is the first extensive example of "compartment
models", similar to those which became very popular in physiology in the 1960's.

Chapters III and IV deal mostly with the problem of glaciations, their
distribution and their periodicity. Chapter III begins with a critique of previous
theories, each of which tended to justify glaciations as general, world-wide
phenomena having one or another single cause, such as variation in solar radiation
due to the earth's orbital geometry, etc. Among them, Wegener's original explanation (1922), in terms of continental drift and migration of the poles, was found
by Kostitzin to be insufficient, especially with regard to the distribution and
periodicity of the Permo-Carboniferous and Quaternary glaciations. Arrhenius's
classical explanation was also found to carry little weight when compared with the
more precise information then available to Kostitzin concerning the circulation of
CO_2 and the changes in its concentration following volcanic activity.

We believe that Kostitzin was the first to employ a precise geophysical model
to explain the local occurence of glaciations and to partially justify their
periodicity. Kostitzin deduced from basic physical principles that the local
uplift of a continental plate could lead to the rapid formation of a thick ice

sheet, whose added weight would be enought to temporarily stop and reverse the
uplift. The block would then sink, and the ice melt, thus allowing the uplift to
resume and give rise to another cycle. The processes of uplifting in the absence of
an ice sheet, its formation, sinking, and subsequent melting are qualitatively
different phenomena which can be described by distinct differential equations, each
holding in a different domain of the state space. Kostitzin's model predicted
suitable conditions for quasi-periodic behavior in the distribution of glacial
sheets. The predictions were in surprisingly good agreement with the known data,
even from a quantitative point of view.

The model itself is of sufficient interest to warrant a slight digression at
this point. Gause and Witt (1934) had applied the same novel technique, at about
the same time and independently of Kostitzin, to a simple model for "relaxation
oscillations" in a predator-prey system. In a further application of this technique to ecological systems, Kostitzin (1937, Chapt. VIII) considered the case of
two competing species (barnacles & mussels) which are eaten by a prey-switching
predator (a whelk). The model describes the discontinuities introduced by the
predator's "functional response" to each of the prey species, and accounts for
their alternate patterns of abundance. For a discussion of this case see Scudo
and Ziegler (1976).

Chapter IV ends with a general discussion of the implications of the
glaciation model in the context of all the other climatic and geophysical factors
which necessarily influence glaciations. Only recently have certain indicators
of global climate, obtained from the examination of ocean sediments, been found
which extend far enough into the past to suggest that climatic changes are substantially correlated with variations in the earth's orbital geometry (cf. Hays,
et al., 1976). Such results are in essential agreement with the classic treatment
of Milankovich. Perhaps more information of this sort would allow a more precise
discrimination among the various interrelated causative agents of glaciations,
much along the line of Kostitzin's thinking. And yet, the 19th century habit of
attempting to justify most aspects of glaciations in terms of a single causative
agent, frequently with global effects, is a slow one to die.

The last chapter begins with a discussion of diffusion of the atmosphere into space. Kostitzin pointed out that a naive interpretation of statistical mechanics had often led to overestimates of the rate of this process. He also pointed out that some of the calculated loss is partially reversible, being rather a form of exchange within a larger, very thin atmosphere which surrounds both the earth and the moon. The monograph concludes with a discussion of the secular changes in the atmosphere caused by life on earth, and the possible effects of secular changes in the atmosphere on the stability of life. In this section there is a remarkable passage in which Kostitzin describes a "multidimensional space which symbolically represents the vital factors: pressure, p; temperature, T; light, l; etc. At any time each living organism is represented by a point in this space, and a species is represented by an ensemble of points." Kostitzin's rigorous conceptual treatment of a species' relation to its surroundings is obviously an important historical antecedent of the n-dimensional niche concept introduced by Hutchinson (1957) some two decades later.

In these concluding remarks Kostitzin again warns against confusing the relative short-term stability of nature with the absolute, but misleading, long-term stability of mechanical systems "which does not, in fact, exist, either in mechanics or in biology." Warnings of this sort have escaped the attention of many ecologists. And yet, such warnings seem all the more pertinent today, as ecologists and non-ecologists alike are responding to a general feeling of impending crisis by constructing naive optimizations and Newtonian models, which project a present without a past into impossible futures.

EVOLUTION OF THE ATMOSPHERE:

Organic circulation,

glacial periods

V. A. KOSTITZIN

TABLE OF CONTENTS

In the present work I study the problem of the evolution of the atmosphere,
and of the organic circulation. This problem has been posed a number of times by
chemists and mineralogists. From the side of chemistry the road has been well paved
by such people as J. Fournet, J. Boussingault, J. B. Dumas, A. Delesse, A. Gauthier,
S. Arrhenius and W. Vernadsky. Here I consider the problem as a mathematician. It
is possible in fact to derive differential equations on the basis of empirical
postulates. In turn, one derives conclusions which are testable either at the
present time, or in the not so distant future. This logical apparatus and the
results one derives from it, are no less valid than empirical conclusions.

I

THE COMPOSITION OF THE ATMOSPHERE AND THE ROLE OF ORGANIC MATTER

1. Oxygen

It is necessary to start by considering a few figures. The atmosphere contains
10^{15} metric tons of free oxygen while there are certainly more than 10^{19} tons
contained in the earth's crust. Taking into account the chemical activity of oxygen,
this fact seems enigmatic. According to R. Perrin (16), in fact: "An increase of
less than one meter in the height of the earth's crust, which has a thickness on the
order of a hundred kilometers, would have fixed all the oxygen in the atmosphere,
the amount of which is almost negligible as compared to that in the crust." How can
it be that there is still some free oxygen? Since there are enormous masses of
magma not saturated with oxygen, how is it possible that one meter more or one less
might make any difference? Would it not be excessive to explain the existence of
free oxygen through the chance that the earth's crust is one meter thinner? To pose
the question in this way is much the same as to solve it. The fact that one still
has free oxygen after a geological history of the order of two billion years must
simply mean that the replacement of oxygen in the atmosphere is not less intense
than the process of oxidation. This replacement must not be sought within the crust,
since any oxygen freed there by any chemical reaction would be immediately absorbed
by oxidation before reaching the atmosphere. Now, among all the chemical processes

at the surface of the earth, those which lead to tertiary substances are certainly accompanied by the liberation of important quantities of oxygen. According to L. Mangin (2), plants assimilate more oxygen than carbon dioxide. The total weight of the oxygen freed in these reactions must amount to a substantial fraction of the total weight of the earth's vegetation. This, in turn, is a non-negligible fraction of the earth's oxygen. One can also point to other processes which are accompanied by the liberation of oxygen. Such are, for example, a number of chemical reactions in the decomposition of organic matter, the activities of certain micro-organisms, etc. These processes which give back free oxygen have as their counterparts the loss of oxygen by oxidation and dispersion:

(1) Oxidation by the respiration of animals and plants;

(2) The diffusion of oxygen from organic decomposition into the earth's crust;

(3) The diffusion of oxygen into the earth's crust through erosion, etc.;

(4) The oxidation of carbon by combustion (man's industrial activities);

(5) The diffusion of oxygen into outer space by the mechanism of collisions between particles.

We will study all of these factors later on.

2. Carbon

The atmosphere contains approximately $2\text{-}3 \times 10^{12}$ metric tons of carbon dioxide. The quantity contained in the oceans is approximately 10^{14} tons. These quantities are minute compared with the weight of carbon contained in the earth's crust, $10^{16}\text{-}10^{17}$ tons. On the other hand, the total weight of oil and other fossil fuels of organic origin is of the order of 2×10^{13} tons. This carbon was formerly atmospheric (†) and it will be transformed back into carbon dioxide by man's industrial activity. One sees then that organic matter is, in the long run, a very important factor. This factor, which we are now going to consider, is capable of producing

(†) This does not mean that all this carbon was dissolved in the atmosphere at any one time. It is more likely that it resulted from the ages-old activity of plants which extracted it from the atmosphere where it was continually introduced by volcanoes and other sources.

basic changes in the atmosphere and in the earth's crust.

It is convenient to start with an important observation on the structure of the organic world. The great French chemists J. B. Dumas and J. Boussingault (4) have already pictured the living world as "an appendix of the atmosphere." Thus, J. B. Dumas (4) states: "Summing up, we see that the primitive atmosphere of the earth has been subdivided into three large components: The present atmosphere, the plants, and the animals.... In this way, what plants get from the atmosphere they pass on to animals; the animals give it back to the atmosphere. This is the eternal circle by which life moves and is manifest, but in which matter is only shifted from place to place. The brute matter of air, which is organized little by little in plants, then comes to function in animals without alteration, to become the material basis for thought. Finally, broken by this effect, it returns as brute matter to the grand reservoir from which it came..." This was the opinion of the great masters of chemistry a century ago. Here is what a contemporary geochemist, W. Vernadsky (21), has to say: "It is certain that more than 97-98 percent of living matter comes from terrestrial gases, and that a smaller quantity of the same order of magnitude, is liberated in the form of gas after the death of an organism." As J. B. Dumas and J. Boussingault (4) have foreseen, the role of plants in this organic circulation is completely different from that in animals. The latter do not synthesize the tertiary substances, which they derive from plants. All carbon compounds in animal tissues come directly or indirectly from plants. We shall see that this important difference gives rise to a sequence of interesting phenomena, both at the biological and at the geological level.

The carbon dioxide liberated by living beings is in part absorbed by the oceans, and in part liberated into the atmosphere, so that the partial pressure of this gas increases. From the observations by de Saussure, Boussingault, and Godlewski (6), the rate at which plants assimilate carbon is roughly proportional to the partial pressure of carbon dioxide. This rate reaches a maximum when the proportion of carbon dioxide is around 10%. The proportion of carbon dioxide in the present atmosphere is far lower, and it seems plausible that such a percentage has never been reached. Therefore, one can consider the plant world as an automatic regulator of the carbon content of the atmosphere. Any time the quantity of this gas increases, the plants respond by increasing their consumption. Notice, again, that by taking into account absorption by oceans, the combustion of all the oil would increase the present content of carbon dioxide in the atmosphere by only a factor of three. Still it would be far from the optimum of 10% referred to above. It is also worth noting that the upper regions of the lithosphere are enriched with carbon to a degree which is 5-10 times that in the lower regions. This remarkable distribution must be attributed to the secular action of living organisms, especially those in the sea.

3. Nitrogen

The atmosphere contains 4×10^{15} metric tons of nitrogen; that is a quantity of the same order as in the lithosphere. This scarcity of nitrogen in the earth's crust is interesting and still poorly understood. Should one conclude that the nitrogen in the soil has an atmospheric origin? In this case, what has been the role of living matter in the transformation? The fact that the total amount of nitrogen in the earth seems to be below 10^{16} tons poses an enigma since the spectroscopic data indicate that nebulae are rich in nitrogen. It is probably in this respect that we can better follow the workings of a kinetic mechanism which we shall consider later on.

Nitrogen is an essential component of living matter, and very special conditions are necessary for the formation of nitrogen compounds and for their assimilation by living beings. Such conditions, which are still poorly understood, are due to the fact that the chemical activity of nitrogen is far less than that of oxygen. On this point, one can still find ridiculous statements in works which are only ten years old. In any case, it seems established that, in general, neither animals nor plants are able to directly assimilate nitrogen. Animals receive their nitrogen from plants, and the plants in turn get it from the soil through certain bacteria. Only the latter are able to fix atmospheric nitrogen. On the other hand, all organisms certainly contribute to the enrichment of the soil with nitrogen compounds.

A few cosmological considerations are necessary to throw some light on the origin of mineralized nitrogen (†). Let us go back to the very distant time when the earth was in a gaseous state. The density of this gas decreased from the center to the periphery. It is very likely that this nebula was composed of a succession of layers, each containing elements of close atomic numbers. Some of these elements must have been in an atomic state, and the chemical composition of different gaseous layers must have been very complex. Any one layer was constantly mixed by ascending and descending currents, similar to those which are now observed in the atmosphere of the sun or of the earth. This mixing facilitated an interplay of chemical combinations, in so far as temperature and other conditions permitted. This can be confirmed by the abundance of light elements in the atmosphere, the hydrosphere, and the earth's crust. In this way chemically active gases tended to be eliminated. Only inert gases were left in the atmosphere, such as argon, neon, krypton, xenon, and, to a large extent, nitrogen. Part of this last gas was able to leave the atmosphere and enter the soil. We can point out some of the chemical reactions through which this movement was facilitated. For example, nitrogen and oxygen in the atmosphere can combine to form nitric acid under the influence of electric discharges and in the presence of water vapor. Nitric acid is likewise produced by the combustion of hydrogen in the air.

Now, the formation of water was one of the most important stages in the chemical history of the earth. Furthermore, nitrogen can combine with boron, silicon, and magnesium, light elements which are very abundant in the earth's crust. In this way nitrogen, although chemically very inert, could easily leave the atmosphere and actually contribute to chemical processes in the soil. The 4×10^{15} tons of mineralized nitrogen are not exclusively due to the activity of living beings. On the other hand, it would also be wrong to think that all the primitive nitrogen was eliminated from the atmosphere before life appeared, and that all the

(†) There is nothing less certain than cosmological considerations. It is enough to notice by the way of an example that the hypothesis that solid bodies develop from nebulae is opposed by some authors who consider the reverse process more likely. I expose here the opinion which I consider the more plausible.

present atmospheric nitrogen resulted from life alone, as is the case for oxygen.
In any event, it is certain that living matter has left its powerful imprint on
every form of nitrogen on earth.

II

EQUATIONS AND THEIR SOLUTIONS

4. Simplifications

Quantitatively, living matter constitutes only a very minute portion of the
external layers of the earth. Yet, its long term effect on the surface of the
earth is enormous. Down to considerable depths, with the exclusion of rocks of
pure magmatic origin, the action of organic factors has affected every fragment of
rock. We have seen in the preceding chapter that the atmosphere itself can be
created and recreated by the action of living beings. Thus, W. Vernadsky (21) says:
"...enormous quantities of gaseous matter must pass each year through living matter.
Such quantities are certainly several times larger than the mass of the entire
atmosphere." The exchange of matter between the organic world and the atmosphere
is, then, very fast and intense. We have already examined the relationships joining
the living world with brute matter. We now have to state this problem in terms of
equations, and work out the various long and short-term effects of this circulation.

In our equation we will neglect the industrial activity of men. In the first
place it is very difficult to attribute numerical values to this factor. In the
second place one can as well neglect the effects of industrial activity, since one
is dealing with time spans at least of the order of centuries, and often much more,
rather than with instantaneous effects.

No matter at what rate the mineral energy sources are destroyed, they will
become exhausted in a few centuries. Mankind will be forced to look for different
sources of energy, and there is no doubt that he will find them. It is clear that
life in general, and human life in particular, will not end at that time. The
carbon dioxide liberated during this time span will be divided between the atmosphere and the oceans. From that moment on, our equations can again be applied
without any need to introduce complementary terms. It will be enough to replace a

few numerical coefficients with others, taking into better account the increase in the partial pressure of carbon dioxide. At that time one would have to choose as initial values of the functions the unknown values which then represented the current situation. The qualitative nature of the equations and their solutions would not be effected.

The difficulty introduced by the effect of the hydrosphere is more serious. We have seen that the total weight of carbon dioxide in the oceans is 30-40 times greater than that in the atmosphere. If this proportion were not to change, no matter what the circumstances, Schloesing (20) would have been correct in saying that the oceanic waters are a great regulator of carbon dioxide. The oceans liberate carbon dioxide when its atmospheric tension decreases, and they absorb carbon dioxide when its atmospheric tension increases. Quoting again from W. Vernadsky (21): "In this way Schloesing (1878) has discovered a great chemical property of the earth's crust, which was later confirmed by independent researchers. The importance of this fact in geology is undoubtedly vast, since the proportion between water and land surfaces has been highly variable on a geological scale. We are still far from recognizing the full geological implications of such facts. In their present manifestation they are more complex than Schloesing had indicated." We will accept this regulatory role for the oceans. This hypothesis allows us to greatly simplify the equations without introducing extra assumptions on the regulatory actions of the oceans.

A third difficulty is introduced by sealife, and by the role of oceanic CO_2 in the organic circulation. If the oceanic circulation is characterized by the same dual scheme (plants $\longrightarrow$ animals) which rules the atmospheric circulation, the mathematical problem is simplified. We will then be in the cases studied in Nos. 5-7 of this work. On the other hand, if the whole body of carbon dioxide in the oceans takes active part in the synthesis of calcium carbonate, the organic circulation in the ocean must be studied separately, as in No. 11.

5. The Circulation of Carbon and Oxygen

The special problem of the circulation of carbon and oxygen will be considered first. Let us set:

x, the total weight of free atmospheric oxygen;

y, the total weight of carbon dioxide in the atmosphere and in the oceans;

v, the total weight of these elements in plants;

u, the total weight of these elements in animals;

s, the total weight of these elements in the earth's crust.

The relationships among the variables can be represented by the scheme which is illustrated in Fig. 1, or by the differential equations:

$$x' = -\alpha_{13}u - \alpha_{14}v + \alpha_{41}v \tag{1}$$

$$y' = \alpha_{32}u - \alpha_{24}v + \alpha_{42}v \tag{2}$$

$$u' = \alpha_{13}u - \alpha_{32}u - \alpha_{35}u + \beta uv \tag{3}$$

$$v' = \alpha_{14}v - \alpha_{41}v + \alpha_{21}v - \alpha_{42}v - \alpha_{45}v - \beta uv \tag{4}$$

$$s' = \alpha_{35}u + \alpha_{45}v . \tag{5}$$

The structure of these equations is very simple. Each term corresponds to one of the arrows in Fig. 1. Equation (1), for instance, shows that atmospheric oxygen is consumed by animal respiration $(-\alpha_{13}u)$, by plant respiration $(-\alpha_{14}v)$, and in the process of assimilation the plants emit free oxygen $(+\alpha_{41}v)$. Equation (2) shows that the atmosphere receives carbon dioxide which is released by animals and by plants through respiration and the decomposition of living matter $(\alpha_{32}u, \alpha_{42}v)$, and that plants assimilate carbon dioxide $(-\alpha_{24}v)$. For simplicity, it is assumed that the intensity of this process does not depend on the atmospheric content of this gas. This hypothesis is certainly wrong if applied over the whole of geologic time, but it is admissable over considerably large intervals of time. For the time being, we will leave out all the other positive and negative sources of CO_2 (volcanism, human industry, erosion, etc.). In equation (3) the terms $(\alpha_{13}u, -\alpha_{32}u)$ represent the amounts animals take in, and respectively give back to the atmosphere. The term (βuv) means that animals live at the expense of plants, and that this process is governed by the so-called principle of encounters between the consumers and their food. The term $(-\alpha_{35}u)$ represents the enrichment of the soil through the decomposition of animal bodies. Equations (4) and (5) are derived in much the same way as the previous ones.

The variable y expresses the total weight of CO_2 which is free in the atmosphere

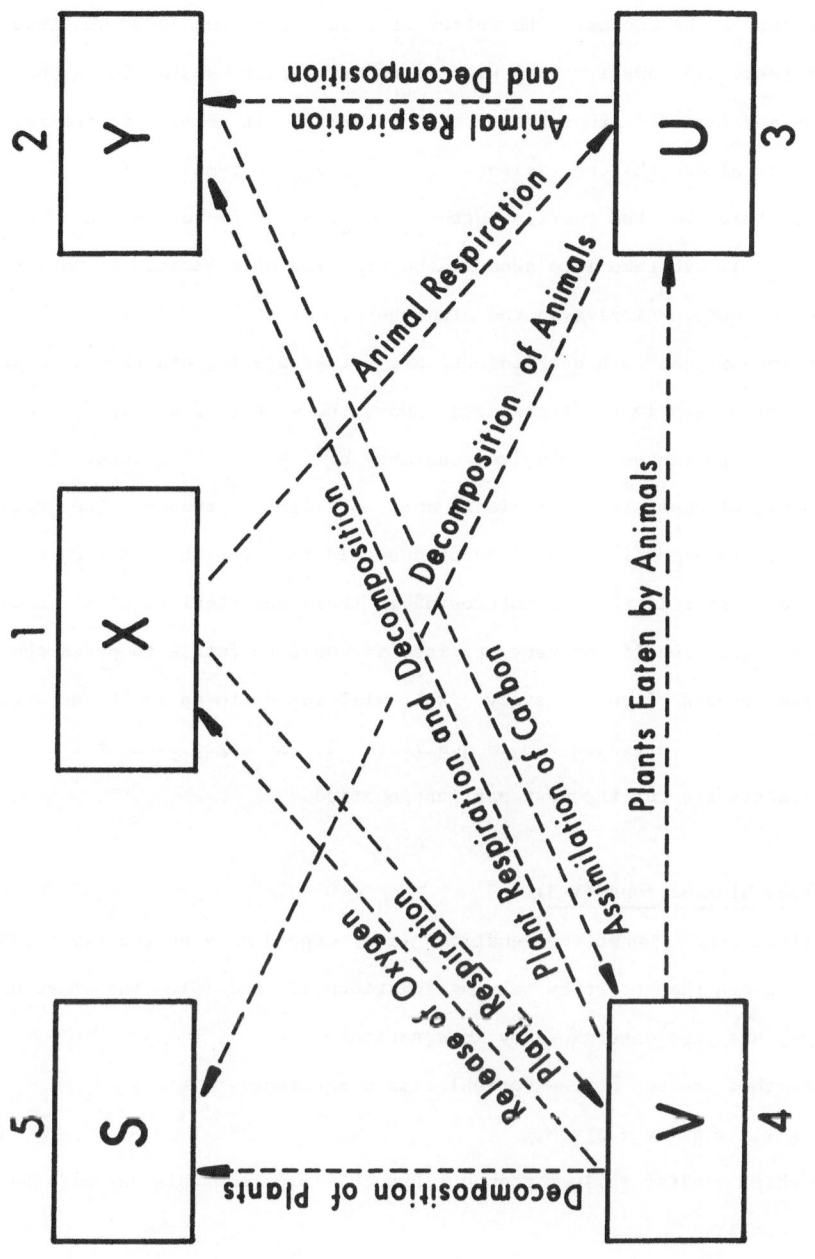

Fig. 1.

or in solution in the oceans. The latter of these two terms is 30-40 times larger than the former. It constitutes a kind of reserve which reestablishes the equilibrium any time the atmospheric content varies. It is easy to realize that this mechanism allows the coefficients α_{32}, α_{24}, α_{42} to remain relatively constant, and the same holds for the coefficients α_{13}, α_{14}, α_{41}. Because of this relative constancy we will not take into account the reaction of organisms to variations in the tension of carbon dioxide in the atmosphere.

Regarding oxygen, such coefficients are not as stable, and they do depend on the tension of oxygen in the atmosphere. Now, the weight of atmospheric oxygen is large enough to guarantee a relative constancy of these coefficients. Therefore, the hypothesis of constant coefficients does not alter appreciably the quality of the equations and simplifies their solutions. In this way, however, we would introduce a serious error if we were interested in their numerical results. Lacking empirical values, even if not very precise, it would be futile to press the quantitative aspects of such a study. What will matter to us is the general behavior of the solutions, which remain qualitatively unchanged even if the values of the coefficients are for the most part unspecified.

6. The Study of Equations (1-5)

The right hand sides of the equations (1-5) depend only on the two variables u and v. One can then start by solving equations (3) and (4). The other unknowns, x, y, and s, are then determined by integration.

Notice that the sum of the variables is a constant:
$$x + y + u + v + s = M, \tag{6}$$
something which clearly follows from our hypothesis. From this one also has the obvious inequality:
$$u + v < M. \tag{6'}$$
Notice also that s is an increasing function, meaning that the process of life leads to the diffusion into the soil of carbon and oxygen from the atmosphere.

Let us give to equations (3) and (4) a simpler form:
$$u' = u(-\lambda + \beta v) \tag{7}$$

$$v' = v(\mu - \beta u) \tag{8}$$

by setting:

$$\lambda = \alpha_{32} + \alpha_{35} - \alpha_{13}$$

$$\mu = \alpha_{14} - \alpha_{41} + \alpha_{24} - \alpha_{42} - \alpha_{45}.$$

The parameter λ is positive since, in the absence of plants, the animals could not exist; the parameter μ is positive since, in the absence of animals, the vegetation would rapidly increase.

From equations (7) and (8) one easily derives a relation between u and v:

$$u^{\mu}v^{\lambda} = H e^{\beta u + \beta v} \tag{9}$$

in which the constant H depends on the initial values u_0 and v_0:

$$H = u_0{}^{\mu}v_0{}^{\lambda}e^{-\beta u_0 - \beta v_0}.$$

The curves (9) are closed (Fig. 3, I) which means that u and v are periodic functions of time t. Let ω be the period and

$$u_c = \frac{\mu}{\beta}, \qquad v_c = \frac{\lambda}{\beta}$$

the coordinates of the common center C of all curves (9). The period ω depends not only on the coefficients λ, μ, β, but also on the constant H, that is, on the initial values u_0 and v_0. The period is zero when $u_0 = u_c$, $v_0 = v_c$, and infinite in the case $H = 0$.

Equations (7) and (8) give:

$$u(t + \omega) - u(t) = -\lambda \int_0^{\omega} u \, dt + \beta \int_0^{\omega} uv \, dt = 0$$

$$v(t + \omega) - v(t) = \mu \int_0^{\omega} v \, dt - \beta \int_0^{\omega} uv \, dt = 0$$

or:

$$\int_0^{\omega} u \, v \, dt = v_c \int_0^{\omega} u \, dt = u_c \int_0^{\omega} v \, dt .$$

One next gets:

$$\log \frac{u(t + \omega)}{u(t)} = 0 = -\lambda\omega + \beta \int_0^{\omega} v \, dt$$

$$\log \frac{v(t + \omega)}{v(t)} = 0 = \mu\omega - \beta \int_0^{\omega} u \, dt$$

so that:

$$u_c = \frac{1}{\omega} \int_0^{\omega} u \, dt, \qquad v_c = \frac{1}{\omega} \int_0^{\omega} v \, dt, \qquad u_c v_c = \frac{1}{\omega} \int_0^{\omega} u \, v \, dt. \tag{10}$$

One then sees that u and v oscillate periodically around the equilibrium point u_c and v_c. What happens to the other variables? The best way to understand their variations is to study the average differences:

$$\frac{x(t + \omega) - x(t)}{\omega}, \qquad \frac{y(t + \omega) - y(t)}{\omega}, \qquad \frac{s(t + \omega) - s(t)}{\omega} .$$

Equations (1), (2), (5) and (10) give:

$$\frac{x(t + \omega) - x(t)}{\omega} = -\alpha_{13}u_c - \alpha_{14}v_c + \alpha_{41}v_c$$

$$\frac{y(t + \omega) - y(t)}{\omega} = \alpha_{32}u_c - \alpha_{24}v_c + \alpha_{42}v_c \qquad\qquad (11)$$

$$\frac{s(t + \omega) - s(t)}{\omega} = \alpha_{35}u_c + \alpha_{45}v_c ,$$

that is:

$$\frac{x(t + \omega) - x(t)}{\omega} = \frac{1}{\beta} [(\alpha_{32} + \alpha_{35})(\alpha_{41} - \alpha_{14}) + \alpha_{13}(\alpha_{45} - \alpha_{24} + \alpha_{42})]$$

$$\frac{y(t + \omega) - y(t)}{\omega} = \frac{1}{\beta} [-\alpha_{32}(\alpha_{45} + \alpha_{41} - \alpha_{14}) - (\alpha_{24} - \alpha_{42})(\alpha_{35} - \alpha_{13})]$$

$$(12)$$

Suppose that:

(1) $\alpha_{14} - \alpha_{41} < 0$,

meaning that plants assimilate less oxygen than they produce.

(2) $\alpha_{42} - \alpha_{24} < 0$,

meaning that plants assimilate more carbon dioxide than they produce.

(3) $\alpha_{45} > \alpha_{24} - \alpha_{42}$,

meaning that plants disperse into the soil more matter than the carbon dioxide they assimilate.

(4) $\alpha_{35} > \alpha_{13}$,

meaning that animals disperse into the soil more matter than the oxygen they assimilate.

Under such conditions the equations (12) show that, apart from periodic oscillations, the free oxygen continues to increase while the carbon dioxide continues to decrease. One has, on the other hand:

$$s(t + \omega) - s(t) = \text{constant} > 0,$$

meaning that the gases lost by the atmosphere disperse into the earth's crust. The living world works as an intermediary in this process.

Suppose instead that:

(1) $\alpha_{14} - \alpha_{41} < 0$

(2) $\alpha_{42} - \alpha_{24} > 0,$

meaning that plants assimilate less carbon dioxide than they produce.

(3) $\alpha_{35} < \alpha_{13},$

meaning that animals contribute to the soil less matter than the oxygen they assimilate. In such circumstances, apart from periodic oscillations, the carbon dioxide continues to decrease while the free oxygen increases.

Finally, suppose that:

(1) $\alpha_{14} - \alpha_{41} > 0$

(2) $\alpha_{24} - \alpha_{42} > \alpha_{45} > \alpha_{14} - \alpha_{41}$

(3) $\alpha_{35} > \alpha_{13}.$

In such conditions both gases decrease. We will not insist on the other possibilities to which the equations (12) give rise.

7. The Circulation of Nitrogen

Let us complete the preceding results by considering the circulation of nitrogen. In this respect, the living world can be divided into two parts:

(1) organisms which are capable of directly binding atmospheric nitrogen;

(2) organisms which can only derive their nitrogen through the intermediary action of organisms in the preceding group.

The precise relationship between these two groups is not yet well known. Suppose that the organisms in group one live as symbiotes of plants. Let us set:

x, the total weight of free atmospheric oxygen;

y, the total weight of carbon dioxide in the atmosphere and in the oceans;

z, the total weight of atmospheric nitrogen;

u, the total weight of these elements in animals;

v, the total weight of these elements in plants;

s, the total weight of these elements in the earth's crust.

We completely neglect the circulation of water; introducing this factor would make our equations bulkier, without modifying their solutions in any essential way.

458

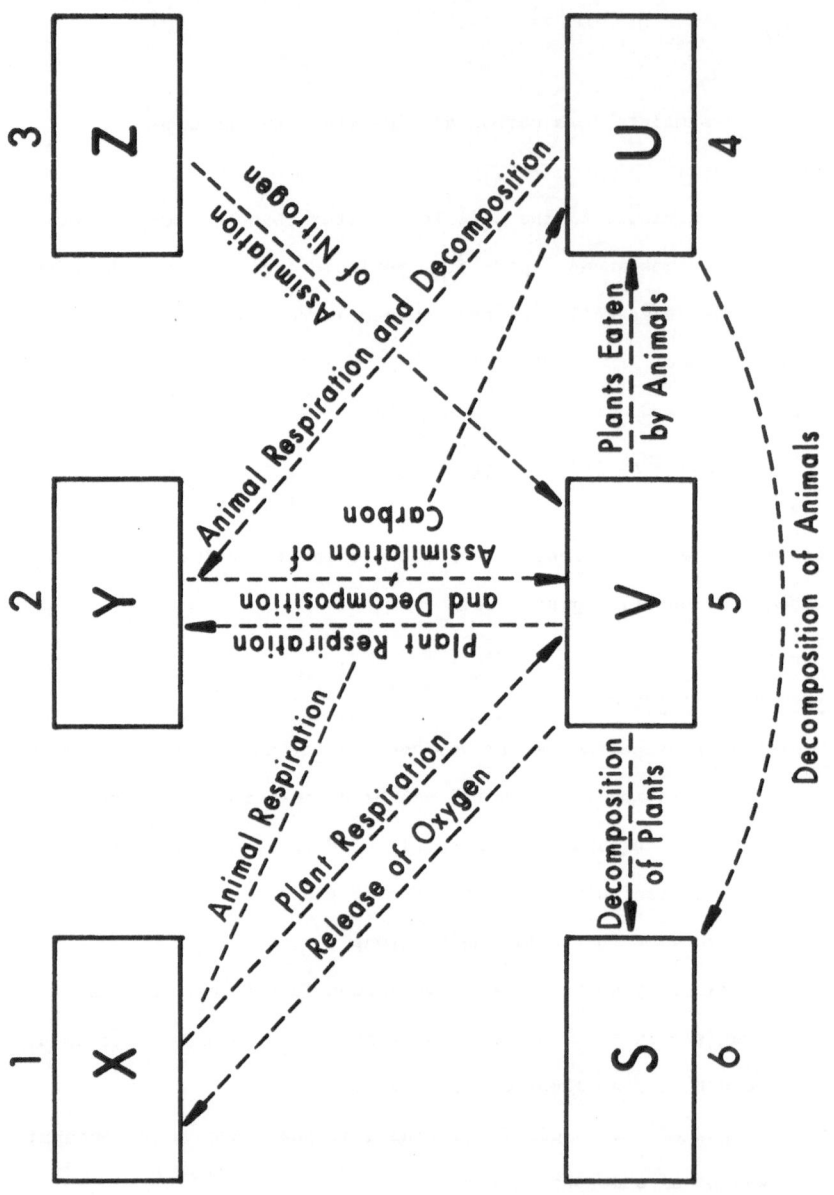

Fig. 2.

The relationships among the variables can be represented by the scheme which is illustrated in Fig. 2, or by the differential equations:

$$x' = -\alpha_{14}u - \alpha_{15}v + \alpha_{51}v \tag{13}$$

$$y' = \alpha_{42}u - \alpha_{25}v + \alpha_{52}v \tag{14}$$

$$z' = -\alpha_{35}v \tag{15}$$

$$u' = \alpha_{14}u - \alpha_{42}u - \alpha_{46}u + \beta uv \tag{16}$$

$$v' = \alpha_{15}v - \alpha_{51}v + \alpha_{25}v - \alpha_{52}v + \alpha_{35}v - \alpha_{56}v - \beta uv \tag{17}$$

$$s' = \alpha_{35}u + \alpha_{45}v \tag{18}$$

One can suppose that:

(1) the vital balance of animals would be negative in the absence of plants:

$$\lambda = \alpha_{42} + \alpha_{46} - \alpha_{14} > 0 \tag{19}$$

(2) the vital balance of plants would be positive in the absence of animals:

$$\mu = \alpha_{15} - \alpha_{51} + \alpha_{25} - \alpha_{52} + \alpha_{35} - \alpha_{56} > 0 \tag{20}$$

In such conditions the solutions of equations (16) and (17) are periodic functions, u, v, which satisfy the relation (9). The variable s always increases, and the variable z always decreases. The relations (12) take the form:

$$
\left.
\begin{aligned}
\frac{x(t + \omega) - x(t)}{\omega} &= \frac{1}{\beta}[(\alpha_{46} + \alpha_{42})(\alpha_{51} - \alpha_{15}) - \\
&\quad - \alpha_{14}(\alpha_{25} - \alpha_{52} + \alpha_{35} - \alpha_{56})] \\
\frac{y(t + \omega) - y(t)}{\omega} &= \frac{1}{\beta}[\alpha_{42}(\alpha_{15} - \alpha_{51} + \alpha_{35} - \alpha_{56}) - \\
&\quad - (\alpha_{25} - \alpha_{52})(\alpha_{46} - \alpha_{14})] \\
\frac{z(t + \omega) - z(t)}{\omega} &= -\frac{\alpha_{35}(\alpha_{42} + \alpha_{46} - \alpha_{14})}{\beta}
\end{aligned}
\right\} \tag{12'}
$$

From this it follows that:

$$\frac{x(t + \omega) + y(t + \omega) + z(t + \omega) - x(t) - y(t) - z(t)}{\omega} =$$

$$= -\frac{\lambda \alpha_{56} + \mu \alpha_{46}}{\beta} < 0 ,$$

that is to say, the atmosphere decreases. We shall not insist on the behavior of the single components of the atmosphere: this behavior depends on inequalities fully analogous to those considered in the previous section.

8. Modifications and Generalizations

These results remain substantially unchanged if one replaces the constant coefficients with others which change slowly, or which depend on unknown functions for x, y, z, u, v, s. The periodic process is changed to an irregular, periodomorphic process. In place of a closed curve (Fig. 3, I) the point representing the organic world (u, v) traces an irregular curve $C_1 C_2 C_3 \ldots$ (Fig. 3, II). The angular points correspond to sudden variations in living conditions, following periods of intense volcanism or glaciations. The decrease in the atmosphere of a gas which is essential to life can entail the rapid extinction of a whole important ensemble of animal species. This disruption of the equilibrium favors the development of other animal species. The result is a sudden variation in the coefficients in our equations, and a segment such as $C_1 C_2$ (Fig. 3, II). If a sufficiently long period of calm follows such perturbations, the equilibrium will be, little by little, reestablished. The oscillations will dampen, and the points (u, v) will tend toward a limit state (u_c, v_c) (Fig. 3, III). This process would appear periodic, if observed throughout a limited amount of time. Eventually, however, a stable state (u_c, v_c) is attained by the organic world. The impoverishment of nitrogen in the atmosphere, and its diffusion into the soil, go on at the same rates.

The positive and negative sources of carbon dioxide, such as volcanic action, erosion, human industry, etc., do not appear in our equations. Suppose that the global contributions of these sources per unit time is given by σ. Equation (14) becomes:

$$y' = \sigma + \alpha_{42} u - \alpha_{25} v + \alpha_{52} v . \tag{14'}$$

It follows that the average change of y(t) is:

$$\frac{y(\omega + t) - y(t)}{\omega} = \alpha_{42} u_c - \alpha_{25} v_c + \alpha_{52} v_c + \frac{1}{\omega} \int_t^{\omega+t} \sigma(s) ds .$$

This expression is, generally, constant and negative. During certain epochs, however, it may change very rapidly, become positive, and even become sufficiently large to cause a rapid enrichment of carbon dioxide in the atmosphere. One can imagine other possibilities, either catastrophic or not, yet with quite large repercussions in the organic world. One can be sure that both of these possibilities

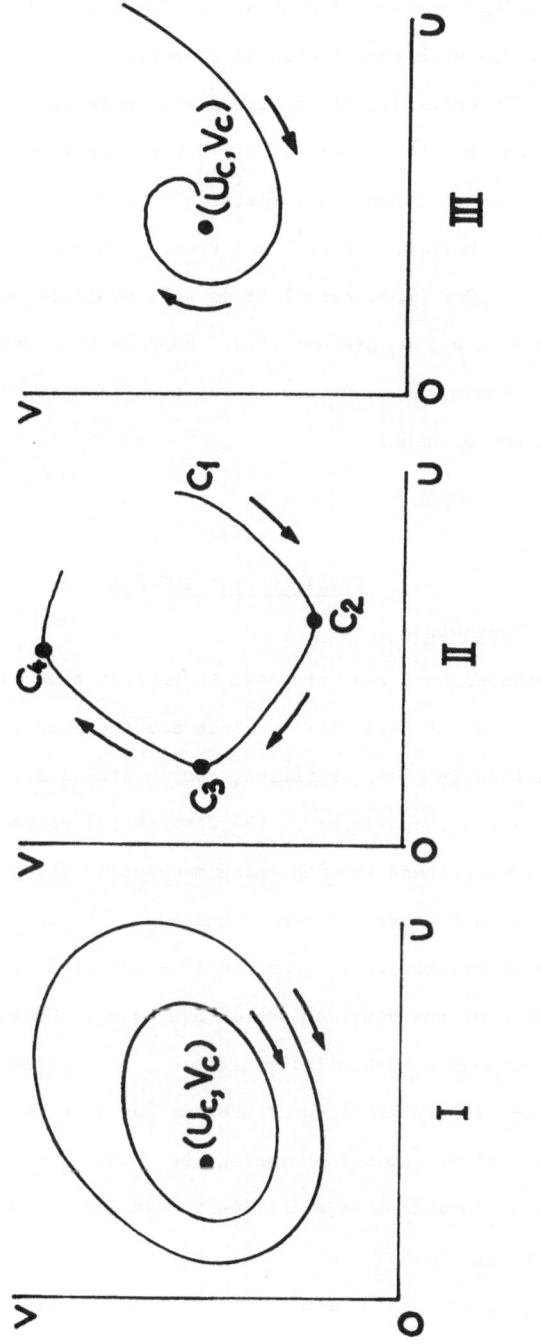

Fig. 3.

have endangered living beings more than once. Therefore, if life exists, it means that life has answered each threat with an adaptation, with an evolutionary novelty. From time to time the point (u, v) in Fig. 3 will come close to the u-axis as well as to the v-axis, and to the point (0, 0). But the resiliency of life allows life itself to respond with an appropriate reaction to each external or internal danger.

Notice also that both the initial and final points of this evolution will escape our analysis. Our equations (1-5) as well as (13-18) will only give non-zero values of u and v when u_o, v_o are non-zero. Suppose that animals and plants originated from a common, unknown stock. Our equations cannot represent the separation of the two branches.

III

ORIGIN OF THE ICE AGES

9. Astronomical Hypotheses

Several hypotheses have been proposed to explain the origin of the ice ages, and their true or apparent periodicity. Such are the precession (3), showers of meteorites, variations in solar radiation, continental drift, wandering of the poles, variation in the composition of the atmosphere, orogenic phenomena, etc.

It is hard to understand through which mechanisms the precession of equinoxes could produce a climatic effect of such magnitude. The earth's axis describes, in fact, a cone, always preserving the same inclination with respect to the ecliptic. It is all a question of the displacement of equinoxes. We know that the ellipticity of the earth's orbit has a weak climatic effect. The largest effect of the precession would amount to the addition or subtraction of a few days (a week at the most) to the length of the summer or the winter. There is such a margin between this effect and the glaciation of a continent, that one is really amazed at the popularity of this hypothesis.

The encounter of the earth with swarms of meteorites, or of large meteorite clouds, can certainly have climatic effects, but it is difficult to see in which direction. It is, in fact, very difficult even to derive the consequences of such an encounter. The problem deserves a special study which is completely outside our present theme.

The changes in solar radiation are real, but, as far as the known solar cycle is concerned, they are insignificant. Are there other cycles of larger period and amplitude? One cannot exclude this possibility, but there are no data for or against it.

10. Migrations of the Continents and the Poles

The hypotheses for the migrations of the continents and the poles (22) are very attractive. At first sight they seem to explain very well all the glaciations beginning with the Proterozoic period. However, a closer examination shows that these hypotheses imply arbitrary elements which are especially introduced for this purpose. In spite of all the effort, it has not been possible to demonstrate the mechanical possibility of such continental motions. Such motions, on the other hand, could not explain the multiplicity of Quaternary glaciations. One can rigorously accept the very slow motion of continents, but we can hardly accept their dancing.

Concerning the Permo-Carboniferous glaciations, Wegener (22) introduces a displacement of the continents which effects only the southern hemisphere and leaves the northern hemisphere unchanged until the Quaternary. The reasons for such an asymmetry in the forces propelling the continents remain unknown. The miracle of the glaciations is explained by a greater, and more mysterious miracle.

In such circumstances it would seem wiser to look for explanations based exclusively on terrestrial factors, which can be studied directly and which, potentially, can be verified. These are the composition of the atmosphere and the vertical movements of the earth's crust.

11. The Atmosphere and Glaciations

Arrhenius (1) was the first to propose the idea that the level of carbon dioxide in the air has a climatic influence. He showed that variations in the content of carbon dioxide can determine temperature variations on the earth's surface. According to his calculations a decrease of 1/3 of the present level of carbon dioxide would be sufficient to lower the average temperature by 3°C. This would determine a new glacial invasion in North America and in the whole of northern Europe. If, on the other hand, the present level of carbon dioxide should double or triple, the warming effect on these same regions would be approximately 8-9°C. Such a warming would correspond to the onset of a climate for these regions analogous to the one which prevailed during the Eocene. Without discussing, for the time being, Arrhenius's calculations, one must warn that their physical basis is doubtful, and that they are in need of revision.

Arrhenius attributed these variations in carbon dioxide levels of the atmosphere to a large increase in volcanic activity in the Tertiary era. He strongly emphasized that assimilation increases with the tension of CO_2, and he implied there would be oscillations of the system:

$$\text{living beings} \xrightleftharpoons{\hspace{2cm}} \text{carbon dioxide}$$

before the equilibrium was reestablished. Several passages in his paper (1) must be read in this way. Particularly the passage stressing the multiplicity of ice ages and the impossibility of explaining it by any hypothesis other than his.

It is very difficult to understand the effect of internally produced gas on the composition of the atmosphere and the climates. According to W. Vernadsky (21): "the quantity of newly produced carbon dioxide which reaches the earth's surface is enormous.... The total amount is of the same order of magnitude as the carbon dioxide in the hydrosphere and the atmosphere." According to Boussingault, Cotopaxi generated in one year over 10^9 cubic meters, or about 2×10^6 metric tons of CO_2. This volcano alone, then, would be able to reconstitute the entire body of atmospheric CO_2 in just two million years.

Let us accept that, following an increase in volcanic activity, the atmosphere receives a considerable quantity of carbon dioxide. What effect will it have on the climate? We do not have any reliable observation. In 1912 the eruption of the volcano Katmai in Alaska was followed by a very rainy and cold summer, which was attributed to the enormous clouds of volcanic ash it had put into circulation. On the contrary, the famous Krakatoa eruption in 1883 was accompanied by a comparable emission of ash which had only optical effects. It is then very difficult to foresee the effect of volcanic eruptions on the climate, especially when taking into account the emission of carbon dioxide which is always accompanied by the emission of water vapor and volcanic ash. The effects might then be very different from those anticipated by Arrhenius.

To start with, let us accept the predictions of Arrhenius, and estimate the quantity of carbon dioxide necessary to modify the climate to the extent he had indicated. Let us again stress that it is not just a matter of carbon dioxide in the atmosphere, but of all the CO_2 contained in the oceans as well as in the atmosphere. If we accept an amount equal to 10^{14} tons, we find that to change the climate of northern Europe by 10°C, a total mass of 3×10^{14} tons of CO_2 must first be added and then removed from the atmosphere. The addition of this quantity would require the action of 150 volcanoes at the level of Cotopaxi for a million years. This quantity must then be removed in 100,000 years by increased plant activity. To give an idea of this activity it is enough to recall that the total weight of fossil fuels is not larger than 2×10^{13} tons. Thus, during this relatively short time span, the vegetation would have had to remove from the atmosphere four times as much coal as was deposited in the ground throughout the entire history of the earth. This conclusion does not at all agree with what we know about Quaternary deposits. It follows that one must either abandon Arrhenius' hypothesis or change all the data on carbon dioxide in the oceans and alter the conditions of gaseous equilibrium between the atmosphere and the oceans. One might try to rescue Arrhenius' hypothesis by assuming that the carbon dioxide in the oceans is rapidly transformed by marine organisms into calcium carbonate, which falls as sediment to the bottom of the sea. This might explain the disappearance of the enormous quantities of carbon dioxide produced by volcanoes.

We shall give a mathematical expression to this hypothesis. Set:

p, the concentration of CO_2 in the oceans;

q, the concentration of living beings in the oceans;

a, the constant, external generation of CO_2 per unit time;

αpq, the relative consumption by organisms of CO_2 per unit time;

$-\beta q$, the negative balance of organisms in the absence of CO_2.

Under such conditions the process is described by the differential equations:

$$p' = a - \alpha pq$$

$$q' = -\beta q + \alpha pq \ . \tag{25}$$

The process tends toward a stationary state:

$$P_c = \frac{\beta}{\alpha} \ , \qquad q_c = \frac{a}{\beta} \ . \tag{26}$$

This state is stable since the characteristic equation:

$$\mu^2 + \mu\frac{a\alpha}{\beta} + a\alpha = 0$$

has negative roots when:

$$a\alpha > 4\beta^2 \ ,$$

and complex roots with negative real parts when:

$$a\alpha < 4\beta^2 \ .$$

Suppose that up to the instant t_o the generation of CO_2, a_o, is weak, so that:

$$a_o\alpha < 4\beta^2 \ ,$$

and suppose that the system is close to the stationary state:

$$P_o = \frac{\beta}{\alpha} \ , \qquad q_o = \frac{a_o}{\beta} \ .$$

There is an increase of volcanic activity from this time to the time t_1, and the average generation of CO_2 becomes a_1:

$$a_1 \gg a_o.$$

The equilibrium is upset, and (p, q) tends toward the new equilibrium state:

$$P_1 = \frac{\beta}{\alpha} \ , \qquad q_1 = \frac{a_1}{\beta} \ .$$

Suppose that this new equilibrium has been nearly attained by the time t_1. One sees that the final concentration of carbon dioxide, p_1, is the same as the initial concentration, p_o, while the concentration of living beings has increased. Assume that the normal conditions are reestablished after this increase of volcanic activity, so that for $t > t_1$ the generation of CO_2, a, again becomes a_o. The equilibrium which was attained is again perturbed, and (p, q) tend toward their previous equilibrium state:

$$P_2 = \frac{\beta}{\alpha} \ , \qquad q_2 = \frac{a_o}{\beta}$$

through decreasing spirals, since the point (p_2, q_2) is a focus. Initially this is a quasi-periodic process. The concentration of carbon dioxide in the oceans, and consequently its atmospheric concentration, pass through a sequence of maxima and minima.

Previously we had studied another biological mechanism which, starting with climatic variations, accounts for oscillations in the level of atmospheric CO_2. Are there any relationships between these two explanations? It is not difficult to write down a system of differential equations comprising the systems (1-5), (13-18), and (25) as special cases. It would be, however, very complex. It is better to study the special systems independently, while assuming that the two mechanisms operate simultaneously. If the amount of living matter in the oceans is much larger than that on the continents, then, for all practical purposes, one can consider only the action of (25).

We had previously considered some of the biological consequences of variations in CO_2 concentration. The two models indicated above allow us to perfect these considerations. The increase in volcanic activity is followed in the seas by the development of marine organisms, and an increased deposition of calcium carbonate. On the continents it is followed by an increase in vegetation, which is also a consumer of carbon dioxide. The reestablishment of a normal CO_2 balance is followed in the oceans and on the continents by a sequence of oscillations which progressively dampen to zero. In particular, the succession of calcareous layers in ocean sediments can be explained by this mechanism. Finally, after having gone through a number of maxima and minima, the process tends to become steady, and it moves towards a new equilibrium state.

One sees that the hypothesis of Arrhenius is quite capable of explaining Quaternary glaciations, provided it is improved by some modifications, and several refinements. Nevertheless one can still bring a few serious objections against this explanatory setup:

(1) A glaciation produced in this manner must occur simultaneously in both hemispheres, and be independent of longitude. Now, the Quaternary glaciations did not effect Europe and Asia in the same way. The Permo-Carboniferous glaciations only left traces in the southern hemisphere.

(2) One cannot find any evident relationship between volcanism and glaciations.

(3) The orogenic and epeirogenic phenomena responsible for variations in

volcanism might have produced glaciations as an indirect consequence. These
problems will be studied in the next chapter.

IV

ORIGIN OF THE ICE AGES (Continued):

12. Orogenic Processes, Volcanic Activity and Glaciations

Relationships among these three types of phenomena are currently established.
Yet, it is very difficult to verify such comparisons with geological data. We know,
in fact, that some orogenic processes were neither followed, nor preceded, by
glaciations. We also know of glaciations separated by millions of years from the
most recent mountain building, and of increases in volcanic activity not accompanied
by any mountain building. In order to convince ourselves of these facts, let us
review some geological information. Starting from the present era, the most recent
orogenic process is the great alpine upfolding of the Late Tertiary. It can be
estimated to have begun 25-30 million years ago. A large proportion of the orogenic
movements which produced the largest mountain chains (Alps, Caucasus, Himalayas,
etc.) go back to the very beginning of the Late Tertiary. These chains were already
formed at the beginning of the Pleiocene. Yet, the first glaciations occurred at
the end of this period. Consequently, these glaciations are separated from the end
of the alpine upfolding by a time span of several million years. If the two were
related by a causal link, one would have to explain why the effect was delayed.

Examining the situation from the point of view of volcanic phenomena, one gets
equally disappointing results. We know that the Mesozoic era enjoyed a relative
stability. Volcanism broke out at the end of the Cretaceous period and the first
half of the Tertiary. The Early Tertiary period is characterized by large eruptions
which can be viewed as forerunners of the Late Tertiary upfoldings. If one assumes
that glaciations are caused by eruptions, it is impossible to understand why the
Late Cretaceous and Early Tertiary volcanisms were not followed by glaciation.
Also, if one considers the Quaternary glaciations as a result of Late Tertiary
volcanism, one must look to this activity for peculiar characteristics which
differentiate it from Early Tertiary volcanism.

Let us now go back to the beginnings of geological history. The two
Precambrian periods, the Archean and the Proterozoic, witnessed several orogenic
processes and as many complete geological cycles. The volcanic activity was very
high, but one looks in vain for a causal relationship between these two kinds of
phenomena. Rather, if such a relationship does exist, we have no idea of its nature.
The only known Precambrian glaciations took place at the beginning of the Proterozoic
period, before the Huronian upfolding, and after an unknown amount of time from the
last Archean upfolding. It is very difficult, then, to extract from these data an
argument for or against any theory whatsoever.

Let us move now to the Caledonian upfolding, which took place at the end of the
Silurian and the beginning of the Devonian. The insignificant volcanic activity
during the Cambrian increased during the Silurian, to become very intense in the
Devonian. Regarding Devonian glaciations, there are traces of them in southern
Africa. On the contrary, there are known morainic deposits in southern Australia
and in the Yangsee region of China which almost certainly date to the Lower Cambrian.
That is to say, they are not at all related with intense upfoldings and eruptions.
It is very difficult to reconcile this fact with Arrhenius' hypothesis. We are
left with examining the very interesting and enigmatic Permo-Carboniferous period.
It alone spans 140 million years, as much as the Cretaceous, the Early Tertiary,
and the Late Tertiary taken together. The Hercynian upfolding begins in the Middle
Carboniferous and ends in the Permian. This means that it is not a matter of a
single upfolding, and that one has to deal with several orogenic processes, probably
independent of each other, and one after the other. This is what emerges from the
study of epeirogenic movements in the Permo-Carboniferous.

Volcanic phenomena played an important role during this time. The Permian
times are particularly remarkable for the intensity of their eruptions. Arrhenius'
hypothesis would lead us to predict multiple and widespread Permo-Carboniferous
glaciations. Multiple and important they were, but not widespread. For some very
strange reason only the southern hemisphere was privileged to be effected by them.
There are extensive glacial deposits in South America (in Brazil, in southern Africa,
in India, and in Australia). It might seem at first sight, that the whole southern

hemisphere was covered with ice, while the northern hemisphere was enjoying a hot climate. In fact, the Brazilian glaciation preceded those in Africa, and the latter preceded the Indo-Australian glaciations, which are clearly Permian. One is, then, rather in the presence of a sequence of local and asynchronous phenomena, which might be periodic or quasi-periodic. What makes the Permo-Carboniferous climate even stranger is the clearly desert-like climate of the northern hemisphere in the Permian.

One sees that it is difficult to reconcile the climatic phenomena in the Permo-Carboniferous with Arrhenius' hypothesis. It does not appear any easier to relate them to the Hercynian upfoldings which primarily affected the northern hemisphere.

From this cursory review we can draw the following general conclusions:

(1) general processes such as the great upfoldings or the variations in the composition of the atmosphere, each acting alone, are not enough to produce glaciations;

(2) glaciations are local phenomena, which can be explained by local causes;

(3) the action of general factors, combined with local causes, can trigger processes of glaciation.

In the following section we will study a mechanism which is local in nature, and which might well be capable of explaining such phenomena.

13. Vertical Movement of the Earth's Crust

Consider a continental mass of cylindrical shape, and depth, h. It is immersed in a magma, and it has some vertical velocity. From the isostatic principle, it is easy to conceive of a succession of vertical oscillations: the formation of glaciers after the initial rise which follows an upfolding, or some other internal phenomenon, overburdens the continental mass, and it begins to sink into the magma; the lower elevation which results allows the glaciers to melt; the lighter continental mass then rises again to an altitude where more abundant precipitation leads to a new glacial invasion. This periodical mechanism has been known to geologists for a long time. We will deal with this hypothesis in a mathematical form.

Let us suppose (Fig. 4) that the surface of the continental block rises by an amount x above the level of the magma, and it supports a layer of ice of thickness y. Let ℓ be an altitude such that the formation of ice is impossible below, and its melting is impossible above. Of course, we are not dealing with seasonal variations, but with the net yearly balance. We suppose that ℓ is very small with respect to the horizontal extent of the block, so that one can neglect both the formation and the melting of ice on the lateral surface of the layer. We can assume that the rate of formation of ice at a level z > ℓ is proportional to the difference in altitude z - ℓ. Similarly, the melting of ice at a level z < ℓ would be proportional to ℓ - z. These are quite plausible working hypotheses.

Suppose at first x > ℓ; the ice does not melt, and y' is proportional to the difference of the altitudes x + y and ℓ:

$$y' = \varepsilon(x + y - \ell) \qquad (x > \ell) \ . \tag{27}$$

For x < ℓ < x + y, ice is formed at the upper surface of the layer and it melts at the lower surface. This gives:

$$y' = \varepsilon(x + y - \ell) - k(\ell - x) \qquad (x < \ell < x + y) \ . \tag{28}$$

For x + y < ℓ, melting takes place at both the upper and the lower surfaces of the layer. One has then:

$$y' = -k(\ell - x) - k(\ell - x - y) \qquad (x + y < \ell) \ . \tag{29}$$

Finally, an elementary, but very important, remark: for x < ℓ, y = 0, the ice layer is not reestablished, and y remains zero as long as x < ℓ. It is indeed this condition which gives rise to the periodicity of the process.

The equations (27-29) describe the variation of the ice layer under the different conditions which may arise. Let us now write down the differential equation for the vertical motion of the continental block. One can neglect its acceleration, but it is necessary to take into account viscosity and other dissipative forces. This is even more necessary since the actual vertical velocity of the earth's crust is well below a meter per century. Let us have a look at what would happen in the absence of dissipative forces. A continental block 24 kilometers thick would take, on the average, five minutes to go through a complete oscillation about an equilibrium position. Such an unrealistic figure compels us to consider

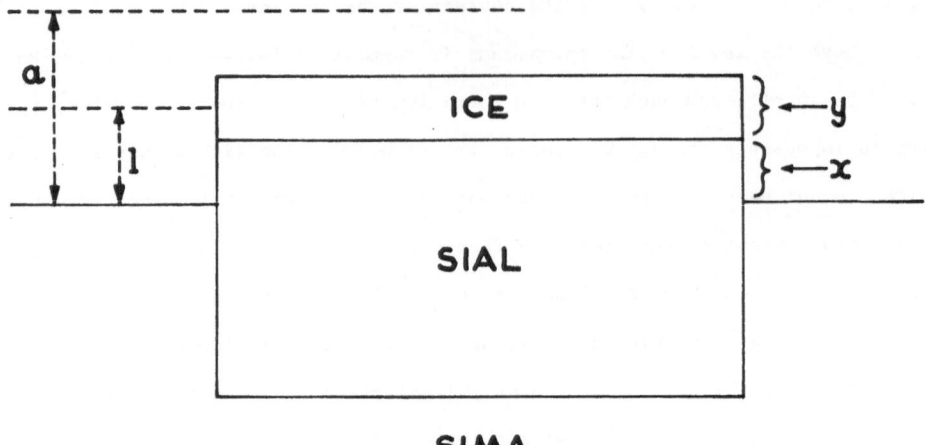

Fig. 4.

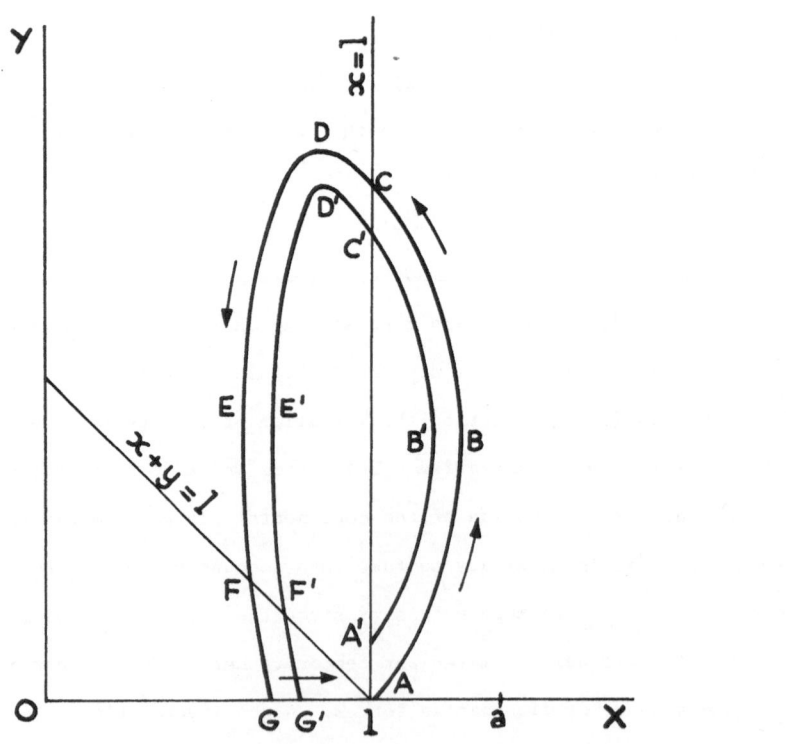

Fig. 5.

the reality of dissipative forces.

Denote by ρ the density of the _sial_, by η the density of ice, by σ the density of the _sima_, by r the coefficient of resistance, by g the acceleration of gravity. Then, $a = h\left(1 - \frac{\rho}{\sigma}\right)$ is the level corresponding to the isostatic equilibrium in the absence of ice, also, $n = \frac{g\sigma}{r}$, $\tau = \frac{g\eta}{r}$. Very simple mechanical considerations allow us to write the required differential equation in the form:

$$x' = na - nx - \tau y .$$ (30)

The density of the _sima_ is approximately three times greater than that of ice, so that $n \approx 3\tau$. On the other hand it is natural to assume that $\ell < a$, since glaciation would be impossible otherwise.

It is very easy to study these differential equations. One can consider them in graphic form, in which any state of our system (x, y) is represented by a point on a plane (Fig. 5). From what has been shown above, the positive quadrant $(x, y > 0)$ is subdivided into three regions corresponding to different regimes. The part of the x-axis corresponding to $x < \ell$ constitutes a fourth region in which:

$$x' = na - nx \qquad (y = 0).$$ (31)

Several cases are now possible. We will examine at first that of a periodic cycle, which is the most interesting. Suppose that the coefficients n, τ, ε, k satisfy the following inequalities:

$$\varepsilon > n, \qquad 4\tau(\varepsilon + k) > (n + \tau)^2 .$$ (32)

The coefficient ε depends on the average annual precipitation. Such inequalitess mean that precipitation is abundant, but at the same time the melting, k, is very intense. Let us choose a starting point $A(\ell, 0)$. We have:

$$x_0' = n(a - \ell) > 0$$

so that x increases. The process is governed by the equations (27) and (30) as long as the point (x, y) remains in the region I $(x > \ell, y > 0)$. In this region the point (x, y) traces the arc A B C. After attaining a maximum, x_{max} (point B), x decreases, and y continues to increase. The point (x, y) then crosses the line $x = \ell$ (point C), passing into the region II $(x < \ell < x + y)$ in which the process is governed by equations (28) and (30). The variable x continues to decrease, and y increase, reaching a maximum, y_{max} (point D), after which y decreases. Proceeding, x passes through a minimum, x_{min} (point E), and then begins to increase.

The point (x, y) crosses the line x + y = ℓ at point F and moves into region III (x + y < ℓ). In this region the process is governed by equations (29) and (30). No more ice is formed, and melting is very rapid, so that in a relatively short time all the ice disappears. The representative point lies on the x-axis (point G), that is, the region IV. From this moment on the process is governed by equation (31). As long as x remains smaller than ℓ, y remains zero: we are in a full interglacial period. After some time x reaches ℓ (point A); one cycle is finished, and a new one is initiated.

It is easy to show that this process is stable. If for any reason one of these cycles is perturbed, it will end with a slightly longer or slightly shorter inter-glacial period. However, a normal cycle will follow. The case in which the deviation in a cycle is such that the interglacial period is eliminated is somewhat more complex. In a finite time, however, the process does return to normal. Suppose, in fact, that the cycle starts at the point A' (Fig. 5). The arc A'B'C' takes place of the arc A B C, and the arcs C'D'E'F' and F'G' replace, respectively, the arcs C D E F and F G. The interglacial period G A is substituted by a shorter one, G'A. From this moment on the cycle continues as normal.

In this way we have a curious periodicity which arises from the fact that:

(1) the thickness of a layer of ice is essentially positive;

(2) the ice melts completely in a finite time;

(3) the interglacial period always begins at the same point A.

We know that in general the periodical behavior of differential equations is due to the fact that the coefficients and the initial values exactly satisfy certain special, and not very probable, relationships. The opposite occurs in this case. One is led to believe that this phenomenon can also occur in a number of other natural processes, which might explain the high incidence of periodical phenomena in nature.

Let us now consider a few figures, as for instance:

$$a - ℓ = 100 \text{ meters}, \qquad n = 3\tau, \qquad ε = 5\tau, \qquad k = 27\tau.$$

To fix the chronological scale suppose that we are in an interglacial period, and

that the actual vertical velocity of the scandinavian plate is approximately 1 meter
per century. Under such conditions the total period will be approximately 150,000
years, and its interglacial fraction about 40,000 years. The maximum difference in
altitude of the continental block is of the order of 400 meters, and the maximum
thickness of the glacial layer is of the order of 1900 meters. This last figure
might appear somewhat excessive, but at least it is not improbable. One can justify
it by supposing, for instance, that the initial ice layer does not occupy the whole
surface of the plate, and that it slowly expands over it. In this case y would
characterize the total load of ice rather than the thickness of the layer. On the
other hand, observations on the thickness of continental glaciers in Greenland
justify, up to a point, our figures.

It is curious that our numerical example predicts an exceedingly fast melting
at the end of a glaciation. This agrees with de Geer's figures for the speed at
which scandinavian glaciers receeded at the end of glacial periods. Finally, the
elevation of the block over the magma, x, always changes at a velocity below 1.5
meters per century, which is within plausible limits.

We consider now the other cases which might arise, and which might be of some
geological interest. Suppose, for instance, that:

$$\epsilon > n, \qquad 4\tau(\epsilon + k) < (n + \tau)^2, \qquad \tau\epsilon + \tau k - \epsilon n > 0 \qquad\qquad (33)$$

Such inequalities mean that precipitation is sufficiently abundant, but that melting,
represented by the coefficient k, is not fast enough to allow the establishment of
a periodic cycle. Under such conditions the process tends toward an equilibrium
point:

$$x_{lim} = \ell - \frac{n\epsilon(a - \ell)}{\tau\epsilon + \tau k - \epsilon n}, \qquad y_{lim} = \frac{n(\epsilon + k)(a - \ell)}{\tau\epsilon + \tau k - \epsilon n}.$$

There is a permanent glacial regime with a considerable thickness of the ice layer.
One can suppose that this case applies to Antarctica and to Greenland. The latter
enjoys much the same maritime climate as Scandanavia, but, being at a higher
latitude, it therefore has a lower rate of melting, k. Another possibility cannot
be excluded regarding Greenland: This island might now be in the initial phase of
a glaciation. In fact, according to the scandinavian sagas, it still had a moderate
climate in historic times. It seems, on the other hand, that this island has a

vertical movement of the same order as that of the scandinavian block.

Suppose, to the contrary, that we have a plate with a harsh and dry continental climate. The coefficient ε in equation (27) is therefore very small. The result would be a permanent regime with no ice, or only a thin layer, as in northern Asia.

Therefore, in spite of its imperfections, our scheme is flexible enough to account for a variety of cases. It illustrates how important is the ensemble of local conditions in determining whether a glacial regime is possible, and, if so, whether it is periodic.

The works by F. Nansen and L. Zenkevitsch (23) partially confirm this hypothesis. According to Nansen the sea level in the glacial period oscillated with an amplitude on the order of 1000 meters. In particular, he considers the bottom of the present Barents sea, with depths of around 400 meters, to have been a system of river valleys which emerged during periods of glaciation, and which immerged during the interglacial periods.

14. Ocean Currents and Other Climatic Factors

Among such local factors we cannot forget the oceanic currents. It is enough to compare the average temperatures on the opposite shores of the Atlantic, to see what would happen if the Gulf Stream were not there. Now, the bathymetric map of the North Atlantic shows the existence of a true oceanic ridge. It passes between Iceland and the Faeroe Islands, and its depth is never more than 600 meters below the surface of the ocean. Suppose that in such places the bottom of the ocean participates in the vertical movements of the scandinavian plate and of Greenland. A decrease in sea level of the order of 600 meters would be enough to stop the passage of the Gulf Stream between Greenland and Scandinavia, and to reroute it west of Greenland to the Davis Straight. This possibility would have important climatic consequences, not the least of which would be the warming up of Labrador, Greenland, and the whole North American archipelago.

The zoogeographical investigations by Zenkevitsch (23) confirm the hypothesis of a periodic isolation of the arctic basin. They show that the present distributions of the large families of fishes (Salmonidae and Gadidae) still clearly reflect

changes in the regime of salinity. Such variations would be due to the periodic
isolation of the arctic basin, as a consequence of the vertical movements of the
earth's crust.

Another locally important factor not included in our equations is represented by the glaciers themselves when they are present. We know that the glaciers of
Mont-Blanc descend almost to Germany (1,039 meters), even though the lower limit
for permanent snowcover is approximately 2,700 meters elevation. This factor
explains the great expansion of glaciers and renders useless those hypotheses which
require a significant decrease in mean temperature as a precondition for their
establishment. A climatic variation of little importance may be sufficient to
trigger the onset of a glaciation, and it is at this point that the first mechanism
we studied -- the rapid variations in the level of atmospheric carbon dioxide --
can intervene. Such trivial atmospheric variations can be enough to alter the
local climatic coefficients ε and k, in such a way that local glaciations
become possible. It is perhaps in this manner that we must explain the Quaternary
glaciations in the two hemispheres. The role of mechanism number 1 is merely to
trigger mechanism number 2 when local conditions are favorable. Now, such conditions were favorable everywhere when the Quaternary glaciations were produced
in the northern hemisphere and in the southern hemisphere (Australia, Tasmania,
New Zealand, Chile, Patagonia, Tierra del Fuego).

Let us further say a few words about the Permo-Carboniferous glaciations.
The geography of this period is rather poorly established. The absence of glaciations in the northern hemisphere would seem to confirm the existence during this
period of important continental masses (Siberian continent, North Atlantic continent) with very poorly defined margins. On the other hand, the glaciations of
the southern hemisphere permit us to assume that the hypothetical continent of
Gondwanaland comprised an assemblage of large and small islands, and that, in
any case, their margins were sufficiently distinct to assure a humid, maritime
climate. Concerning the enigmatic displacement of the glaciers from west to east,
one can relate it to the very important epeirogenic movements of the time. One
can also assume that there existed a kind of give and take between the different

fragments of the earth's crust, such that the descent of one region resulted in the
elevation of another. This hypothesis does not enjoy the sympathies of the
geodesists, but it nevertheless has been invoked by such eminent geologists as
Karpinsky (8) in order to account for the transgressions and regressions of the
oceans. Moreover, nothing prevents us from assuming the existence of connected
regions which are subject to alternating vertical movements and periodic glaciations with a phase difference of one-half period. In any event, there is no need
to assume that enormous glaciations extended over entire hemispheres, or to
invoke extraordinary factors. Small variations in well known factors would be
enough to have produced the spread of Quaternary and Permo-Carboniferous glaciations.

V

THE FUTURE OF THE ATMOSPHERE AND OF LIFE:

15. Kinetic Diffusion and the Atmosphere

We have not yet considered the kinetic diffusion of the atmosphere resulting
from the collision of particles. The basis of this phenomenon is known. Through a
succession of shocks a particle can acquire a velocity which is higher than a
certain critical value. This value is 11.2 Km/sec at the surface of the earth,
and 2.4 Km/sec at the surface of the moon. By attaining such critical velocities,
a particle can escape the attraction of a body, and move in an open orbit. The
probability of escape from the terrestrial atmosphere is small, but it is not zero,
and such escapes do take place. Each escape decreases the probability of further
escapes so that the dispersion of the atmosphere declines exponentially (†).

The reality of this factor cannot be denied. However, there are reasons to
believe that its importance has been somewhat exaggerated. It is, in fact,
unjustified to apply to a free and thin atmosphere the reasoning which only holds

---

(†) It is difficult to understand how one can find in certain works imprecise
considerations on the relation between the critical velocity and the average
velocity of molecules in a gas. The average velocity can be less than the critical
velocity, but this will not prevent the atmosphere from dissipating into space.

for dense gases in rigid containers. There is an enormous difference between dense gases and rarified gases; between gases with random heat movement and gases with regular molecular motion. In the lower layers of the atmosphere a proportion of the molecules can attain velocities higher than 11.2 Km/sec. However, since the average free path there is very small, their velocity is also rapidly diminished by collisions. Under such conditions it would be a mistake to assume that all particles which attain speeds greater than 11.2 Km/sec will be lost from the earth's atmosphere. Conversely, in the higher layers of the atmosphere the mean free path is very long, but the directions of movement are not at all randomly distributed. In this case Kiveliovitch's theorem, on the existence of an upper limit for the number of collisions in an n-body system (9), has to be taken into account. It then becomes very improbable that a molecule could attain in the upper atmosphere a velocity higher than the critical one through shocks alone. Therefore, the diffusion of the atmosphere into outer space is probably so slow that it can be neglected, especially with respect to diffusion into the earth's crust.

This conclusion, however, is valid only for the stated conditions. It does not take into account one very important factor: namely, the ionization of the upper layers of the atmosphere by solar radiation, which can accelerate the loss of the atmosphere into outer space. On the other hand, the gaseous contribution of cosmic matter into the earth's atmosphere cannot be considered as negligible.

One final remark: of all the possible directions of escape from the earth, that toward the moon is particularly favorable. But, the molecules which have so escaped are not all lost. A considerable fraction of them simply move from the terrestrial atmosphere to the lunar atmosphere. Since the reverse movement is also possible, we have, in effect, two communicating atmospheres with exchanges between them going on all the time. The two atmospheres constitute, in some way, a pear-shaped gas body containing the earth and the moon. The density of this body decreases rapidly with distance, and very slowly with time. Because of such mechanisms, this pear-shaped distribution is completely different from the distribution proposed by Poincaré.

16. The Diffusion of the Atmosphere into the Soil

All atmospheric gases, with the exception of the noble gases, disperse into
the soil. Our equations show that the rate of this process is relatively constant,
as long as the coefficients do not vary. Naturally, such coefficients change in
the long run with the physico-chemical characteristics of the atmosphere and the
organic processes.

The maximum rate of diffusion of nitrogen can be inferred from simple
considerations. Certainly, the weight of atmospheric nitrogen which diffuses in
this way is less than the weight of nitrogen contained in the earth's crust. If
we suppose that this process has already continued for more than two billion years,
we find that the average loss of nitrogen per year is below 2×10^6 metric tons.
At this rate the atmospheric nitrogen would not be depleted for two billion years.

An analogous calculation for oxygen would be difficult. The fact that it is
still present in the atmosphere after two billion years of geological history
supports the view that both the atmosphere and life could persist for a practically
indefinite period of time.

Concerning carbon dioxide, suppose that volcanos, and all the other interior
sources, liberate on the average 10^8 metric tons of CO_2 per year. This figure
corresponds to the annual emission of fifty volcanos of the magnitude of Cotopaxi.
In this way 2×10^{17} tons of CO_2, or equivalently, 6×10^{16} tons of carbon, would
have been emitted since the beginning of geologic history. Now, the content of
carbon in the earth's crust is approximately double this figure. It seems plausible
that the carbon dioxide of volcanic origin was subtracted from the atmosphere by the
activity of marine organisms, which transformed it into calcarious rocks. The
carbon dioxide in solution in the ocean would only be the intermediate link in this
process.

Let us return to the equations (25). They do not change if the concentrations
are replaced by the total weights of factors taking part in the process. According
to W. Vernadsky (21), the total weight of marine organisms is of the order of 10^{14}
metric tons. This figure is almost certainly too large. We will assume, then that
$q_c = 10^{13}$ tons. We have already seen that $p_c = 10^{14}$ tons, and $a = 10^8$ tons. Sup-

pose that at the present moment we are at a stable equilibrium. The coefficients of the equations (25) are easily derived as:

$$\beta = 10^{-5}, \qquad \alpha = 10^{-19}, \qquad a\alpha = 10^{-11}, \qquad \beta^2 = 10^{-10}$$

It follows that:

$$a\alpha \ll 4\beta^2 . \tag{35}$$

Under these conditions, for each sudden increase of CO^2 the system, oceans -- atmosphere -- organic world, responds by quasi-periodic oscillations which rapidly damp out as the normal level of CO_2 is reestablished. Consequently, there is no risk that the continuing diffusion of carbon into the soil will deplete the carbon in the oceans or in the atmosphere as long as the earth exists.

17. The Stability of Life

Let us imagine a multidimentional space which symbolically represents the vital factors: pressure, p; temperature, T; light, ℓ; etc. At any time each living organism is represented by a point in this space, and a species is represented by an ensemble of points. An infinitesimal volume (dp dT dℓ...) around a point (p, T, ℓ,...) characterizes an infinitesimal ensemble of life conditions. Let us denote by n the number of species living in such conditions at the instant t:

$$n = F(t; p, T, \ell,...) . \tag{36}$$

We know that by changing (p, T, ℓ,...) within very large limits we can always find species which will adjust themselves to such conditions. The function F will have a certain number of maxima which correspond to optimal conditions for existence:

$$N_1(t) , \qquad N_2(t) ,...$$

The absolute minimum n = 0 is only attained for exceptionally large or small values of the variables. In fact, we do not know of any spots on earth which are absolutely uninhabitable, or devoid of life (17).

At a certain time t_o the function:

$$n_o = F(t_o; p, T, \ell,...) \tag{37}$$

is defined in a certain domain D_o of the (p, T, ℓ,...) space. In this domain there are a few privilaged regions which favor a very high number of species, $N(t_o)$:

$$N_1(t_o); \; p_1(t_o), \; T_1(t_o), \; \ell_1(t_o), \ldots$$
$$N_2(t_o); \; p_2(t_o), \; T_2(t_o), \; \ell_2(t_o), \ldots$$
$$\cdot \; \cdot \; \cdot \; \cdot \; \cdot \; \cdot \; \cdot \; \cdot \; \cdot \; \cdot \; \cdot \; \cdot \; \cdot \; \cdot$$
$$\left. \right\} \quad (38)$$

At another time t_1 the function:

$$n_1 = F(t_1; \; p_1, \; T_1, \; \ell_1, \ldots) \tag{39}$$

can differ considerably from n_o. The maxima $N(t_1)$ and the favorable regions:

$$N_1(t_1); \; p_1(t_1), \; T_1(t_1), \; \ell_1(t_1), \ldots$$
$$N_2(t_1); \; p_2(t_1), \; T_2(t_1), \; \ell_2(t_1), \ldots$$
$$\cdot \; \cdot \; \cdot \; \cdot \; \cdot \; \cdot \; \cdot \; \cdot \; \cdot \; \cdot \; \cdot \; \cdot \; \cdot \; \cdot$$
$$\left. \right\} \quad (40)$$

may have nothing in common with the region (38), although they are connected to it by a continuous passage of time. On the other hand, each unit considered in isolation (species, individual, etc.) is not stable, and, for a sufficiently large t_1-t_o, the ensemble (39-40) will certainly be composed of different units than the ensemble (37-38). Nevertheless, one can still talk of the relative stability of life. In mechanics the stability of a system is determined by its reaction to small displacements, to small variations of the variables and the parameters. One can define in like fashion the relative stability of life.

One must not confuse this stability with the secular stability, which does not in fact exist, either in mechanics or in biology. Certainly, by neglecting dissipation and passive resistances, etc., one can construe ideal mechanical systems which are theoretically capable of working for indefinite periods of time. To achieve this, however, one pays the price of a most dangerous departure from reality.

In the same way, it is possible to consider certain processes in biology as either stationary or periodical. Such models can be useful for studying the different details of vital phenomena. One can, for example, consider a healthy adult animal as a mechanism which functions in a manner which is in part stationary and in part periodic. One can also treat certain simplified problems in the struggle for life such as those of the well known type: Carnivore - Herbivore - Plant. It is clear, nevertheless, that such procedures are insufficient in mechanics, and even more so in biology.

But, along with the instability of individuals, species, etc., one can talk of the general stability of vital processes. In this regard, one can be sure that the survival of living beings is guaraneeed for an unlimited span of time, barring of course the sudden annihilation of the earth by a cosmic catastrophe. The slow but sure diffusion of the atmosphere will slowly displace the domain D toward other regions of the symbolic space (p, T, ℓ,...). Most likely, not even the disappearance of the present atmosphere would be enough to terminate life. Given the enormous resources contained in the lithosphere, life would go on, although possibly in more primitive forms.

BIBLIOGRAPHY

1. ARRHENIUS (S.). - Revue Gén. des Sc., 10, p. 337-342 (1899).

2. BONNIER (G.) et MANGIN (L.). - Ann. Soc. Nat. Bot. (7), 3, p. 5 (1886).

3. CROLL (J). - Geol. Mag. (n. s.), 2, I, p. 306-314, 346-353 (1874).

4. DUMAS (J. B.) et BOUSSINGAULT (J.). - Essai de statique chimique des êtres organisés. Paris (1844).

5. GEER (G. de). - Sverig. Geol. Und. (C), no 161 (1896).

6. GODLEWSKI (E.). - Arb. d. bot. Inst. Würzburg, I, p. 343 (1873).

7. HAUG (E.). - Traité de géologie. Paris, Colin.

8. KARPINSKY (A.). - Ann. de Géogr., 5, p. 179-192 (1896).

9. KIVELIOVITCH (M.). - Thèse Paris (1932).

10. KOSTITZIN (V. A.). - C. R., 195, p. 1219-1222 (1932).

11. KOSTITZIN (V. A.). - C. R., 198, p. 326-328 (1934).

12. KOSTITZIN (V. A.). - Revue Cén. des Sc., 45, p. 239-243 (1934).

13. KOSTITZIN (V. A.). - Le Mois, no 42, p. 264-270 (1934).

14. LOTKA (A. J.). - Elements of physical biology (1925).

15. LOTKA (A. J.). - Théorie anal. des associations biologiques, I. Paris (1934).

16. PERRIN (R.). - C. R., 198, p. 105-107 (1934).

17. PRENANT (M.). - Géographie des animaux. Paris (1933).

18. PRENANT (M.). - Adaptation, écologie et biocoenotique. Paris (1934).

19. RUSSELL (H. N.). - Nature, 135, no 3406, suppl. (1935).

20. SCHLOESING (T.). - C. R., 81, p. 82, p. 1852, etc (1875).

21. VERNADSKY (W.). - Géochimie. Paris (1924).

22. WEGENER (A.). - Die Entstehung d. Kontinente und Ozeane (1922).

23. ZENKEVITSCH (L.). - Zool. J. Moscou, 12, f. 4, p. 17-34, 1933.

GENERAL BIBLIOGRAPHY

Contains original references for all papers reprinted in this volume,
plus all papers cited by the editors in their introductory comments.

BRELOT, M. 1931. Sur le problème biblogique héréditaire de deux espèces
dévorante et dévorée. Ann. Matem. Ser. 4, 9: 57–74.

BROWNLEE, J. 1906. Studies in immunity: theory of an epidemic. Proc. R.
Soc. Edinb. 26: 484–521.

BROWNLEE, J. 1910. The mathematical theory of random migration and
epidemic distribution. Proc. R. Soc. Edinb. 31: 262–289.

CHAPMAN, R. N. 1931. Animal ecology with especial reference to insects.
Appendix by V. VOLTERRA. McGraw-Hill. New York & London.

D'ANCONA, U. 1942. La Lotta per L'Esistenza. Giulio Einaudi. Torino.

D'ANCONA, U. 1954. The struggle for existence. E. J. Brill. Leiden.

DARWIN, C. 1859. The origin of species by means of natural selection.
Reprinted by The Modern Library, Random House. New York.

GAUSE, G. F. 1934. The struggle for existence. Williams & Wilkins.
Baltimore.

GAUSE, G. F. and A. A. WITT. 1934. On the periodic fluctuations in the
numbers of animals. A mathematical theory of the relaxation interaction
between predators and prey and its application to a population of
Protozoa. Bull. Acad. Sci. U.R.S.S.: 1551–1559.

HAYS, J. D., J. IMBRIE, and N. J. SHACKLETON. 1976. Variations in the
earth's orbit: pacemaker of the ice ages. Science 194: 1121–1132.

HUTCHINSON, G. E. 1957. Concluding remarks. In Cold Spring Harbor
Symposium Quant. Biol. 22: 415–427.

KERMACK, W. O. and A. G. McKENDRICK. 1927. A contribution to the mathematical theory of epidemics. Proc. R. Soc., Ser. A. 115: 700–721.

KERMACK, W. O. and A. G. McKENDRICK. 1932. Contributions to the mathematical theory of epidemics. II: The problem of endemicity. Proc.
R. Soc., Ser. A. 138: 55–83.

KERMACK, W. O. and A. G. McKENDRICK. 1933. Contributions to the mathematical theory of epidemics. III: Further studies of the problem of endemicity. Proc. R. Soc., Ser. A. 141: 94-122.

KERMACK, W. O. and A. G. McKENDRICK. 1937. Contributions to the mathematical theory of epidemics. IV: Analysis of experimental epidemics of the virus disease mouse ectromelia. J. Hyg., Camb. 37: 172-187.

KOLMOGOROFF, A. N. 1936. Sulla teoria di Volterra della lotta per l'esistenza. Giorn. Instituto Ital. Attuari. 7: 74-80.

KOLMOGOROFF, A. N. 1972. The quantitative measurement of mathematical models in the dynamics of populations. (In Russian). Problems of Cybernetics 25: 100-106.

KOSTITZIN, V. A. 1932. Sur une application géologique des équations différentielles. Comptes rendus de l'Ac. des Sciences 195: 1219-1222.

KOSTITZIN, V. A. 1933. Sur quelques phénomènes quasi-périodiques dans les bassins fermés. Comptes rendus de l'Ac. des Sciences 196: 214-215.

KOSTITZIN, V. A. 1934. Symbiose, parasitisme et évolution (étude mathématique). Hermann. Paris.

KOSTITZIN, V. A. 1935. Évolution de l'atmosphère: Circulation organique, époques glaciaires. Hermann. Paris.

KOSTITZIN, V. A. 1936a. Sur les équations différentielles du problème de la sélection mendélienne. Comptes rendus de l'Ac. des Sciences 203: 156-157.

KOSTITZIN, V. A. 1936b. Sur les solutions asymptotiques d'équations différentielles biologiques. Comptes rendus de l'Ac. des Sciences 203: 1124-1126.

KOSTITZIN, V. A. 1937. Biologie Mathématique. A. Colin. Paris.

KOSTITZIN, V. A. 1938a. Équations différentielles générales du problème de sélection naturelle. Comptes rendus de l'Ac. des Sciences 206: 570-572.

KOSTITZIN, V. A. 1938b. Sur les coefficients medéliens d'hérédité. Comptes rendus de l'Ac. des Sciences 206: 883-885.

KOSTITZIN, V. A. 1938c. Sur les points singuliers des équations différentielles du problème de la sélection naturelle. Comptes rendus de l'Ac. des Sciences 206: 976-978.

KOSTITZIN, V. A. 1938d. Sur les équations différentielles du problème de la sélection naturelle dans le cas de mutation d'un chromosome sexuel. Comptes rendus de l'Ac. des Sciences 206: 1273-1275.

KOSTITZIN, V. A. 1938e. Sélection naturelle et transformation des espèces du point de vue analytique, statisque et biologique. Comptes rendus de l'Ac. des Sciences 206: 1442-1444.

KOSTITZIN, V. A. 1939a. Sur les équations intégrodifférentielles de la théorie de l'action toxique du milieu. Comptes rendus de l'Ac. des Sciences 208: 1545-1547.

KOSTITZIN, V. A. 1939b. Mathematical biology. Harrap and Co. London.

KOSTITZIN, V. A. 1940a. Sur la loi logistique et ses généralisations. Acta Biotheoretica. 5: 155-159.

KOSTITZIN, V. A. 1940b. Sur la ségrégation physiologique et la variation des espèces. Acta Biotheoretica. 5: 160-168.

KOSTITZIN, V. A. 1956. Sur le développement des populations bactériennes. Comptes rendus de l'Ac. des Sciences 242: 611-612.

KOSTITZIN, J. and V. A. KOSTITZIN. 1932. Sur la statistique d'infestation des Pagures par les Chlorogaster. Comptes rendus de l'Ac. des Sciences 193: 86-88.

LOTKA, A. J. 1907. Relation between birth rates and death rates. Science 26: 21-22.

LOTKA, A. J. (with F. R. SHARPE) 1911a. A problem in age-distribution. Philosophical Magazine, Series 6, 21: 435-438

LOTKA, A. J. 1911b. Die Evolution vom Standpunkte der Physik. Ostwalds Annalen der Naturphilosophie 10: 59.

LOTKA, A. J. 1912. Quantitative studies in epidemiology. Nature 88: 497.

LOTKA, A. J. 1915. Efficiency as a factor in organic evolution. J. Washington Acad. Sciences 5: 360-397.

LOTKA, A. J. 1919. A contribution to quantitative epidemiology. J. Washington Acad. Sciences 9: 73.

LOTKA, A. J. 1923a. Contribution to quantitative parasitology. J. Washington Acad. Sciences 13: 152-158.

LOTKA, A. J. 1923b. Contribution to the analysis of malaria epidemiology. I. General Part. Supplement to the Amer. J. Hygiene 3: 1-37.

LOTKA, A. J. 1923c. Contribution to the analysis of malaria epidemiology. II. General Part (continued). Comparison of two formulae given by Sir Ronald Ross. Supplement to the Amer. J. Hygiene 3: 38-54.

LOTKA, A. J. 1923d. Contribution to the analysis of malaria epidemiology. III. Numerical Part. Supplement to the Amer. J. Hygiene 3: 55-95.

LOTKA, A. J. 1923e. Contribution to the analysis of malaria epidemiology. V. Summary. Supplement to the Amer. J. Hygiene 3: 113-121.

LOTKA, A. J. 1924. Elements of physical biology. Williams and Wilkins. Baltimore, Md. Reprinted 1956, Dover, New York.

LOTKA, A. J. 1932. The growth of mixed populations: Two species competing for a common food supply. J. Washington Acad. Sciences 21: 461-469.

MARTINI, E. 1921. Berechnungen und Beobachtungen zur Epidemiologie und Bekämpfung der Malaria auf Grund von Balkanerfahrungen. Gente. Hamburg.

MARTINI, E. 1941. Lehrbuch der medizinischen Entomologie. 2nd. edition, Fischer. Jena.

NÄGELI, C. 1874. Verdrängung der Planzenformen durch ihre Mitbewerber. Sitz. Akad. Wiss. München. 11: 109-164.

NICHOLSON, A. J. and V. A. BAILEY. 1935. The balance of animal populations. Part I. Proc. Zool. Soc. Lond. 3: 551-598.

PEREZ, C. 1928. Sur le cycle évolutif des Rhyzocéphales du genre Chlorogaster. Comptes rendus de l'Ac. des Sciences 187: 771-773.

PERES, C. 1931. Statistique d'infestation des Pagures par les Chlorogaster. Comptes rendus de l'Ac. des Sciences 192: 1274-1276.

POLUEKTOV, et al. 1974. The dynamical theory of biological populations. Nauka. Moscow.

PUMA, M. 1939. Elementi per una teoria matematica del contagio. Edit. Aeron. Rome.

RESCIGNO, A., and I. W. RICHARDSON. 1967. The struggle for life: I. Two species. Bull. Math. Biophys. 29: 377-388.

RESCIGNO, A., and I. W. RICHARDSON. 1973. The deterministic theory of population dynamics. In Foundations of mathematical biology. Vol. III, pp. 283-360. Ed: R. ROSEN. Academic Press. New York.

ROSS, R. 1911. The prevention of malaria, 2nd. edition. Murray. London.

SCUDO, F. M. 1971. Vito Volterra and theoretical ecology. Theor. Popula. Biol. 2: 1-23.

SCUDO, F. M. and J. R. ZIEGLER. 1976. Vladimir Aleksandrovich Kostitzin and theoretical ecology. Theor. Popula. Biol. 10: 395-412.

SHARPE, F. R. and A. J. LOTKA. 1923. Contribution to the analysis of malaria epidemiology. IV. Incubation lag. Supplement to the Amer. J. Hygiene 3: 96-112.

VOLTERRA, V. 1901. Sui tentativi di applicazione della matematiche alle scienze biologicha e sociali. Giornale degli economisti 23: 436-458.

VOLTERRA, V. 1906. Les mathematiques dans les sciences biologiques et sociales. La Revue du Mois, Paris 1: 1-20.

VOLTERRA, V. 1926. Variazioni e fluttuazioni del numero d'individui in specie animali conviventi. Mem. accad. Lincei, (6) 2: 31-113.

VOLTERRA, V. 1927. Variazioni e fluttuazioni del numero d'individui in specie animali conviventi. R. Comitato Talassografico Italiano, Memoria 131: 1-142.

VOLTERRA, V. 1931a. Variations and fluctuations of the number of individuals in animal species living together. In CHAPMAN, R. N., Animal ecology with especial reference to insects, McGraw-Hill, New York.

Translated from a French version which appeared in Journal du Conseil international pour l'exploration de la mer, III. Vol. 1, 1928; a translation from the original Italian text (Volterra, 1926).

VOLTERRA, V. 1931b. Leçons sur la théorie mathematique de la lutte pour la vie. (Redigees par M. BRELOT). Gauthier-Villars. Paris.

VOLTERRA, V. 1934. Remarques sur la note de M. Régnier et Mlle. Lambin (Étude d'un cas d'antagonisme microbien). Comptes rendus de l'Ac. des Sciences 199: 1684-1686.

VOLTERRA, V. 1937a. Applications des mathématiques a la biologie. Enseign. math. 36: 297-330.

VOLTERRA, V. 1937b. Principes de biologie mathématique. Acta Biotheoretica. 3: 1-36.

VOLTERRA, V. 1938. Population growth, equilibria and extinction under specified breeding conditions: A development and extension of the theory of the logistic curve. Human Biology 10: 1-11.

VOLTERRA, V. 1939a. The general equations of biological strife in the case of historical actions. Proc. Edinburgh Math. Soc. 6: 4-10.

VOLTERRA, V. 1939b. Calculus of variations and the logistic curve. Human Biology 11: 173-178.

VOLTERRA, V. 1962. Opere matematiche di Vito Volterra. Accademia Nazionale dei Lincei. Rome. 6 vols.

VOLTERRA, V. and U. D'ANCONA. 1935. Les associations biologiques étudiées au point de vue mathématique. (Act. Sci. industr. nr. 243). Hermann. Paris.

VOLTERRA, V. and V. A. KOSTITZIN. 1938. Remarques sur l'action toxique du milieu à propos de la Note de M. Régnier, Mlle. Lambin (Étude sur le croit microbien en fonction de la quantité de substance nutritive des milieux de culture). Comptes rendus de l'Ac. des Sciences 207: 1146-1148.

WEGENER, A. 1922. Die Entstehung den Kontinente und Ozeane.

Bio—mathematics

Managing Editors: K. Krickeberg, S. A. Levin

Editorial Board: H. J. Bremermann, J. Cowan,
W. M. Hirsch, S. Karlin, J. Keller, R. C. Lewontin,
R. M. May, J. Neyman, S. I. Rubinow, M. Schreiber,
L. A. Segel

Volume 1:
Mathematical Topics in Population Genetics
Edited by K. Kojima
1970. 55 figures. IX, 400 pages
ISBN 3-540-05054-X

"...It is far and away the most solid product I have
ever seen labelled biomathematics."
American Scientist

Volume 2: E. Batschelet
Introduction to Mathematics for Life Scientists
2nd edition. 1975. 227 figures. XV, 643 pages
ISBN 3-540-07293-4

"A sincere attempt to relate basic mathematics to the
needs of the student of life sciences."
Mathematics Teacher

M. Iosifescu, P. Tăutu
**Stochastic Processes and Applications in Biology
and Medicine**

Volume 3
Part 1: **Theory**
1973. 331 pages.
ISBN 3-540-06270-X

Volume 4
Part 2: **Models**
1973. 337 pages
ISBN 3-540-06271-8

Distributions Rights for the Socialist Countries:
Romlibri, Bucharest

"... the two-volume set, with its very extensive bibliography, is a survey of recent work as well as a textbook. It is highly recommended by the reviewer."
American Scientist

Volume 5: A. Jacquard
The Genetic Structure of Populations
Translated by B. Charlesworth, D. Charlesworth
1974. 92 figures. XVIII, 569 pages
ISBN 3-540-06329-3

"...should take its place as a major reference work.."
Science

Volume 6: D. Smith, N. Keyfitz
Mathematical Demography
Selected Papers
1977. 31 figures. XI, 515 pages
ISBN 3-540-07899-1

This collection of readings brings together the major
historical contributions that form the base of current
population mathematics tracing the development of
the field from the early explorations of Graunt and
Halley in the seventeenth century to Lotka and his
successors in the twentieth. The volume includes
55 articles and excerpts with introductory histories
and mathematical notes by the editors.

Volume 7: E. R. Lewis
Network Models in Population Biology
1977. 187 figures. XII, 402 pages
ISBN 3-540-08214-X

Directed toward biologists who are looking for an
introduction to biologically motivated systems
theory, this book provides a simple, heuristic
approach to quantitative and theoretical population
biology.

Springer-Verlag
Berlin
Heidelberg
New York

A
Springer
Journal

Journal of

Mathematical Biology

Ecology and Population Biology
Epidemiology
Immunology
Neurobiology
Physiology

Artificial Intelligence
Developmental Biology
Chemical Kinetics

Edited by H.J. Bremermann, Berkeley, CA; F.A. Dodge, Yorktown Heights, NY; K.P. Hadeler, Tübingen; S.A. Levin, Ithaca, NY; D. Varjú, Tübingen.

Advisory Board: M.A. Arbib, Amherst, MA; E. Batschelet, Zürich; W. Bühler, Mainz; B.D. Coleman, Pittsburgh, PA; K. Dietz, Tübingen; W. Fleming, Providence, RI; D. Glaser, Berkeley, CA; N.S. Goel, Binghamton, NY; J.N.R. Grainger, Dublin; F. Heinmets, Natick, MA; H. Holzer, Freiburg i. Br.; W. Jäger, Heidelberg; K. Jänich, Regensburg; S. Karlin, Rehovot/Stanford CA; S. Kauffman, Philadelphia, PA; D.G. Kendall, Cambridge; N. Keyfitz, Cambridge, MA; B. Khodorov, Moscow; E.R. Lewis, Berkeley, CA; D. Ludwig, Vancouver; H. Mel, Berkeley, CA; H. Mohr, Freiburg i. Br.; E.W. Montroll, Rochester, NY; A. Oaten, Santa Barbara, CA; G.M. Odell, Troy, NY; G. Oster, Berkeley, CA; A.S. Perelson, Los Alamos, NM; T. Poggio, Tübingen; K.H. Pribram, Stanford, CA; S.I. Rubinow, New York, NY; W. v. Seelen, Mainz; L.A. Segel, Rehovot; W. Seyffert, Tübingen; H. Spekreijse, Amsterdam; R.B. Stein, Edmonton; R. Thom, Bures-sur-Yvette; Jun-ichi Toyoda, Tokyo; J.J. Tyson, Blacks-bough, VA; J. Vandermeer, Ann Arbor, MI.

Springer-Verlag
Berlin
Heidelberg
New York

Journal of Mathematical Biology publishes papers in which mathematics leads to a better understanding of biological pheno-mena, mathematical papers inspired by biological research and papers which yield new experimental data bearing on mathema-tical models. The scope is broad, both mathematically and biolo-gically and extends to relevant interfaces with medicine, chemistry, physics and sociology. The editors aim to reach an audience of both mathematicians and biologists.